Rheology and Deformation of the Lithosphere at Continental Margins

MARGINS Theoretical and Experimental Earth Science Series

Series Editors:

Garry D. Karner, *Lamont-Doherty Earth Observatory, Columbia University*

Julie D. Morris, *Washington University*

Neal W. Driscoll, *Scripps Institution of Oceanography*

Eli A. Silver, *University of California, Santa Cruz*

Continental margins are the Earth's principle loci for producing hydrocarbon and metal resources, for earthquake, landslide, volcanic, and climatic hazards, and for the greatest population density. Despite the societal and economic importance of margins, many of the mechanical, fluid, chemical, and biological processes that shape them are poorly understood. Progress is hindered by the sheer scope of the problems and by the spatial-temporal scale and complexities of the processes.

The MARGINS Program (a research initiative supported by the U.S. National Science Foundation) seeks to understand the complex interplay of processes that govern continental margin evolution. The objective is to develop a self-consistent understanding of the processes that are fundamental to margin formation and evolution. The books in the MARGINS series investigate aspects of these active systems as a whole, viewing a margin not so much as a geological entity of divergent, translational, or convergent types but more in terms of a complex physical, chemical, and biological system subject to a variety of influences.

Rheology and Deformation of the Lithosphere at Continental Margins

Edited by

GARRY D. KARNER

BRIAN TAYLOR

NEAL W. DRISCOLL

DAVID L. KOHLSTEDT

Columbia University Press / *New York*

Columbia University Press
Publishers Since 1893
New York   Chichester, West Sussex

Copyright © Columbia University Press 2004

Library of Congress Cataloging-in-Publication Data

Rheology and deformation of the lithosphere at continental margins / edited by Garry D. Karner
. . . [et al.].
    p. cm. — (MARGINS theoretical and experimental earth science series)
Includes bibliographical references and index.
    ISBN 0-231-12738-3 (cl.); 0-231-12739-1 (pbk.)
1. Rock deformation.   2. Continental margins.   3. Earth—Crust.   4. Earth—Mantle.   I. Karner,
Garry D., 1953–   II. Title.   III. Series.
    QE604.R45   2004
    551.8—dc22

                                                                2003017649

∞

Columbia University Press books are printed on permanent and durable acid-free paper.
Printed in the United States of America
c   10   9   8   7   6   5   4   3   2   1
p   10   9   8   7   6   5   4   3   2   1

CONTENTS

CONTRIBUTORS

Dr. Gary Axen
Department of Earth & Space
 Sciences
University of California
595 Circle Drive East
Los Angeles, CA 90095–1567

Dr. Yves Bernabé
Institut de Physique du Globe
Université de Louis Pasteur, Centre
 National de la Recherche
 Scientifique
Strasbourg
France

Dr. W. Roger Buck
Lamont-Doherty Earth Observatory
P.O. Box 1000
Palisades, New York 10964

Dr. F. M. Chester
Geology & Geophysics
Texas A&M University
Mail Stop 3115,
College Station, TX 77843

Dr. J. S. Chester
Geology & Geophysics
Texas A&M University
Mail Stop 3115
College Station, TX 77843

Dr. Mark Davis
Shell E&P
Carel van Bylandtlaan 23
PO Box 663, 2501 CR

The Hague
The Netherlands

Dr. J. P. Evans
Department of Geology
Utah State University
Logan, UT 84322-4505

Dr. Brian Evans
Department of Earth, Atmospheric &
 Planetary Sciences
Massachusetts Institute of Technology
77 Massachusetts Avenue, Room 54–
 718
Cambridge, MA 02139

Dr. Thomas Hanks
U.S. Geological Survey
Menlo Park, CA 94025

Dr. Greg Hirth
Geology and Geophysics
Woods Hole Oceanographic Institution
Woods Hole, MA 02543

Dr. R. D. Hyndman
Pacific Geoscience Centre
Geological Survey of Canada
8960 W. Saanich Road
Sidney, BC V8L 4B2
Canada

Dr. James Jackson
Bullard Laboratories
University of Cambridge
Madingley Road

Cambridge CB3 0EZ
United Kingdom

Dr. D. L. Kirschner
Department of Earth and Atmospheric
  Sciences
Saint Louis University
Saint Louis, MO 63103

Dr. D. L. Kohlstedt
Department of Geology & Geophysics
University of Minnesota
310 Pillsbury Drive SE
Minneapolis, MN 55455

Dr. Nick Kusznir
Department of Earth Sciences
University of Liverpool
PO Box 147
Liverpool L69 3BX
United Kingdom

Dr. Daniel C. Pope
Department of Earth and Space
  Sciences
University of Washington
Seattle WA 98195

Dr. Larry J. Ruff
Department of Geological Sciences
University of Michigan

2534 C.C. Little Building, 425 E.
  University Avenue
Ann Arbor, MI 48109–1063

Dr. Christopher Scholz
Lamont-Doherty Earth Observatory
P.O. Box 1000
Palisades, NY 10964

Dr. S. E. Schulz
Department of Geology
Utah State University
Logan, UT 84322-4505

Dr. Sean D. Willett
Department of Earth and Space
  Sciences
University of Washington
Seattle WA 98195

Dr. Yaqin Xu
Chorum Technology
1303 E. Arapaho Road
Richardson TX 75081

Dr. M. E. Zimmerman
Department of Geology and
  Geophysics
University of Minnesota
Minneapolis MN 55455

PREFACE

"Rheology and deformation of the lithosphere at continental margins"

Garry D. Karner, Brian Taylor, Neal W. Driscoll and David L. Kohlstedt

This volume is a collection of papers resulting from presentations made during a four-day short course at the first U.S. MARGINS Theoretical and Experimental Institute (TEI) held January 23–26, 2000. The institute was funded by the National Science Foundation and examined field, laboratory, and modeling constraints on how lithosphere rheology and deformation evolve throughout continental margin evolution. Traditionally, investigations of the rheology and deformation of the lithosphere have taken place at one scale in the laboratory and at entirely different scale in the field; development of an understanding of large-scale processes requires an integrated approach. The long-term objective of the short course and its ensuing publication is to stimulate cross-disciplinary inquiry into the rheology and deformation of the lithosphere. The first day of the short course provided an overview of the setting and nature of deformation at extensional and compressional continental margins. Day two concentrated on: (1) observations supporting, and models explaining, strain partitioning within the crust and lithosphere and (2) numerical and analogue modeling experiments that address the scaling problem of comparing physical experiments with natural systems. Day three focused on laboratory observations related to frictional sliding and crack healing along fault surfaces. Day four was centered on experimental studies of the rheology of crustal and mantle rocks.

The institute significantly influenced the subsequent research objectives and directions of the MARGINS Rupturing Continental Lithosphere (RCL) initiative, which were examined during a two-day workshop that followed the short course. The RCL initiative had as its basic tenet that the mechanisms allowing continental lithosphere to be deformed by weak tectonic forces were not understood, and neither was the manner in which strain was partitioned and magma distributed. These problems were encapsulated by the following themes: (1) the low-stress paradox of lithospheric deformation and (2) strain partitioning of the lithosphere during deformation. A series of papers verified the existence and complexities of the spatial and temporal distribution of strain within deforming lithosphere (chapters 1, 4, and 7: Buck, Davis and Kusznir, and Willett and Pope).

However, the low-stress paradox of lithospheric deformation that figured so prominently in all MARGINS planning documents prior to the TEI was significantly challenged. This paradox relates to the fact that large fault structures (subduction thrusts, major transforms, and perhaps normal detachments) accommodate a major component of strain but move at resolved shear stresses far smaller than those expected to cause failure. In turn, this apparent low-strength property of large faults may be corollary to an even more fundamental issue; namely, the tectonic forces available are insufficient to rupture the continental lithosphere as defined by the integrated yield–stress envelope of the continental lithosphere. Buck (chapter 1) elegantly showed that dike intrusion could reduce the amount of tectonic force required to rift normal continental lithosphere by an order of magnitude below that needed to stretch lithosphere in the absence of dykes.

Active low-angle normal detachments are the extreme case of the weak fault/low-strength paradox. The present debate revolves around whether low-angle faults mapped in such regions as the Whipple and Mormon Mountains of the western United States actually moved at low fault dip angles or moved on high-dipping faults whose footwalls were rotated into the observed field relationships, either in a domino style or by a rolling hinge mechanism. Continental intraplate earthquake focal mechanisms are predominantly related to high-dipping faults. Nevertheless, the megamullion structures of seafloor spreading centers and the geological reconstructions summarized by Axen (chapter 3) for the fault systems of southeastern Nevada, southwestern Utah, and southeastern California appear to require an active period of low-angle normal faulting. The controversy continues.

This same weak fault/low-strength paradox issue was the rationale behind the Ocean Drilling Program drilling (Leg 180) of the Moresby detachment zone in Papua New Guinea, one of the few examples of an active, low-angle ($\sim 30°$) normal fault (Taylor and Huchon 2002). Studies there showed the existence of many meters of talc-chlorite-serpentinite gouge with low coefficients of friction (0.21–0.3; Kopf et al. 2003) within a permeable, porous, and anisotropic fault zone at greater than hydostatic fluid pressures. Scholz and Hanks (chapter 9) effectively dismiss the paradox of the Moresby Detachment in demonstrating that its lock-up angle is consistent with Andersonian failure theory.

The weak fault/low-strength paradox has become entwined with the elastic thickness controversy in which earthquakes in midplate settings rarely occur below 40 km depth, indicating that the physical and chemical conditions prevailing in deeper rocks do not permit them to deform by brittle failure. In support of this observation, the elastic thickness of the continents inferred from free-air gravity, Bouguer gravity, and topography data is typically less than 40 km and less than the local depth to the Moho. In contrast, estimates of flexural loading of the lithosphere require elastic conditions to prevail to depths of 40 to 100 km over time periods of many millions of years. Hence the controversy: how is it possible for the Earth to support loads elastically at great depth and over long periods when the crust fails seismically at shallow depth and at short periods? Directly linked to this controversy is the viability of the yield-stress envelope for continental lithosphere. For many years, laboratory measurements of high-temperature creep of rock-forming minerals has been used to infer that crustal minerals should deform more readily than olivine at the same temperature. This led to the "jelly sandwich" image of a brittle upper crust, a potentially weak ductile lower crust, and a stronger upper mantle. Topography and the distribution of deformation near the Earth's surface concur with this image, at least for regions like the Basin and Range Province and Tibet. To what extent does a jelly sandwich simulate the rheology of continental lithosphere? Jackson (chapter 2) introduces a contentious idea suggesting that the strength of the continental lithosphere resides in its seismogenic layer, which is contained wholly within the crust, and that the continental lithospheric mantle is characterized by a wet rheology and thus is relatively weak. Willett and Pope (chapter 7), via a series of finite-element modeling experiments for the regional and intensive compressional deformation of continental lithosphere (bivergent orogenic edges and orogenic plateaus), offer important insights into the actual rheological behavior of the lithosphere.

In a set of related papers, Ruff and Hyndman (chapters 5 and 6, respectively) characterize the rheology of the zone between interacting converging plates, the seismogenic zone, which is defined by the spatial extent of earthquakes. Their intent is to define the processes controlling the updip and downdip rupture limits of the seismogenic zone. In this environment, the updip fault rheology appears to be dominated by temperature, which in turn controls the onset of seismic behavior via the dehydration of stable sliding smectite clay to stick-slip chlorite/illite, either in overlying sediments or within the fault zone gouge. The downdip limit of the seismogenic zone appears to be a function of the temperature dependence of the slip characteristics, for example, from stick-slip to stable sliding, in the fault zone material and the composition and thus rheology of the material in the overriding plate. Ruff (chapter 5) also attempts to define the controls on the various depths to the seismogenic limit within continental interiors, which seems to require more than just a temperature control.

Having a weak lithospheric mantle appears to be consistent with the laboratory studies reported by Xu et al. (chapter 10) and Evans et al. (chapter 11). Xu et al. investigated the role of melt on the anelastic and plastic properties of partially molten rocks as well as the effect of deformation on the distribution of the melt phase. The melt phase provides short-circuit diffusion paths or melt-rich bands, which aid in the relaxation of stress concentrations. The melt-rich bands are zones of low viscosity and high permeability, which act on a geologic scale to produce a marked anisotropy in seismic properties in addition to profoundly influencing the style of deformation. The link between magmatic processes and lithospheric strength is further explored by Evans et al., who show that rock strength decreases significantly when even a small amount of melt is present. In contrast, Chester et al. (chapter 8) describe the details of the porosity and permeability structure of large-displacement, strike-slip fault zones of the San Andreas system. The damaged zone and fault core are composed of very fine-grained, altered fault rocks in which the relatively permeable damage zone acts as a conduit for fluid flow along the fault and the low-permeability fault core serves as a barrier for cross-fault flow; the fault zone at least within the upper crust is conducive to fluid flow.

As with any professional publication, the quality of the final product is a strong function of the expertise and dedication of reviewers within the earth science community. The volume editors would like to acknowledge the following people for their unselfish donation of time and critical reviews of the various chapters: Rick Allmendinger, Atilla Aydin, Chris Beaumont, Roger Buck, Fred Chester, Reid Cooper, Tim Dixon, Rebecca Dorsey, Georg Dresen, Bob Engdahl, Laurel Goodwin, Greg Hirth, Bill Holt, John Hopper, Roy Hyndman, Barbara John, Dan Lizarralde, Steve Mackwell, Simon Peacock, John Platt, Leigh Royden, Carolyn Ruppel, Ernie Rutter, Dale Sawyer, Chris Scholz, Rick Sibson, Eli Silver, Joann Stock, Olaf Svenningsen, Uri ten Brink, Harold Tobin, Doug Wiens, Colin Williams, and Teng-fong Wong. The volume editors would also like to acknowledge the efforts of Andreas Aichinger and Steffi Rausch of the Hawaii MARGINS Office in organizing the Snowbird MTEI and the dedication, tenacity, and patience of Joan Basher of the Lamont MARGINS Office in bringing this publication to closure. The efforts and contribution from all these people are much appreciated.

References

Kopf, A., J. B. Behrmann, A. Deyhle, S. Roller, and H. Erlenkeuser. 2003. "Isotopic Evidence (B, C, O) of Deep Fluid Processes in Fault Rocks from the Active Woodlark Basin Detachment Zone." *Earth Planet. Earth Sci.* 2082: 51–68.

Taylor, B., and P. Huchon. 2002. "Active Continental Extension in the Western Woodlark Basin: A Synthesis of Leg 180 Results." In P. Huchon, B. Taylor, and A. Klaus, eds., 3*Proc. ODP, Sci. Results* 180 [online]. Available at: <http://www-odp.tamu.edu/publications/180_SR/synth/synth.htm>.

Rheology and Deformation of the Lithosphere at Continental Margins

CHAPTER ONE

Consequences of Asthenospheric Variability on Continental Rifting

W. Roger Buck

Introduction

The earliest ideas about continental drift (Wegener 1929) were based on the observation that the eastern coasts of North and South America matched the shape of the western coasts of Europe and Africa. This implies that the continents somehow split apart. Plate tectonics holds that continental breakup involves rifting the entire lithosphere, the cold outer layer of the earth that is too strong to flow along with the deeper interior.

During the thirty plus years since the acceptance of plate tectonics, much effort has been made to characterize rifts and rifted margins and understand the processes affecting them. One of the clearest messages from such studies is that continental rifts form with a variety of geometries, faulting patterns, and subsidence histories. For example, some rifts are wide, like the Basin and Range Province, and some are narrow, like the Red Sea, (e.g., England 1983). Some areas of apparently narrow rifting, such as metamorphic core complexes, do not subside locally (e.g., Coney and Harms 1979; Davis and Lister 1988), whereas some rifts, like those in East Africa, form deep basins even with modest amounts of extension (e.g., Rosendahl 1987; Ebinger et al. 1989). It has become accepted that the condition of the lithosphere at the time of rifting, its thermal structure and crustal thickness, can have a profound effect on the tectonic development of a rift (e.g., Sonder et al. 1987; Braun and Beaumont 1989; Dunbar and Sawyer 1989; Buck 1991, Bassi 1991).

There has been far more work on the effect of variations in lithospheric, as opposed to asthenospheric conditions, on the evolution of continental rifts. There are several good reasons for this lithospheric emphasis. It is easier to constrain lithospheric conditions by characterizing the geologic history and the geophysical structure of a rift, and the heat flow in adjacent areas. Further, it has taken time to work out the physics of lithospheric stretching and how processes can vary for different initial conditions and rates of extension. Also, lithospheric stretching models have been very successful at explaining many features of rifts and passive margins (e.g., McKenzie 1978).

A growing number of observations have not been explained by lithospheric stretching models. The most obvious observation involves magmatism. Many, possibly most, margins seem to be affected by massive magmatic intrusion and volcanic outflows, even before the onset of faulting and subsidence that mark stretching (e.g., Sengor and Burke 1978). As more data are collected, more margins appear to be "volcanic" (see White and McKenzie 1989; van Wijk and Cloetingh 2002). For example, the North American East Coast was once regarded as a prime example of nonvolcanic passive margin (e.g., Steckler and Watts 1981). Now, seismic data for the offshore area and geologic mapping onshore indicate that as much magma and lava was emplaced along the East Coast as for any margin (Holbrook and Keleman 1993). Models of the opening of the South Atlantic emphasize the effect of lithospheric stretching and detachment faulting (e.g., Etheridge et al. 1989; Lister et al. 1991), but massive piles of volcanic flows are inferred for the South American margin (Hinz 1981, White and McKenzie 1989). Thick volcanic layers are also seen on the 2,000-km-long Greenland margin (e.g., Mutter et al. 1988). The earliest rifting stage of the Red Sea is marked by massive flood basalts at the southern end of the Red Sea (e.g., Menzies et al. 1992) and dike intrusion in the north (Pallister 1987), yet models of Red Sea rifting usually ignore magmatic effects (e.g., Steckler 1985; Wernicke 1985; Martinez and Cochran 1988; Buck et al. 1988; Chery et al. 1992).

There are at least three major problems with "tectonic extension" models that ignore the effects of magmatism and flow of melt-depleted asthenospheric. They can be described in terms of three paradoxes.

- The "Tectonic Force" Paradox. It may take more force to extend thick lithosphere than is available. Stretching models imply faulting of cold upper mantle under rifts in normal lithosphere, but deep earthquakes are not observed in such settings (e.g., Maggi et al. 2000). Magmatic accommodation of extension may be needed to explain rifting in areas of low-to-normal heat flow.
- The "Extra Subsidence" Paradox, also known as the "Upper Plate" Paradox (Driscoll and Karner 1998). Observations indicate that some margins subside more than is predicted by simple models, prompting development of kinematic models with fairly complex patterns of assumed lithospheric stretching (Royden and Keen 1980; Wernicke 1985; Driscoll and Karner 1998). Dynamical models produce such strain patterns only when special patterns of preexisting weakness are assumed (e.g., Dunbar and Sawyer 1989).
- The "No Magma" Paradox. Some margins are amagmatic, even where the crust is highly attenuated or mantle is present at the ocean floor (e.g., at the Iberia Margin, Whitmarsh et al. 1990). This is surprising because stretching implies lithospheric thinning, causing vertical advection and pressure release melting of normal asthenosphere. The melt should be emplaced at or near the surface.

This chapter makes a case that, in addition to variability in lithospheric properties, asthenospheric variability may also be needed to explain observed features of rifts and rifted margins. Specifically, this chapter considers ways that melting of extremely hot (possibly plume related) mantle may resolve the paradoxes of continental rifting listed previously. The three major sections of the chapter deal with a suggested resolution for each paradox.

Magma-Assisted Rifting of Thick Continental Lithosphere (The "Tectonic Force" Paradox)

Both simple analytic or semianalytic rifting models (e.g., McKenzie 1978; Buck 1991) and more complex numerical simulations (e.g., Braun and Beaumont 1989; Bassi 1991; Dunbar and Sawyer 1989) assume that the average stress or tectonic force required to initiate rifting is available. This may not be true, however, for rifting of thick, strong lithosphere in the absence of basaltic magmatism. Several authors have estimated that the tectonic forces likely to be available for rifting is in the range of 3–5 TeraNt/m (Forsyth and Uyeda 1975; Solomon et al. 1980). The tectonic force needed for amagmatic extension of initially thick lithosphere may be up to an order of magnitude greater than that available (Kusznir and Park 1987; Hopper and Buck 1993; and discussion in the next section). Intrusion of basaltic dikes can allow lithosphere to separate at much lower levels of tectonic force than possible without the dikes.

Areas of initially thin lithosphere should rift at relatively low levels of tectonic force. Consistent with this, some areas of high heat flow and initially thick crust, such as the North American Basin and Range Province, did seem to start extending with little or no basaltic volcanism. Models neglecting magmatism do predict the general patterns of observed extensional strain inferred for such areas (Buck 1991). It should be noted that these "hot" weak areas are not typical of continents. The effects of orogenesis, especially thickening of radiogenic crust, have been suggested as a way of heating regions such as the Basin and Range and the Aegean Sea extensional provinces (e.g., Sonder et al. 1987). But, in areas of low-to-normal heat flow, the earliest phase of rifting is often accompanied by basaltic magmatism.

Morgan (1971) noted that huge volumes of flood basalt extruded in many areas of continental breakup. He suggested that such magmatism was associated with active mantle plumes, which somehow triggered rifting. Since that time much work has been done on estimating how plumes could produce extensional stresses by causing regional uplift (Sengor and Burke 1978; Spohn and Schubert 1982; Bott 1991). Great effort has also been made modeling plume temperatures and the magma volumes that result from active plumes or "plume head" upwelling (Richards et al. 1989; Griffiths and Campbell 1990; Hill 1991), and from passive upwelling related to the stretching and thinning of lithosphere (White and McKenzie 1989). However, the mechanical effects of large-scale magma intrusion on lithosphere-scale rifting remain largely unquantified.

This section will suggest the association of magmatism and rifting in areas of low-to-normal heat flow is not coincidental, but that magmatic intrusion may allow rifting to proceed given available tectonic forces. Specifically, it is suggested that: (1) considerable extensional tectonic force may be needed to cause lithosphere-cutting dike intrusion rather than extrusion of large magma volumes; (2) such dike intrusion would accommodate extension plate separation at as little as one tenth the tectonic force needed for amagmatic rifting; (3) after a short period of magma-assisted rifting (from 1 to 10 million years [m.y.] for a reasonable range of extension rates), initially thick lithosphere could be heated and weakened sufficiently to continue extending at a low-force level even without continued intrusion; and (4) the uplift and subsidence patterns of several major rifts and continental margins are more consistent with magma-assisted rifting than with simple lithospheric stretching models. A simple numerical model for estimating temperature and strength changes during rifting with dike intrusion is described.

Tectonic Force for Extension

Separation of lithospheric plates requires extensional stresses at a rift. At any depth those stresses can cause yielding by fault slip, ductile flow, or dike intrusion, whichever takes the least stress (figure 1.1). The extensional state of stress is approximated using the usual assumption that the vertical or z-direction is the largest principal stress and equals the lithostatic stress (Anderson 1951) given by

$$\sigma_1(z) = g \int_0^z \rho_r(z')\,dz' \qquad (1.1)$$

where g is the acceleration of gravity and ρ_r is the density of rock in the lithosphere. In the crust the assumed density is 2,800 kg/m^3; and in the mantle, density is 3,300 kg/m^3.

Dikes are magma intrusions with a thickness much smaller than their width or length. Molten basalt is assumed to be the material filling rift-related dikes, because mantle melting can produce basaltic magma and more felsic dikes might be too high in viscosity to propagate easily. Dikes should form in planes perpendicular to the least principal stress, σ_3; for a rift, this should be in vertical planes parallel to the rift (Anderson 1951). It is assumed that preexisting vertical fractures are prevalent to avoid the complications of fracture mechanisms (e.g., Rubin and Pollard 1987). However, the extra stress needed to break open dikes in unbroken rock should be limited by the rock tensile strength, which would make a small contribution to the tectonic forces estimated here. Neglected also are the viscous stresses associated with the flow of magma in a dike, since the goal is to estimate the minimum stress difference (defined as $\sigma_1 - \sigma_3$, where σ_1 is the maximum principal stress) required to have magma stop and freeze at a given depth in a dike.

Before freezing, magma in a dike can cease moving up or down when the static pressure in the dike equals the horizontal stress at the dike wall (Lister and

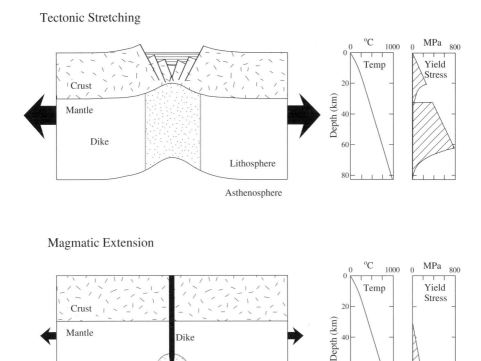

Figure 1.1 Schematic of the difference between extension of thick lithosphere with and without magmatic intrusion. Note the large difference in the yield stress, the stress difference needed to get extensional separation of two lithospheric blocks.

Kerr 1991). The vertical pressure variation in a static column of magma is related to its specific weight, ρ_m, so for magma emplacement: $\partial \sigma_3 / \partial z = \rho_m g$. To specify the level of magma pressure, it is assumed that dikes always cut to the surface, where the pressure is zero. In that case the stress difference required for dike emplacement is

$$\sigma_m(z) = \sigma_1(z) - g\rho_m z \qquad (1.2)$$

where ρ_m is the magma density, taken to equal 2,700 kg/m^3. Clearly, with these simplifications, the stress difference for magma to allow extensional separation between blocks of lithosphere depends only on the density difference between the lithosphere and magma (figure 1.2). Since the crustal density and magma density are taken to be equal, the stress difference for crustal diking is zero. In a mantle of density 3,200 kg/m^3, the stress difference required for intrusion increases at a

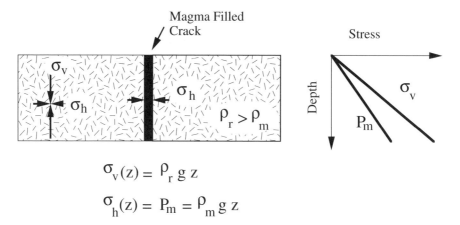

$$\sigma_v(z) = \rho_r \, g \, z$$

$$\sigma_h(z) = P_m = \rho_m \, g \, z$$

Figure 1.2 The stress distribution when extensional separation of two lithospheric blocks is accommodated by magmatic intrusion. Here the lithospheric rock density is ρ_r, greater than the magma density ρ_m. The horizontal stress σ_m equals the pressure in the dike, while the vertical stress is taken to be the overburden.

rate of 5 MPa/km of depth into the mantle. If the mantle is too weak to maintain such stresses, then the magma cannot be emplaced at depth and will be extruded.

At high temperature, rocks can flow in response to stress differences without forming macroscopic fractures. For such ductile flow the stress difference, σ_d, and strain rate, $\dot{\varepsilon}$, are found to be related through a flow law:

$$\sigma_d = (\dot{\varepsilon}/A)^{1/n} \, \exp(E/nRT) \qquad (1.3)$$

where T is absolute temperature, R is the universal gas constant, E is activation energy (e.g., Goetze and Evans 1979), and A is a constant for given material. The ductile yield stress depends on the composition of rock as well as temperature. Dry anorthite rheology is assumed for the crust and a dry olivine rheology for the mantle. For anorthite, $E = 238$ kJ mol^{-1}, $A = 5.6 \times 10^{-23}$ Pa^{-n} s^{-1}, and $n = 3.2$; for olivine, $E = 500$ kJ mol^{-1}, $A = 1.0 \times 10^{-15}$ Pa^{-n} s^{-1}, and $n = 3$ (Kirby and Kronenberg 1987).

Following Brace and Kohlstedt (1980), the stress difference needed for normal faulting is estimated under the assumption that cohesionless fractures exist in all directions to accommodate fault slip. The yield stress for faulting is

$$\sigma_f(z) = B(\sigma_1(z) - P_P(z)) \qquad (1.4)$$

where $B = 2f/[(1 + f^2)^{1/2} + f]$, where f is the coefficient of friction. Assuming $f = 0.85$, which is the average friction coefficient for a wide range of rocks (Byerlee 1978), makes the constant $B = 0.8$. The pore pressure in the rock, P_p, is taken to be hydrostatic.

To estimate the stress difference for extension (the yield stress) as a function of depth, z, we must specify the temperature profile through the lithosphere. This is done by assuming temperatures are in steady state with a constant heat flow from below, radioactive heat production within the crust, and a given heat flow at the surface. The thermal conductivity is set to 2.5 W m^{-1} °C^{-1} for the crust and 3.0 W m^{-1} °C^{-1} for the mantle. The crustal heat production is set to 3.3 $\times 10^{-7}$ W m^{-3}, which contributes 10 mW m^{-2} to the surface heat flow for a 30-km-thick crust. The mantle heat flow is adjusted to provide a given surface heat flow for a specific crustal thickness.

Figure 1.3 shows yield stress profiles for a moderate heat-flow temperature profile, assuming a 30-km-thick crust. The dashed lines show the model yield stress if no magma is available to accommodate extension; the solid line shows the situation if enough magma is available to just reach the surface. If an intermediate amount of magma were supplied in this conceptual model, it would be emplaced at depth, while extension near the surface was accomplished by faulting (e.g., Rubin and Pollard 1987). In some sense the dashed profile for amagmatic stretching can be seen as an upper limit on the yield stresses and the solid line as a lower limit.

The horizontal force per unit length required to cause extensional yielding of the entire model lithosphere, F_{ys}, is estimated by integrating yield stress over depth (figure 1.4). This force depends strongly on the temperature profile and, thus, on the surface heat flow as well as the magma supply. To extend continental lithosphere with a heat flow of about 40 mW/m^2, as is seen adjacent to some rifts like the Red Sea (Martinez and Cochran 1988), may require as much as 30 TeraNt/m of tectonic force if no magma were intruded. Extending the same lithosphere with copious magma may take almost an order of magnitude less force.

Two situations lead to extrusion of magma. First, if the tectonic force is too small to allow lithosphere-cutting dikes, then magma should be extruded along with dikes of small lateral extent. In that case, magma-assisted rifting cannot occur. This

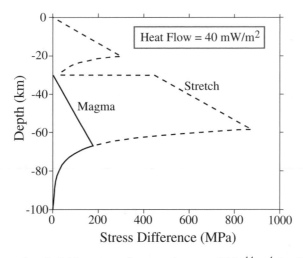

Figure 1.3 Example of yield stresses for a strain rate of 10^{-14} s^{-1} for 30-km-thick crust with a thermal profile derived (as described in the text) for a surface heat flow of 40 mW/m^2. The solid line shows the stress difference for magmatic rifting and the dashed line shows the yield stress for tectonic stretching.

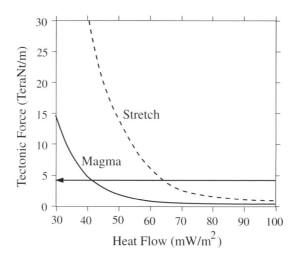

Figure 1.4   Tectonic force for extension either with or without magma as a function of the surface heat flow for a crustal thickness of 30 km. The tectonic force is the result of integrating yield stress envelopes such as those shown in figure 3. The horizontal bold line is the estimated value of plate extensional driving forces.

may be the situation for most ocean island basalts and some continental flood basalts, such as the Columbia River flood basalts. The other interesting case is if the tectonic force is great enough for large-scale diking, but the rate of extension requires less magma than is supplied. In this case, extrusion should occur on top of an area of rifting, as may have occurred in many areas discussed by White and McKenzie (1989), such as the rifting of Greenland from Norway, Madagascar from India, the East Coast of North America from Africa, South Africa from Antarctica.

The amount of basaltic magma available to facilitate rifting may vary with distance along some rifts. There is strong seismic and geodetic evidence from the active extensional plate boundary in Iceland that dikes propagate at least 60 km from central volcanoes (Einarsson and Brandsdottir 1980). These dikes can be intruded at depth with surface normal faulting and no accompanying extrusion of lava (Trygvasson 1984). During dike intrusion sequences there appears to be significant extrusion near the central volcano, while there is no extrusion far from the volcano (Trygvasson 1984). It is possible that a similar pattern occurs on a larger scale for some continental rifts. For example, the rifting of Arabia from Africa coincides with copious volcanism along the southern Red Sea coast (Menzies et al. 1992) while there is on-land evidence of a few dikes and little volcanism in the northern Red Sea region (Pallister 1987). More dikes may have intruded at depth in the northern Red Sea than made it to the surface.

Magma-Assisted Rift Evolution

To begin to relate these ideas about magma emplacement to observed characteristics of rifts and margins, a simple model is used to estimate the time evolution of rift-zone temperature, crustal composition, and lithospheric strength given a large flux of basaltic magma. The key assumption is that magma is intruded as dikes only at depths where the lithosphere is strong enough to hold the magma down for a given density structure. Magma is intruded at a temperature of 1,200°C and the latent heat of fusion adds another 300°C effective initial temperature. Also, it is assumed that magma is intruded at the rift center where the lithosphere is thinnest. Thus, the models differ from those of Royden et al. (1980), who investigated the thermal effect of an arbitrary distribution of intruded magma into stretching lithosphere and did not consider evolution of strength.

In these simple two-dimensional thermal models we are not concerned with whether the magma comes from below the rift or by lateral flow along the rift. The axis of the rift is considered to be a line of divergence so that lithosphere moves horizontally away from the line of dike intrusion (figure 1.5). Here, the lithosphere is defined as any material at a depth where initial rift temperatures were less than 1,200°C. Between the depth where lithospheric stresses are large enough for magma emplacement and the base of the lithosphere, we assume that plate separation occurs by distributed pure shear. The width of the pure shearing region is taken to equal its thickness.

The initial temperature field varies only in the vertical direction, with the profile derived from the same steady-state model parameters described earlier. Given the flow field and temperature structure we compute the time evolution of the temperature field using a standard finite difference scheme (e.g., Buck et al. 1988). As the temperature field changes, the ductile yield stresses should change, and so should the depth range where magma can be emplaced.

Specifying the depth of the transition from dike intrusion to ductile flow depends on estimating the ductile yield stress. Estimating this stress is not straightforward, since it depends on the temperature and strain rate fields. The common assumption used to estimate yield stresses is that the strain rate is uniform with depth and over the region of extension. This situation may never really be obtained in a rift and certainly should not be the case when there are lateral temperature

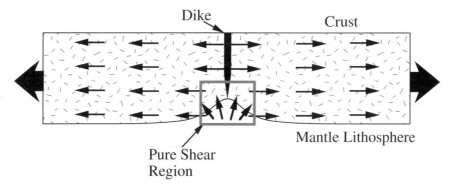

Figure 1.5 Illustration of the velocity field assumed for a thermal model of magma-assisted rifting. Pure shear is assumed in the region where ductile stresses are too low to allow magma emplacement.

variations; instead the hottest area should be the weakest and should thus strain the fastest.

Short of solving the full two-dimensional equilibrium equations for elastic, plastic, and viscous deformation, the following approximations are made. The stress at any depth is taken to be constant. For thermally activated creep and laterally varying temperatures, the strain rates must be greatest where the temperatures are greatest. For simplicity, all components of the strain rate tensor are neglected except the horizontal normal one, and so $\dot{\varepsilon}$ in equation (1.2) is replaced with $\dot{\varepsilon}_{xx}$. The rifting velocity must equal the integral of $\dot{\varepsilon}_{xx}$ across the rift, so the stress difference required to give the assigned rifting velocity u_r for a given temperature field $T(x,z)$ is

$$\sigma_d(z) = \left| \frac{u_r}{2A \int_0^{W/2} \exp\left(\frac{E}{RT(x,z)}\right) dx} \right|^{\frac{1}{n}} \tag{1.5}$$

where W, the width for integration, equals the lithospheric thickness.

Using the stress difference from equation (1.5) to calculate the maximum depth of dike intrusion, the flow field is adjusted as described previously. Temperature changes due to extensional flow in the lower lithosphere and dike intrusion at shallower levels are computed at every time step. At any time, the yield stress can be integrated over depth to estimate the tectonic force needed to continue extension.

Results for Magma-Assisted Rifting

We are interested in how rifting and magma intrusion weakens the lithosphere, since we want to consider whether a rift may continue to extend for a given regional tectonic force even if the magma supply is reduced. Therefore, through a calculation in which magma supply is sufficient to be intruded at all possible depths, we calculate the yield stress as if the magma were suddenly "shut off."

Figures 1.6–1.8 show results from one calculation of the evolution of magma-assisted rifting. Here, the half-rifting velocity was taken to be 1 mm/yr, the initial surface heat flow, $Q_s = 40$ mW m^{-2}, and the crustal thickness was 30 km. Figure 1.6 shows the model isotherms after 10 m.y. of extension. At the start of rifting, magma was emplaced down to ~60 km depth; by the end of the calculation magma was being emplaced only within the crust, because the mantle was too hot and weak to retain magma.

Figure 1.7 shows yield-stress profiles calculated at million-year intervals. The integrated yield stresses through time, both with and without magma supply, are shown in Figure 1.8. After approximately 5 m.y. into the calculation, the tectonic force for amagmatic extension has dropped to the level required to initially achieve

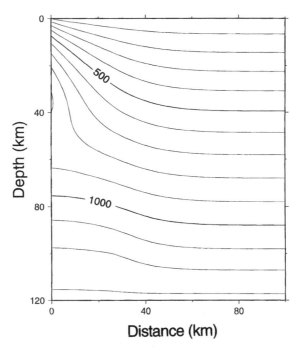

Distance (km)

Figure 1.6 Contours of temperature across half a rift zone after 10 m.y. of magmatic extension at a rate of 0.1 cm/yr. The initial crustal thickness was 30 km and the initial surface heat flow was 40 mW/m^2.

magmatic extension. The amagmatic (stretching) force is only weakly dependent on rifting velocity, so when an area has sufficiently weakened, the rate of extension might well increase. In contrast, the rate of magmatic extension depends mainly on the rate of magma supply.

It should be noted that only end-member models have been considered: either no magmatic intrusion, or enough to intrude at all depths where the lithosphere is strong. Other possibilities clearly exist. If the tectonic force is intermediate between the amount needed for these end members, then magma should be intruded over a reduced depth range, with strain at shallow depths accommodated by either elastic deformation, fissure opening, or fault offset (Rubin and Pollard 1987). Also, the portion of the lithosphere stretching tectonically might increase as the lithosphere thins.

The main potential objection to the idea that magma intrusion may be needed to allow rifting thick lithosphere is that our estimate of the stretching strength, based on the approach of Brace and Kohlstedt (1980), may be too high. There is evidence that for large-offset thrust faults, and possibly strike-slip faults, the stress for faulting the shallow lithosphere may be much less than the frictional stress assumed here. Usually, high pore fluid pressures are taken to be the cause of low stress brittle deformation (e.g., Hubbert and Rubey 1959). However, it is not clear whether areas of stretching can be characterized by low-stress brittle deformation, since pore pressures should be decreased by extension.

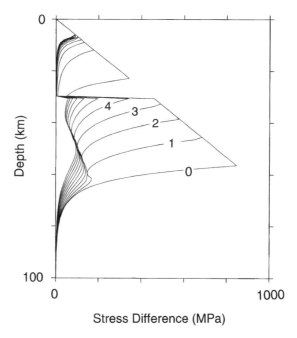

Figure 1.7 The tectonic yield stress as a function of depth for 1 m.y. time steps in the calculation illustrated in figure 6. The curves show the yield stresses needed for continued extension if the magma supply were suddenly cut off at the given time.

Comparison with Observations

Various effects of magmatic intrusion into rifts may be observable. The most direct observation would involve seismic imaging of basaltic bodies at depth. This is particularly challenging if there was little volcanism and most magma was intruded at tens of kilometers depth. A related challenge is that the seismic velocity of basalts may be only slightly greater than that for typical continental crust (see Kelemen and Holbrook 1995).

Another important effect to consider is the pattern of subsidence and uplift across rifts. Magmatic accommodation of extension should result in less subsidence than tectonic stretching of continental lithosphere, as discussed next.

The most promising places to test magmatic rifting models may be at the distal ends of young rifts and margins that are clearly affected by magmatism. The younger the rift or margin the better the chance that the early magmatic and tectonic history can be resolved. Therefore, we focus most of the present discussion on the northern sections of the 2,000-km-long Red Sea rift.

The Gulf of Suez is one of the best-characterized recent continental rifts in an area of low ($\sim$40 mW/m^2) heat flow. This rift is a part of the Red Sea rift system that ceased most extension when the Aquaba-Dead Sea oblique rift/transform developed 12–13 Ma (LePichon and Gaulier 1988). Dikes intruded this region beginning about 35 Ma (Pallister 1987; Dixon et al. 1989). However, rapid subsidence and rift shoulder uplift did not begin until after 25 Ma (Jarrige et al. 1990;

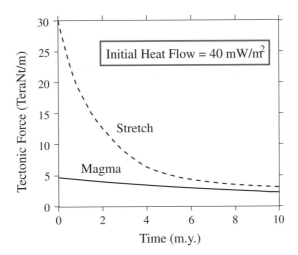

Figure 1.8 Evolution of the extensional force needed for tectonic and magmatic extension in the model of magmatic extension shown in figure 6. The curve labeled tectonic can be thought of as the force needed for continued extension if the magma supply were suddenly cut off. The curve labeled magmatic assumes that enough magma to form dikes that reach the surface is present at all times.

Omar et al. 1989; Omar and Steckler 1995). The magma-assisted rifting model may explain some observed features for the northern Red Sea and the Gulf of Suez rifts, long considered good examples of passive, essentially amagmatic, rifting (Steckler 1985; Martinez and Cochran 1988).

Seismic Structure

One would expect that a similar amount of basalt might have been intruded during the early phase of northern Red Sea rifting as may have intruded into the Gulf of Suez. The northern Red Sea has undergone much more extensional widening than the Gulf of Suez and so shows far greater average subsidence (Martinez and Cochran 1988). Any intrusives may be harder to image in the Red Sea because of the greater bathymetric relief and the greater thickness of salt in that region (Gaulier et al. 1988). Also, one would be searching for a rather subtle difference in seismic velocity and crustal thickness structure. The Gulf of Suez may never have reached the phase of large-magnitude tectonic subsidence that shaped the present day Red Sea. For the less extended Gulf of Suez, the difference in the crustal structure predicted by a pure tectonic stretching model, as opposed to a magma-assisted model, might be more readily resolved.

The calculation illustrated in figures 1.6 through 1.9, constructed with the early history of the Gulf of Suez and northern Red Sea in mind, allows rifting at low initial tectonic stresses. According to the model presented here, the average crustal thickness might have changed little during the rifting that produced the Gulf of Suez. This may seem contradictory to the observation of large tectonic faults that

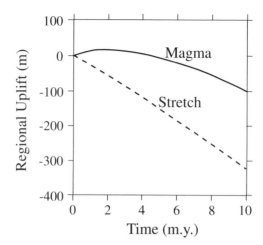

Figure 1.9 Comparison of the predicted average regional isostatic elevation changes with time for two rift models. Density changes are taken to affect the elevation over a region of width equal to the initial thermal lithospheric thickness. The solid line is for the model of magma-assisted rifting described in the text, and the dashed line is for a model of pure shear necking over a region as wide as the initial thickness of the lithosphere.

account for kilometers of near-surface brittle stretching. However, the intrusion of basalt may have occurred at greater depths, even well into the mantle at the start of rifting. If we consider the intruded basalt to be part of the crust, then during the early phase of magma-assisted rifting the average regional crustal thickness could increase. As the lithosphere is weakened by intrusive heating, a greater proportion of extension might have been accommodated by tectonic stretching, producing the observed slip on faults.

Subsidence/Uplift

Tectonic stretching (without magmatism) also may change the average density and so the elevation of a region. Continental lithosphere can be thought of as hot mantle replacing crust and cold mantle. Thinning the compositionally low-density crust causes subsidence, whereas thinning the thermally dense lithosphere causes initial uplift. The total initial and long-term effect of stretching typical continental lithosphere should be regional subsidence (McKenzie 1978). Regional, rather than local, elevation must be considered, since the lithosphere maintains finite strength during rifting.

Solidified basaltic magma is less dense than mantle, so basalt intrusion can affect the average density and the isostatic elevation of a region. The emplacement of large quantities of basalt into a rift can accommodate extension with little or no crustal thinning. In fact the intrusion of basalt into the mantle can effectively thicken the crust. So dike intrusion can lessen the initial amount of subsidence or even lead to regional uplift.

Figure 1.9 compares the average isostatic elevation through time for magma-assisted rifting with that predicted by a standard stretching model. The uplift or subsidence is calculated assuming that at 0°C crust and basalt both have a density of 2,800 kg/m³, while mantle has a density of 3,300 kg/m³ at the same temperature. The temperature field computed during rifting is related to the density field using a thermal expansion coefficient of 3.5×10^{-5} °C^{-1}. Density changes related to crustal thinning, basalt intrusion, and temperature changes are integrated over depth, $D = 150$ km, and over a 100-km-wide region of the center of the rift. Decreases or increases in the weight of the rift region cause uplift or subsidence, respectively, because the region is taken to float on hot mantle asthenosphere with a density, ρ_a, of 3,285 kg/m³. Formally, the elevation change equals $D\Delta\rho/\rho_a$ where $\Delta\rho$ is the average density change of the rift region. Figure 1.9 shows potentially observable differences between the tectonic stretching and the magma-assisted rifting models.

Some continental margins such as the Bay of Biscay seem to fit the general predictions of the stretching model (LePichon and Sibuet 1981). However, Royden and Keen (1980) showed that the subsidence history recorded in wells on the passive margin off the Canadian East Coast do not fit the McKenzie (1978) stretching predictions. These data require less initial tectonic subsidence (related to crustal thinning) relative to the long-term thermal subsidence. The magma-intrusion model gives less tectonic subsidence than the stretching model, as shown in figure 1.9. The subsequent thermal subsidence (not shown here) for comparable amounts of extension is affected little by the magma intrusion. Thus, this model predicts subsidence patterns that are consistent with the general trend of the data analyzed by Royden and Keen (1980).

Steckler (1985) showed that the Gulf of Suez does not match the predictions of the tectonic stretching model. He analyzed the tectonic subsidence in the rift and the surrounding rift-shoulder uplift and found that the average present-day regional elevation is close to zero: the volume of the uplifted rift shoulders approximately equals the volume of the subsided gulf basin, after corrections are made for loading of basin sediments (see figure 1.10). Tectonic stretching should produce long-term average regional subsidence. Steckler (1985) explained the lack of such subsidence in terms of a convective input of heat. However, the lack of large magnitude regional subsidence across the Gulf of Suez is consistent with the injection of significant quantities of magma into the rift in the early stages of extension.

Other failed rifts may have extended when basaltic magma was being intruded. One candidate for magma-assisted rifting is the Mesozoic Dnieper-Donets Basin of southern Ukraine, where well data indicate very little tectonic phase subsidence, but large-magnitude thermal subsidence (Starostenko et al. 1999)

Magmatic input may be necessary for the active rifting seen in several areas of presumed thick lithosphere, including the Rhinegraben, the Baikal Rift, the Rio Grande Rift, and along parts of the East African Rift. The Rhinegraben cuts a region of northwest Europe characterized by normal heat flow, averaging about 40 mW m^{-2} (Illies and Greiner 1978). As argued previously, such heat flow may indicate very large lithospheric strength in extension. Volcanism is contempora-

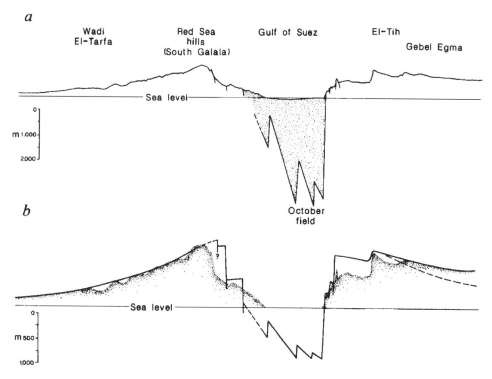

Figure 1.10 (a) Topography and basement relief for a transect across the central part of the Gulf of Suez Rift. (b) Shows the topography modified by the effect of flexural sediment unloading (from Steckler 1985). This shows that the net subsidence averaged across the rift is close to zero, since the uplifted flanks nearly balance the down-dropped center of the rift. The average elevation of the rift plus flanks is positive, but this may reflect the fact that the region outside the flanks has a positive elevation of ~500 m.

neous with the rifting along the Rhinegraben (Illies and Greiner 1978). The region around Baikal, another Cenozoic rift, is also about 40 mW m^{-2} (Morgan 1982). Seismic surveys across this Siberian rift show evidence for a ~10-km-thick layer at the base of the rifted crust with a seismic velocity consistent with basaltic "underplating" (Zorin 1981). The East African rift and the Rio Grande Rift also cut areas with near-normal heat flow (Morgan 1982) and parts of these rifts are characterized by recently active volcanoes (Ebinger et al. 1989, Mohr 1992).

Dynamic Subsidence of Passive Margins (The "Extra Subsidence" or "Upper Plate" Paradox)

Many margins show more subsidence after the early "tectonic" phase than is predicted by uniform pure shear stretching of typical crust and mantle lithosphere (see Lister et al. 1986; Driscoll and Karner 1998). For example, analysis of deep-well data for the Atlantic margin of Canada shows extra "thermal phase" subsidence after a phase of assumed tectonic subsidence (Royden and Keen 1980).

Observations similar to those described here led several workers to suggest that the geometric pattern of lithospheric extension is significantly more complex than uniform pure shear. Royden and Keen (1980) proposed a "two-layer" stretching model in which the mantle lithosphere stretched more than the crust. An alternative model to explain subsidence with little near-surface extension (stretching) is the "simple shear" model (e.g., Wernicke 1985). The idea is that a lithosphere cutting low-angle fault or shear zone accommodates much of the strain during rifting. Vertical sections through parts of the side of the rift above the shear zone, called the upper plate, would experience little crustal thinning but large amounts of mantle lithosphere thinning. The simple shear model had the added appeal that it could explain the topographic asymmetry seen across many conjugate margins (e.g., Lister et al. 1986). To explain a subsidence event for the Exmouth Plateau, off N.W. Australia, that does not seem to involve upper crustal extension, Driscoll and Karner (1998) proposed an extreme variant on the simple shear model. To explain the Exmouth subsidence they called for several hundred kilometers of offset between upper crustal thinning and lower crustal/mantle lithospheric thinning.

Numerical models of lithospheric stretching that treat the evolution of mechanical strength during rifting (e.g., Braun and Beaumont 1989; Bassi 1991; Chery et al. 1992) tend to show fairly symmetric patterns of deformation that are similar to the necking pattern seen for laboratory necking of metal rods. The necking strain predicted by such dynamical numerical models is similar to that predicted by kinematic pure shear models, if the width of pure shear necking equals the thickness of strong lithosphere. Dynamical numerical models that predict very asymmetric strain patterns generally assume preexisting laterally offset regions of crust and mantle strength (e.g., Dunbar and Sawyer 1989). Dynamical models with strain weakening applied to extension of initially symmetric lithosphere do produce asymmetric fault patterns, but typically only on the scale of the upper crust (e.g., Buck and Poliakov 1998; Lavier et al. 2000). If highly asymmetric lithospheric deformation is common during rifting, then one must assume large-scale and large-magnitude prerift asymmetric weak zones in the lithosphere.

In at least one site where simple shear lithospheric stretching was suggested to explain rift asymmetries, subsequently collected data contradicted that suggestion. The strong topographic asymmetry across the Red Sea rift, with the Arabian side ~500 m higher than the Egyptian side, had made this rift system a prime example of a possible simple shear rift (Wernicke 1985). Cochran et al. (1993) did detailed transects of heat-flow measurements across the northern Red Sea, an ~20 million year old rift that has opened ~100 km. Thermal models of pure shear stretching and simple shear extension were compared with the data (Buck et al. 1988; Martinez and Cochran 1988). This showed that the pure shear models fit the heat-flow and subsidence data in the rift. The simple shear models fit neither the heat flow nor the asymmetric topography of that rift.

Another potential problem with the simple shear model is that in some cases both sides of a conjugate margin look like "upper plate margins" in that they show more long-term subsidence than can be explained in terms of the estimated local crustal stretching (Driscoll and Karner 1998). This has been dubbed the "upper plate paradox" by Driscoll and Karner (1998).

Driscoll and Karner (1998) conducted a detailed study of a particularly clear example of subsidence that cannot be explained by uniform pure shear lithospheric stretching: the Exmouth Plateau. They used seismic reflection lines and deep-well data to look at the tectonic and subsidence history of the ~300-km-wide Exmouth Plateau margin. The plateau was tectonically extended and faulted during the Middle Jurassic. This rifting event did not result in seafloor spreading but did produce observable subsidence, because sediment filled in the area to about sea level. Another rifting event affected the plateau during the early Cretaceous, and seafloor spreading commenced adjacent to the western side of the plateau. The interior of the plateau shows no evidence of tectonic extension, because the sediments deposited since the first rifting event are not faulted. However, the plateau subsided at least a kilometer and, where loaded by new sediments, subsided even further.

To explain the large plateau subsidence that begins during seafloor spreading adjacent to the plateau, Driscoll and Karner (1998) assumed upper crustal thinning on the west side of the plateau and lower crustal thinning under the entire plateau. In their kinematic model, strain was distributed through the shear flow within the presumably weak lower crust. Many workers have argued that weak lower crust can flow under areas of high heat flow like the Basin and Range Province of western North America (e.g., Gans 1987; Block and Royden 1990; Bird 1991). In the model of Buck (1991) the change from core complex style extension to Basin and Range style extension is due to a decrease in the rate of lower crustal flow. The rate of lower crustal flow is related to its viscosity, which should be a function of temperature. High-temperature crust should have a low viscosity and would easily flow.

Other areas where the lower crust may flow easily are high-elevation plateaus like the Altiplano and Tibet. The correlation between elevation and crustal flow is probably not a coincidence. These plateaus are high because the crust is thick. The base of thick crust can be very hot even with moderate heat flow and so temperature gradients. The active part of the Basin and Range has an average elevation of nearly 2 km while the crustal thickness there is only about 30 km (~average or below average for continents). The higher-than-normal elevation is likely to reflect very high temperatures below the region. This is consistent with the very high surface heat flow (~100 mW/m^2) (e.g., Lachenbruch and Sass 1978). Thus, it is not surprising that the lower crust in the Basin and Range is hot and therefore can flow.

It is much harder to explain how the lower crust of the Exmouth Plateau was hot enough to shear easily when rifting and seafloor spreading occurred. The main problem is that the region was close to sea level at that time (Driscoll and Karner 1998). The crustal thickness then, according to these authors, would have been

~5 km thicker than its present average thickness of 20–25 km. To explain the sea level elevation would require very low temperatures below the plateau and so very low thermal gradients. It is therefore hard to understand how the lower crust under the Exmouth Plateau would have been hot enough to shear at high rates during a rifting event. In the following section, it is hypothesized that the lateral flow of low-density mantle asthenosphere may cause the "extra" subsidence seen on some margins.

Assume that a local area of thin lithosphere exists in the vicinity of a large volume of anomalously hot mantle (figure 1.11). The hot mantle might have

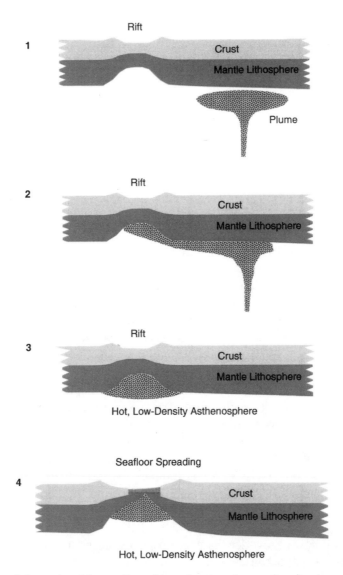

Figure 1.11 Schematic of the pooling of low-density, plume-related asthenosphere under an area of thin lithosphere.

been delivered to the shallow asthenosphere by a plume or plume head (e.g., Campbell and Griffith 1990), though its genesis is of little consequence to this discussion. The hot mantle should be much lower in density than surrounding "normal" asthenospheric mantle both due to thermal expansion and possibly due to depletion on partial melting (Oxburgh and Parmentier 1977). This hot asthenosphere should then pond beneath the area of thin lithosphere. The hot, low-density asthenosphere displaces the dense asthenosphere below the thin lithosphere, resulting in uplift of that area.

To estimate the amount of uplift produced by such ponding we need to know the density and thickness of the hot mantle. The elevation, e_0, will be related to the normal mantle density, ρ_{NM}, the anomalously hot mantle density, ρ_{AM}, as

$$e_0 = (H_L - H_R)(\rho_{NM} - \rho_{AM})/(\rho_{AM} - \rho_W) \qquad (1.6)$$

(1.6) assumes the elevation is submarine with a water density, ρ_W, and that $(H_L - H_R)$ is the thickness of the region of anomalous low-density mantle. H_L equals the thickness of the lithosphere around a region with thinner lithosphere of thickness H_R (figure 1.12).

One may get some idea of the uplift produced by hot asthenosphere by looking at the anomalous depth of midocean ridges that are affected by mantle plumes. A prime example is the Reykjanes Ridge south of Iceland. The crustal thickness is about 8 km (Ritzert and Jacoby 1985), about average for midocean ridges, but the water depth is only about 1000 m (e.g., Talwani et al. 1971). The usual depth to ridges is about 3,000 m (e.g., Small 1998). Thus, we can attribute 2,000 m of elevation to low-density asthenosphere below the northern Reykjanes Ridge. If the hot, depleted mantle layer were 200 km thick, then it would have to be just 23 kg/m^3 less dense than normal mantle to explain the anomalous depth of the Reykjanes Ridge. This assumes the density of normal mantle $\rho_{NM} = 3,300$ kg/m^3 and $\rho_W = 1,000$ kg/m^3. Such a layer thickness is not inconsistent with recent models of plume-ridge interactions (e.g., Sleep 1990; Ribe et al. 1995; Ito et al. 1996). If the layer of hot asthenosphere is thinner, its density has to be lower to explain the ridge depth.

Part of the density anomaly is likely to be due to depletion of the hot mantle caused by pressure release melting on ascent. The material density will become progressively smaller as more melt is extracted at shallower depths of melting (e.g., Oxburgh and Parmentier 1977; Klein and Langmuir 1987). Thus, even if material pooled beneath an area of thin lithosphere cools, it can still remain lower in density, and so positively buoyant, compared with normal mantle. If the area of thin lithosphere is rifted, the buoyant asthenosphere could pour into the area of extremely thin lithosphere where seafloor spreading is beginning (figures 1.11 and 1.12).

The lateral flow of low-density asthenosphere causes subsidence in the region where it was pooled, because it is replaced by mantle of normal density. The rate of subsidence is related to the velocity at which rifting or seafloor spreading occurs, u_p, and on the width of the lithosphere (figure 1.12). It also depends on the

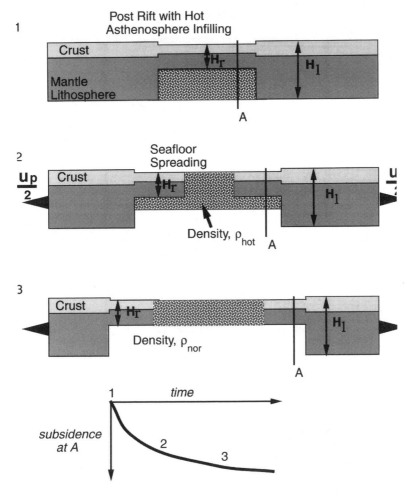

Figure 1.12 Geometry assumed to calculate the effect of flow of a low-density asthenospheric layer on subsidence of a rift.

depth to the base of the ponded layer, H_L, and of course, on the density contrast with normal mantle. The local isostatic subsidence (the change in elevation of a point in the region of pooled lithosphere) due to outflow is given by

$$\frac{e(t)}{e_0} = \left[\left(\frac{1}{1+u_p t/W}\right) - \frac{H_r}{H_L}\right]\bigg/\left(1 - \frac{H_R}{H_L}\right) \qquad (1.7)$$

Figure 1.13 shows the predicted "extra subsidence" due to flow of low-density asthenosphere. The maximum subsidence equals e_0 and depends on the initial thickness of the depleted layer and on the density contrast. The time needed for the entire depleted layer to flow out from under the initially thin lithosphere, and for the related subsidence to cease, equals $(W/u_p)[(H_L/H_R) - 1]$. For $(H_L/H_R) = 3.33$, $W = 100$ km, and $u_p = 1$ cm/yr, the time for total outflow is about 23 m.y. This case is one of the two shown in figure 1.13. The other case had an initially

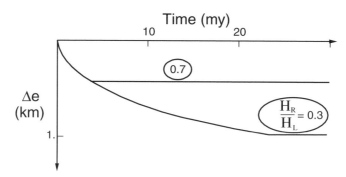

Figure 1.13 Results of analytic model calculation of dynamic subsidence using equation (1.7). The two curves correspond to cases with the same model parameters except that the ratio of the initial depth to top and bottom of the layer, H_R/H_L, are 0.3 and 0.7, as labeled. The initial width of thin lithosphere, W, is 100 km; the plate-spreading velocity, u_p, is 1 cm/yr; $H_L = 150$ km; and the density difference between anomalous and normal mantle is 25 kg/m^3.

thinner depleted layer and so subsided less and took less time to have all depleted material flow out.

Viscous Flow of Depleted Layer

Up to now, it has been assumed that flow in the depleted layer is fast enough to keep the base of the layer flat. One would expect that the flow rate is limited by the viscosity of the layer. If the viscosity were extremely great then the layer would not flow out in a geologically observable time. The pressure-driven flow in a broad, relatively thin viscous layer can be approximated by one-dimensional channel flow, making it easy to relate asthenospheric viscosity to the time for layer thinning.

The viscous layer thins as it flows into the space created by seafloor spreading (figure 1.12). To first order, the thickness with time and distance from the site of seafloor spreading obeys a diffusion relation. The effective diffusivity for the thinning of the layer can be estimated in the way often done for flow-related thinning of viscous lower crust (see Bird 1991; Buck 1991). It is assumed that the top and bottom boundaries of the layer can be described as "no slip." This is probably reasonable for the top boundary, but may not describe the bottom boundary where underlying mantle may flow easily. Changing the bottom boundary condition to free slip would decrease the estimated time for flow and thinning by a factor of four. Since an order-of-magnitude estimate of parameters is needed here, more complex boundary conditions will not be considered.

The flow is taken to be driven by pressure gradients that arise due to local isostatic compensation of the lateral density variations associated with layer thickness variations. Then the effective flow diffusivity is

$$\kappa_f = \frac{g\Delta\rho^*(H_L - H_R)^3}{\mu} \tag{1.8}$$

where μ is the layer viscosity and the effective density contrast $\Delta\rho^* = (\rho_{NM} - \rho_{AM})\rho_{AM}/\rho_{NM}$. For diffusive processes, the characteristic time for significant thinning over a horizontal distance d is:

$$t_c = \frac{d^2}{4\kappa_f} = \frac{ud^2}{4g\Delta\rho^*(H_L - H_R)^3} \tag{1.9}$$

If we assume that $d = 300$ km, the initial layer thickness $(H_L - H_R) = 50$ km, $\Delta\rho^* = 20$ kg/m^3, then for $\mu = 10^{20}$ Pa s, the flow time $t_c = 3$ m.y.

Most estimates of shallow mantle viscosity based on observations of glacial and lake desiccation loads are about 10^{19} Pa s (e.g. Cathles 1975). Therefore, it seems likely that flow of a low-density asthenospheric layer could occur on the timescale of the observed postrift subsidence of passive margins.

One might well ask how initially hot, pooled asthenosphere could remain warm enough to flow if the region cools for many tens of millions of years. Even conductive cooling is expected to turn oceanic asthenosphere into lithosphere to a depth of more than 70 km in ~100 m.y. (e.g., Parsons and Sclater 1977). Two things may allow some significant thickness of warm pooled asthenosphere to remain after geologically long time intervals. First, if the layer bottom (H_L in figure 1.12) is deep, say 125 km, then the layer top could be cooled to 75 km and still leave 50 km of warm, low-density material to flow. Second, the presence of low-conductivity sediments and radioactive continental crust over the pooled asthenosphere should slow its cooling compared with the oceanic lithosphere.

Possible Application of Model to the Exmouth Plateau

A possible scenario for applying the asthenospheric depletion and flow model to the Exmouth Plateau would involve an early phase of emplacement of depleted asthenosphere during or after the Jurassic rifting phase. Some of that low-density asthenosphere might then flow out from under the plateau during Cretaceous seafloor spreading. Karner and Driscoll (1999) showed the subsidence with time for a theoretical well in the central part of the Exmouth Plateau (figure 1.14). The amount of sediment loaded subsidence was ~2 km starting at ~120 m.y. before the present. This was during the rifting event that led to seafloor spreading to the west of the margin, but did not tectonically stretch the upper crust of the plateau margin. Were there no sediment load, the water-loaded subsidence would have been about 700 m. This is in the range predicted by lateral flow of a moderately thick, low-density layer of depleted asthenosphere.

Post-rift subsidence form

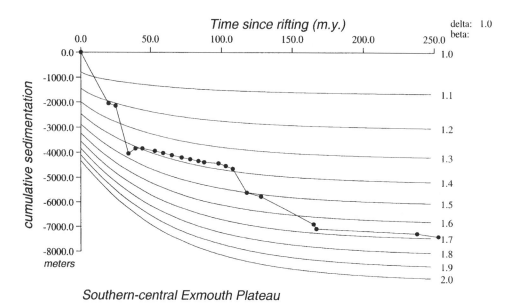

Southern-central Exmouth Plateau

Figure 1.14 Subsidence versus time since rifting for a site on the Central Exmouth Plateau (Karner and Driscoll 1999). Note the subsidence event beginning at about 120 m.y. after the start of rifting. This is associated with no tectonic stretching of the plateau, but breakup to the west of the plateau occurs at about this time.

Convection in the Depleted Layer and Amagmatic Rifting (The "No Magma" Paradox)

This section deals briefly with another possible effect of the pooling of the residue of hot asthenospheric melting. Namely, it might contribute to later amagmatic rifting. The key to this process is the convective cooling of the hot asthenosphere after it is first emplaced beneath thin lithosphere (as in figure 1.11). If the layer did not convect and cool after pooling, then it could melt further during a subsequent rifting event, assuming that event caused asthenospheric upwelling.

Convection can inhibit later rift-related melting of the asthenosphere in two ways. First, it could cool the layer, taking the material well below its solidus. Second, it could homogenize the layer by carrying very depleted material down and bringing less depleted material up to a depth where it would melt, and be depleted further. If convection were very efficient, then the layer would become so cold that its viscosity would increase significantly and the layer could not flow laterally; thus neither allowing the kind of dynamic subsidence discussed previously nor the suppression of later melting discussed here. What is needed is a little convection, but not too much.

To consider convection of a depleted layer due to cooling from above we have to consider its density structure. As hot asthenosphere first upwells beneath thin lithosphere, an increasing fraction of it may melt. The melt can segregate from the residue at depth and be emplaced at shallow depths. The residue becomes increasingly depleted at shallower depths. Melting tends to first remove the dense aluminous silicate phases, like garnet, and the more iron-rich olivines and pyroxenes. This means that the more depleted the mantle becomes, the lower is its density (e.g., Oxburgh and Parmentier 1977).

The density changes caused by cooling a layer from above result in cold, dense material overlying hotter, less dense mantle. It is this density structure that drives thermal convection. The depletion-related density structure is opposed to that caused by cooling, and so it may suppress thermal convection.

The gradient of depletion-related density with depth depends on the amount and composition of melt as a function of depth during the upwelling of the hot asthenosphere. Neglecting complications in the melting process, a linear profile of depletion-related density with depth is assumed. The following thought experiment illustrates how such a density structure might allow convection only over a limited depth range of the layer. Assume that the difference in depletion-related density from the top to the bottom of a 100-km-thick hot depleted layer is 40 kg/m^3. If the layer top cools by 200°C, its local density increases by about 20 kg/m^3. This means that a blob of cooled material could descend, at most, 50 km before it was surrounded by less depleted but warmer material of the same density. The cold downwelling could descend no further. Thus, the convection should be strongly depth limited within a layer or layers of the depleted asthenosphere. Since the vigor and cooling efficiency of convection depends strongly on the thickness of the convecting layer, this would lead to less vigorous cooling.

A layer within the pooled depleted asthenosphere might convect until cooling increased the viscosity of the layer. It is interesting to estimate the viscosity at which such convection would cease to determine if that layer would be weak enough to later flow in response to lateral pressure gradients. The Rayleigh number offers a measure of whether a layer of thickness, l, and viscosity, μ, can convect, and for thermal convection can be expressed as

$$Ra = \frac{\alpha g \rho \Delta T l^3}{\mu \kappa_t} \tag{1.10}$$

where α is the thermal expansion coefficient, ΔT is the temperature difference across the layer driving the convection, and κ_t is the thermal diffusivity. It is assumed that compositional density variations mainly limit the convecting layer thickness and the effective temperature drop driving convection. If the Rayleigh number is larger than a critical value (~1,000), then the layer convects. If it is less than the critical value, then the layer does not convect. Taking $l = 50$ m, $\alpha = 3 \times 10^{-5}$, $\kappa_t = 10^{-6}$ m^2/s, and $= 100°C$ means that the viscosity, μ, would be about 10^{19} Pa s when the Rayleigh number was just critical. Convection should cease if the layer cooled further and viscosity increased.

Therefore, convection in a hot, pooled, depleted layer of asthenosphere could homogenize and cool the layer so that later rifting might lead to the kind of dynamic flow pictured in figure 1.11. During rifting, the shallowest part of the asthenosphere under the active rift would be the highly depleted asthenosphere that would not melt for moderately large stretching-induced upwelling. This would suppress melting of assumed "normal" asthenosphere, which would be pulled up below the depleted asthenosphere. The underlying, undepleted asthenosphere would not reach a shallow enough depth for significant melting. If the rifting and initial seafloor spreading were moderately slow, then depleted asthenosphere at shallow levels under the new rift could be frozen in place. This would prevent later melting of "normal" asthenosphere as the depleted layer thinned by flow.

The kind of amagmatic rifting seen for the Iberia margin may have been a result of two stages of rifting. In the first stage, the region might have been affected by the magmatic rifting of North America from Africa in the late Jurassic and early Triassic (e.g., Keleman and Holbrook 1995). The melting associated with the 201 Ma basalts associated with that rifting were emplaced mainly south of the Iberia margin, but depleted asthenosphere might have flowed north beneath the later Iberian margin. The Iberian rifting, beginning at roughly 135 Ma (Whitmarsh et al. 1990), would then have involved rifting of lithosphere overlying highly depleted mantle asthenosphere that did not melt as it flowed to shallower depths under new seafloor.

Summary

This chapter considered some possible effects on rifts of upwelling of hotter-than-normal (possibly plume-generated) mantle asthenosphere. Such upwelling could promote rifting by generating large volumes of magma that split the lithosphere much more easily than it could have been tectonically stretched. More speculative still are the possible effects of the residue of such massive melting events. Flow of that depleted and cooled residue could contribute to the observed "extra" subsidence seen at many continental margins and it may even contribute to rifting with very little magma production. One hopes that the simple calculations described here will encourage development of more rigorous and general rifting models and more careful comparison of model predictions to observed features of rifts and margins.

Acknowledgments

Thanks to Dale Sawyer, Carolyn Ruppel, Leigh Royden, Garry Karner, Neal Driscoll, and David Kohlstedt for many helpful comments that greatly improved this chapter. Support for this work came from National Science Foundation grants EAR 98–14576 and OCE 98–19866. Lamont Contribution 6459.

References

Anderson, E. M. 1951. *The Dynamics of Faulting*. Edinburgh: Oliver and Boyd.

Bassi, G. 1991. Factors controlling the style of continental rifting; insights from numerical modelling. *Earth Planet. Sci. Lett.* 105(4):430–452.

Bird, P. 1991. Lateral extrusion of lower crust from under high topography, in the isostatic limit. *J. Geophys. Res.* 96(6):10275–10286.

Block, L. and L. Royden. 1990. Core complex geometries and regional scale flow in the lower crust. *Tectonics* 9(4):557–567.

Bott, M. H. P. 1991. Ridge push and associated plate interior stress in normal and hot spot regions. *Tectonophysics* 200:17–32.

Brace, W. F. and D. L. Kohlstedt. 1980. Limits on lithospheric stress imposed by laboratory experiments. *J. Geophys. Res.* 85:6248–6252.

Braun, J. and C. Beaumont. 1989. A physical explanation of the relation between flank uplifts and the breakup unconformity at rifted continental margins. *Geology* 17:760–764.

Buck, W. R. 1991. Modes of continental lithospheric extension. *J. Geophys. Res.* 96(12):20161–20178.

Buck, W. R., F. Martinez, M. S. Steckler, and J. R. Cochran. 1988. Thermal consequences of lithospheric extension: Pure and simple. *Tectonics* 7:213–234.

Buck, W. R. and A.N.B. Poliakov. 1998. Abyssal hills formed by stretching oceanic lithosphere. *Nature* 392:272–275.

Byerlee, J. D. 1978. Friction of rocks. *Pure Appl. Geophys.* 116:615–626.

Campbell, I. H. and R. W. Griffiths. 1990. Implications of mantle plume structure for the evolution of flood basalts. *Earth Planet. Sci. Lett.* 99:79–93.

Cathles, L. M. 1975. *The Viscosity of the Earth's Mantle*. Princeton, NJ: Princeton University Press.

Chery, J., F. Lucazeau, M. Daignieres, and J. P. Vilotte. 1992. Large uplift of rift flanks; a genetic link with lithospheric rigidity? *Earth Planet. Sci. Lett.* 112(1–4):195–211.

Cochran, J. R., J. A. Goff, A. Malinverno, D. J. Fornari, C. Keeley, and X. Wang. 1993. Morphology of a "superfast" mid-ocean ridge crest and flanks: The East Pacific Rise 7°–9°S. *Mar. Geophys. Res.* 15:65–75.

Coney, P. J. and T. A. Harms. 1984. Cordilleran metamorphic core complexes: Cenozoic extensional relics of Mesozoic compression. *Geology* 12(9):550–554.

Davis, G. A. and G. A. Lister, eds. 1988. Detachment faulting in continental extension; Perspectives from the southwestern U.S. codillera. Processes in Continental Lithospheric Deformation. Boulder, CO: Geological Society of America.

Dixon, T. H., E. R. Ivins, and B. J. Franklin. 1989. Topographic and volcanic asymmetry around the Red Sea: Constraints on rift models. *Tectonics* 8:1193–1216.

Driscoll, N. W. and G. D. Karner. 1998. Lower crustal extension across the Northern Carnarvon basin, Australia: Evidence for an eastward dipping detachment. *J. Geophys. Res.* 103(B3): 4975–4991.

Dunbar, J. A. and D. S. Sawyer. 1989. How pre-existing weaknesses control the style of continental breakup. *J. Geophys. Res.* 94:7278–7292.

Ebinger, C. J., A. L. Dieno, R. E. Drake, and A. L. Tesha. 1989. Chronology of volcanism and rift basin propagation: Rungwe volcanic province, East Africa. *J. Geophys. Res.* 94:15585–15803.

Einarsson, P. and B. Brandsdottir. 1980. Seismological evidence for lateral magma intrusion during the July 1978 deflation of the Krafla volcano in NE-Iceland. *J. Geophys.* 47:160–165.

England, P. 1983. Constraints on extension of continental lithosphere. *J. Geophys. Res.* 88:1145–1152.

Etheridge, M. A., P. A. Symonds, and G. A. Lister. 1989. Application of the detachment model to reconstruction of conjugate passive margins. In A. J. Tankard and H. R. Balkwill, eds., *Extensional Tectonics and Stratigraphy of the North Atlantic Margins*. Memoir, vol. 46, pp. 23–40. Tulsa, OK: American Association of Petroleum Geologists.

Forsyth, D. W. and S. Uyeda. 1975. On the relative importance of the driving forces of plate motion. *Geophys. J. R. Astron. Soc.* 43(1):163–200.

Gans, P. 1987. An open-system, two layer crustal stretching model for the eastern Great Basin. *Tectonics* 6(1–12):1987.

Gaulier, J. M., X. LePichon, N. Lyberis, F. Avedik, L. Geli, I. Moretti, A. Deschamps, and S. Hafez. 1988. Seismic study of the crustal thickness, Northern Red Sea and Gulf of Suez. *Tectonophysics.* 153:55–88.

Goetze, C. and B. Evans. 1979. Stress and temperature in the bending lithosphere as constrained by experimental rock mechanics. *Geophys. J. R. Astron. Soc.* 59:463–478.

Griffiths, R. W. and I. H. Campbell. 1990. Stirring and structure in starting mantle plumes. *Earth Plant. Sci. Lett.* 99:66–78.

Hill, R. I. 1991. Starting plumes and continental break-up. *Earth Planet. Sci. Lett.* 104:398–416.

Hinz, K. 1981. A hypothesis on terrestrial catastrophes: Wedges of very thick oceanward dipping layers beneath passive margins. *Geol. Jahrb. Reihe E.* 22:3–28.

Holbrook, W. S. and P. B. Kelemen. 1993. Large igneous province on the US Atlantic margin and implications for magmatism during continental breakup. *Nature* 364:433–436.

Hopper, J. and W. R. Buck. 1993. The initiation of rifting at constant tectonic force: The role of diffusion creep. *J. Geophys. Res.* 98:16213–16221.

Hubbert, M. K. and W. Rubey. 1959. Role of fluid pressure in mechanics of over-thrust faulting. Pts. I and II. *Geol. Soc. Am. Bull.* 70:115–205.

Illies, J. H. and G. Greiner. 1978. Rhinegraben and the Alpine system. *Geol. Soc. Am. Bull.* 89:770–782.

Ito, J., J. Lin, and C. W. Gable. 1996. Dynamics of mantle flow and melting at a ridge-centered hotspot: Iceland and the Mid-Atlantic Ridge. *Earth Planet. Sci. Lett.* 144:53–74.

Jarrige, J.-J., P. O. D'Estevou, P. F. Burollet, C. Montenat, P. Prat, J.-P. Richert, and J. P. Thiriet. 1990. The multistage tectonic evolution of the Gulf of Suez and the Northern Red Sea continental rift from field observations. *Tectonics* 9(3):441–465.

Karner, G. D. and N. W. Driscoll. 1999. Style, timing and distribution of tectonic deformation across the Exmouth Plateau, northwest Australia, determined from stratal architecture and quantitative basin modeling. In C. Mac Niocaill and P. D. Ryan, eds., *Continental Tectonics*, Geological Society Special Publication 164, pp. 271–311. London, UK: Geological Society of London.

Kelemen, P. B. and W. S. Holbrook. 1995. Origin of thick, high-velocity igneous crust along the U.S. East-Coast Margin. *J. Geophys. Res.* 100(B6):10077–10094.

Kirby, S. H. and A. K. Kronenberg. 1987. Rheology of the lithosphere: Selected topics. *Rev. Geophys.* 25:1219–1244.

Klein, E. M. and C. H. Langmuir. 1987. Global correlations of ocean ridge basalt chemistry with axial depth and crustal thickness. *J. Geophys. Res.* 92:8089–8115.

Kusznir, N. J. and R. G. Park. 1987. The extensional strength of the continental lithosphere: Its dependence on geothermal gradient, and crustal composition and thickness. In M. P. Coward, J. F. Dewey, and P. L. Hancock, eds., *Continental Extensional Tectonics,* Geological Society Special Publication 28, pp. 35–52. London, UK: Geological Society of London.

Lachenbruch, A. H. and J. H. Sass. 1978. Models of an extending lithosphere and heat flow in the Basin and Range Province. In R. B. Smith and G. P. Eaton, eds., *Cenozoic Tectonics and Regional Geophysics of the Western Cordillera*, vol. 152, pp. 209–250. Boulder, CO: Geological Society of America.

Lavier, L., W. R. Buck, and A. N. B. Poliakov. 2000. Factors controlling normal fault offset in an ideal brittle layer. *J. Geophys. Res.* 105:23431–23442.

LePichon, X. and J.-M. Gaulier. 1988. The rotation of Arabia and the Levant fault system. *Tectonophysics* 153:271–294.

LePichon, X. and J.-C. Sibuet. 1981. Passive margins: A model of formation. *J. Geophys. Res.* 86:3708–3720.

Lister, G. A., M. A. Etheridge, and P. A. Symonds. 1991. Detachment models for the formation of passive continental margins. *Tectonics* 10: 1038–1064.

Lister, G. S., M. A. Etheridge, and P. A. Symonds. 1986. Detachment faulting and the evolution of passive continental margins. *Geology* 14: 246–250.

Lister, J. R. and R. C. Kerr. 1991. Fluid-mechanical models of crack propagation and their application to magma transport in dykes. *J. Geophys. Res.* 96:10049–10077.

Maggi, A., J. A. Jackson, D. McKenzie, and K. Priestley. 2000. Earthquake focal depths, effective elastic thickness, and the strength of the continental lithosphere. *Geology (Boulder)* 28:495–498.

Martinez, F. and J. R. Cochran. 1988. Structure and tectonics of the northern Red Sea: Catching a continental margin between rifting and drifting. *Tectonophysics* 150:1–32.

McKenzie, D. P. 1978. Some remarks on the development of sedimentary basins. *Earth Planet. Sci. Lett.* 40:25–32.

Menzies, M. A., J. Baker, D. Bosence, C. Dart, I. Davison, A. Hurford, M. Al'Kadasi, K. McClay, A. Al'Kadasi, and A. Yelland. 1992. The timing of magmatism, uplift and crustal extension: Preliminary observations from Yemen. In B. C. Storey, T. Alabaster, and R. J. Pankhurst, eds., *Magmatism and the Causes of Continental Break-up*. Geological Society Special Publication 68, pp. 293–304. London, UK: Geological Society of London.

Mohr, P. 1992. Nature of the crust beneath magmatically active continental rifts. *Tectonophysics* 213:269–284.

Morgan, P. 1982. Heat flow in rift zones in continental and oceanic rifts. In G. Palmason, ed., AGU Geodynamics Series, no. 8, pp. 107–122. Washington, DC: American Geophysical Union.

Morgan, W. J. 1971. Convection plumes in the lower mantle. *Nature* 230:42–43.

Mutter, J. C., G. A. Barth, P. Buhl, R. S. Detrick, J. Orcutt, and A. Harding. 1988. Magma distribution across ridge-axis discontinuities on the East Pacific Rise from multichannel seismic images. *Nature* 336:156–158.

Omar, G. I. and M. S. Steckler. 1995. Fission track evidence on the initial rifting of the Red Sea: Two pulses, no propagation. *Science* 270:1341–1344.

Omar, G. I., M. S. Steckler, W. R. Buck, and B. P. Kohn. 1989. Fission-track analysis of basement apatites at the western margin of the Gulf of Suez rift, Egypt: Evidence for sychroneity of uplift and subsidence. *Earth Planet. Sci. Lett.* 94:316–328.

Oxburgh, E. R., and E. M. Parmentier. 1977. Compositional and density stratification in oceanic lithosphere: Causes and consequences. *J. Geol. Soc.* 133:343–355.

Pallister, J. S. 1987. Magmatic history of Red Sea rifting: Perspective from the central Saudi Arabia coastal plain. *Geol. Soc. Am. Bull.* 98:400–417.

Parsons, B. and J. G. Sclater. 1977. Ocean floor bathymetry and heat flow. *J. Geophys. Res.* 82:803–827.

Ribe, N., U. R. Christensen, and J. Theissing. 1995. The dynamics of plume-ridge interaction, 1: Ridge-centered plumes. *Earth Plant. Sci. Lett.* 134:155–168.

Richards, M. A., R. A. Duncan, and V. E. Courillot. 1989. Flood basalts and hot-spot tracks: Plume heads and tails. *Science* 246:103–107.

Ritzert, M. and W. R. Jacoby. 1985. On the lithospheric seismic structure of Reykjanes Ridge at 62.5 degrees N. *J. Geophys. Res.* 90(12):10117–10128.

Rosendahl, B. R. 1987. Architecture of continental rifts with special reference to East Africa. *Annu. Rev. Earth Sci.* 15:443–503.

Royden, L. and C. E. Keen. 1980. Rifting process and thermal evolution of the continental margin of eastern Canada determined from subsidence curves. Earth Planet. Sci. Lett. 51:343–361.

Royden, L., J. G. Sclater, and R. P. Von Herzen. 1980. Continental margin subsidence and heat flow: Important parameters in formation of petroleum hydrocarbons. *Am. Assoc. Petrol. Geol. Bull.* 64:173–187.

Rubin, A.M. and D. D. Pollard. 1987. *Origins of Blake-Like Dikes in Volcanic Rift Zones*, U.S. Geological Survey Professional Paper 1350, pp. 1449–1470.

Sengor, A.M.C. and K. Burke. 1978. Relative timing of rifting and volcanism on Earth and its tectonic implications. *Geophys. Res. Lett.* 5:419–421.

Sleep, N. H. 1990. Hotspots and mantle plumes: Some phenomenology. *J. Geophys. Res.* 95:6715–6736.

Small, C., ed. 1998. *Global Systematics of Mid-Ocean Ridge Morphology.* AGU Monograph: Faulting and Magmatism at Mid-Ocean Ridges. Washington, DC:American Geophysical Union.

Solomon, S. C., R. M. Richardson, and E. A. Bergman. 1980. Tectonic stresses: Models and magnitudes. *J. Geophys. Res.* 85:6086–6092.

Sonder, L.J.P.C.E., B. P. Wernicke, and R. L. Christiansen. 1987. *A Physical Model for Cenozoic Extension of Western North America.* Conference on Continental extensional tectonics, Durham, UK, April 18–20, 1985, Geological Society Special Publication 28, pp.187–201. London, UK: Geological Society of London.

Spohn, T. and G. Schubert. 1982. Convective thinning of the lithosphere: A mechanism for this initiation of continental rifting. *J. Geophys. Res.* 87:4669–4681.

Starostenko, V. I., V. A. Danilenko, D. B. Vengrovitch, R. I. Kutas, S. M. Stovba, R. A. Stephenson, and O. M. Kharitonov. 1999. A new geodynamical-thermal model of rift evolution, with application to the Dnieper-Donets Basin, Ukraine. *Tectonophysics* 313(1–2):29–40.

Steckler, M. S. 1985. Uplift and extension in the Gulf of Suez, indications of induced mantle convection. *Nature* 317:135–139.

Steckler, M. S. and A. B. Watts. 1981. Subsidence history and tectonic evolution of Atlantic-type continental margins. In R. A. Scrutten, ed., *Dynamics of Passive Margins,* AGU Geodynamics Series, no. 6, pp. 184–196. Washington, DC: American Geophysical Union.

Talwani, M., C. C. Windisch, and M. G. Langseth. 1971. Reykjanes Ridge Crest: A detailed geophysical study. *J. Geophys. Res.* 76:473–517.

Tryggvason, E. 1984. Widening of the Kafka Fissure Swarm during the 1975–1981 volcano-tectonic episode. *Bull. Volcanol.* 47(1):47–69.

van Wijk, J. W. and S. A. P. L. Cloetingh. 2002. Basin migration caused by slow lithospheric extension. *Earth Planet. Sci. Lett.* 198(3–4):275–288.

Wegener, A. 1929. *The Origin of Continents and Oceans.* London, UK: Methuen.

Wernicke, B. 1985. Uniform-sense normal simple shear of the continental lithosphere. *Can. J. Earth Sci.* 22:108–125.

White, R. S. and D. McKenzie. 1989. Magmatism at rift zones: The generation of volcanic continental margins and flood basalts. *J. Geophys. Res.* 94:7685–7729.

Whitmarsh, R. B., P. R. Miles, and A. Mauffret. 1990. The ocean-continent boundary off the western continental margin of Iberia: Part 1, Crustal structure at 40 degrees 30′ N. *Geophys. J. Int.* 103(2):509–531.

Zorin, Y. A. 1981. The Baikal Rift: An example of the intrusion of asthenospheric material into the lithosphere as the cause of the disruption of lithospheric plates. *Tectonophysics* 73:91–104.

CHAPTER TWO

Velocity Fields, Faulting, and Strength on the Continents

James Jackson

Problems in Continental Tectonics

The simple concepts of plate tectonics, in which the deformation of the ocean basins is adequately described by the relative motions of rigid blocks, are not easily applicable in continental tectonics, where the deformation is usually much more diffuse than in the oceans and is not restricted to narrow plate boundaries (McKenzie 1972; Molnar and Tapponnier 1975). A different framework is therefore needed within which to view continental deformation. Within the broad deforming belts on the continents some large, flat, aseismic regions such as central Turkey, central Iran, and the Tarim basin appear to be rigid and can usefully be thought of as "microplates" (McKenzie and Jackson 1984; Avouac and Tapponnier 1993; McClusky et al. 2000). But in most continental areas the scale on which the active deformation and its consequent topographic features, such as mountain belts, plateaus, and basins, are distributed makes it more practical to describe the overall characteristics of that deformation by a velocity field, rather than by the relative motions of rigid blocks (England and Jackson 1989). An important problem is then to obtain this velocity field and understand its relation to the motions of the rigid plates that bound the deforming region. A major advance of the past decade has been in estimating such velocity fields, either directly from GPS measurements (Clarke et al. 1998; McClusky et al. 2000), from spatial variations in strain rates estimated from seismicity (Holt et al. 1991; Jackson et al. 1992), or from fault slip rates (England and Molnar 1997a).

Most of the velocity fields obtained for large areas look as though they vary smoothly, either because the techniques used to acquire them involve smoothing the earthquake or fault slip data (Holt et al. 1991; England and Molnar 1997a), or because the geodetic stations are widely spaced (McClusky et al. 2000) and usually span time intervals that include few major earthquakes, so that motion on faults is represented by their smooth interseismic elastic strain fields. Thus, regional velocity fields rarely show discontinuities associated with faults and can be described by continuous functions such as polynomials or splines, which in turn opens their scrutiny and analysis to the language of continuum mechanics (Holt and Haines 1993). These continuous velocity fields are often assumed to represent the average deformation of the whole lithosphere, which consists of a relatively thin (typically 10–20 km) seismogenic layer that deforms by faulting above a much thicker (80–100 km) layer that deforms by more distributed creep. This assumption is, to some extent, an act of faith and it is important to realize that a scale is imposed by this reasoning: the continuum description is only likely to be a reasonable approximation to the behavior of the whole lithosphere at length scales much larger than the thickness of the seismogenic upper crust (10–20 km), within which long-term deformation is discontinuous across faults.

So a velocity field that describes the average, or long-wavelength, deformation is only a partial description of what is happening because it does not describe the detailed and, in the long-term, discontinuous deformation of the seismogenic layer. An additional problem is then to understand how this faulting in the seismogenic layer is able to accommodate the velocity field that describes the deformation of the lithosphere as a whole.

The two fundamental questions in continental kinematics are therefore: (1) What is the continuous velocity field that describes the average deformation at large length scales?, and (2) How is that velocity field achieved by discontinuous slip on faults? If we can answer these questions, we can pose a third: (3) What is the relation between the two? Because all our observations of faulting are restricted to the top 15–20 km of the lithosphere, and because the lower 100 km of the lithosphere is assumed to deform by a more distributed ductile flow, any attempt to address this question leads to another: (4) What controls the deformation of the continents—the strength of the upper crustal blocks and their interactions, or the flow in the lower lithosphere? This is one of the issues discussed by this chapter, but it has proved difficult to grasp and is not yet properly understood. It calls into question our view of where long-term strength resides within the lithosphere and what controls it. This too is a controversial issue and is briefly reviewed. The concepts discussed here are illustrated with examples from the active tectonics of central Greece, where many of them originated. (See Jackson 1994 and Goldsworthy et al. 2002 for reviews.)

Faulting, Flow, and Rotations

Distributed Faulting and Velocity Fields

The way in which discontinuous faulting in the upper crust accommodates regional velocity fields is not always straightforward, as illustrated by the case of central Greece. The earthquakes in figure 2.1 show how northeast-southwest right-lateral strike-slip faulting in the northern Aegean Sea changes into a system of east-west normal faults in central Greece. The strike-slip faulting represents a continuation of the right-lateral shear on the North Anatolian fault system of Turkey, allowing the southwest motion of the southern Aegean relative to Eurasia. The strike-slip

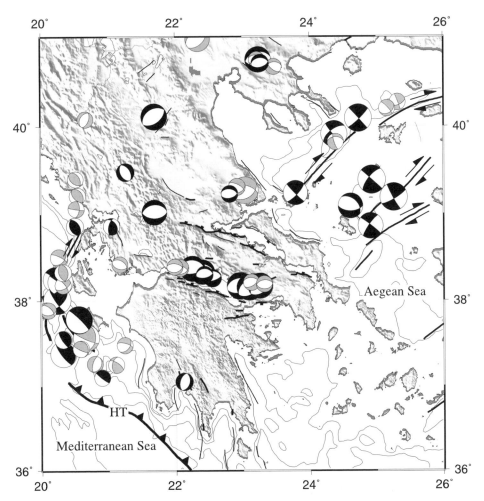

Figure 2.1 Fault plane solutions in central Greece and the western Aegean. Black focal spheres are those constrained by body wave modeling. Gray focal spheres are Harvard CMT solutions for additional earthquakes with $M_w \geq 5.3$ and with more than 70% double-couple component. Plio-Quaternary and active fault trends are marked, with those thought to be currently active shown in thicker lines. HT is the Hellenic Trench. Bathymetric contours are shown at 500, 1,000, 2,000, and 3,000 m. Note how the northeast-southwest right-lateral strike-slip faults in the northern Aegean change abruptly to east-west normal faults in central Greece.

motion in the northern Aegean must connect with the thrusting in the Hellenic Trench to the southwest, or central Greece would be shortening, not extending (McKenzie 1972). Thus, the overall deformation in central Greece must involve a distributed northeast-southwest right-lateral shear, which can be seen in the velocity fields obtained from the seismicity (Jackson et al. 1992) and, more accurately, from GPS measurements (figure 2.2b; Clarke et al. 1998; McClusky et al. 2000). But this shear is not achieved by northeast-southwest strike-slip faults

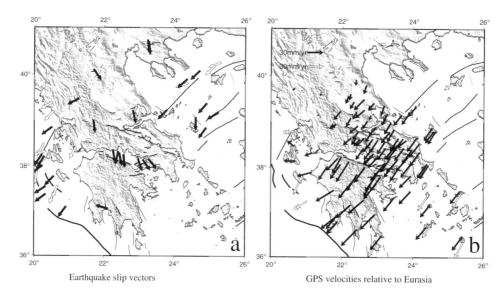

Earthquake slip vectors   GPS velocities relative to Eurasia

Figure 2.2 (a) Slip vectors for earthquakes shown in (a), with black arrows from body wave solutions and white arrows from CMT solutions. Note the abrupt change in slip-vector direction where the strike-slip faulting changes to normal faulting in central Greece. (b) Velocities relative to Eurasia, determined by GPS. Black arrows are from Clarke et al. (1998), white arrows are from McClusky et al. (2000). Note that the change in slip-vector direction in central Greece is not seen in the velocity azimuths.

crossing central Greece; instead, it is achieved by normal faulting. In a conventional transform fault setting, the slip vectors on the strike-slip and normal faults would be in the same direction, but in mainland Greece they are very different (figure 2.2a), with those on the active normal faults having azimuths in the range south to south-southeast rather than southwest (Taymaz et al. 1991). How can slip vectors to the south achieve an overall motion to the southwest? The answer must be that the fault-bounded blocks rotate clockwise as they move, an inference that is supported by paleomagnetic measurements on Late Miocene and Pliocene rocks in the region (Kissel and Laj 1988). Various simplistic block models can be used to illustrate the geometric relations between the faulting and the velocity field (McKenzie and Jackson 1983, 1986; Taymaz et al. 1991; Goldsworthy et al. 2002). The point to emphasize here is that it is the slip vectors on the normal faults, apparently in the wrong direction to achieve the overall motion, which indicate that rotations about vertical axes must occur.

Similar arguments have been made elsewhere. In eastern Iran and in eastern Tibet the presence of east-west left-lateral faults in regions of probable north-south right-lateral shear strongly suggests that the fault-bounded blocks rotate clockwise (Jackson and McKenzie 1984; England and Molnar 1990). In the Western Transverse Ranges of California east-west thrust faults have slip vectors directed north to northeast, in the wrong direction to accommodate the northwest-southeast right-lateral shear that dominates the regional velocity field (Jackson and Molnar 1990;

Molnar and Gipson 1994). Once again, clockwise rotations, which in the Western Transverse Ranges are well documented by paleomagnetic declinations (Hornafius et al. 1986), can reconcile the faulting with the velocity field.

No-Length-Change Directions and Faulting

We can pursue the relationship between faulting and the velocity field further. If we know the velocity field, we know the distribution of strain rates ($\dot{\varepsilon}$) everywhere. (If the velocities have been measured by GPS, there is an intermediate step, which is to fit a continuous surface to the points where the velocities have been measured, to allow interpolation.) We can then ask the question: What orientation and type of faulting can produce this strain-rate tensor? If the faulting occurs in regular subparallel sets in which along-strike variations in the magnitude of slip on the faults is unimportant, then Holt and Haines (1993) showed that there are two possible strike directions of faulting that can produce the horizontal components of the strain rate tensor. These are directions of zero-length change in the velocity field, and they correspond to the strikes of the two nodal planes in an earthquake fault plane solution. In terms of the elements of the strain rate tensor, these strikes are

$$\tan \theta_f = \frac{-\dot{\varepsilon}_{xy} \pm \sqrt{\dot{\varepsilon}_{xy}^2 - \dot{\varepsilon}_{xx}\dot{\varepsilon}_{yy}}}{\dot{\varepsilon}_{yy}} \tag{2.1}$$

where θ_f is the strike of the fault, and x and y are orthogonal horizontal axes. This result is correct only if

$$\dot{\varepsilon}_{xy}^2 \geq \dot{\varepsilon}_{xx}\dot{\varepsilon}_{yy},$$

which is equivalent to requiring that the principal horizontal strain rates have opposite signs, or that one is zero. Pure contraction and pure extension require two sets of faults to be active, but other strain fields require only one set.

So, as long as the faulting is locally organized into subparallel sets, which is one of the fundamental characteristics of most regions of distributed continental tectonics (McKenzie 1972; Molnar and Tapponnier 1975), there will be two different orientations of faulting that can achieve the observed strain-rate field. This result is illustrated by the cases of eastern Iran and eastern Tibet, where north-south right-lateral shear could be accommodated by north-south right-lateral strike-slip faults but is instead accommodated by east-west left-lateral faults that rotate clockwise.

Analyses of this sort are helpful in understanding how the active faulting accommodates the present-day (instantaneous) velocity field but, as we shall see, run into problems when confronted with large finite strains.

Faulting, Rotations, and Flow

In the late 1970s it became apparent from paleomagnetic studies that large systematic rotations of crustal blocks about vertical axes, of the sort described in the previous section, were a common feature of distributed deformation on the continents (Beck 1976; Luyendyk et al. 1980). The way in which such rotations are achieved by faulting became a subject of great interest, because it potentially contains information about whether the deformation of the upper crust in such regions is controlled by the strength of the rigid fault-bounded blocks within it or by distributed flow in the aseismic lithosphere beneath (McKenzie and Jackson 1983; England and Jackson 1989). This is clearly a fundamental question in continental tectonics, but one that has proved remarkably difficult to address.

We would like to know how (or if) the rotation of these blocks is related to the velocity field that describes the 'flow' of the lithosphere at long wavelengths. The simplest scheme to imagine is that of a rigid circular disc floating alone in a fluid that is sheared. If the velocity gradient across the deforming zone of width a is U/a, then a circular disc rotates at a rate of $-U/2a$ about a vertical axis, where U/a is the vorticity ($\dot{\omega}$) of the fluid (McKenzie and Jackson 1983). The rotation rates of blocks with a more general shape depend on their shape and on their orientation (Lamb 1987); for example, a stick aligned parallel with the shearing flow will not rotate, whereas one oriented across the flow will rotate. To examine whether fault-bounded blocks behave as if they were equi-dimensional or elongated inclusions in a fluid, or whether their motions are dominated by their interaction with other blocks, we need to know their present-day rotation rates to better than a factor of two, even in very simplistic models (McKenzie and Jackson 1983). Such accuracy is rarely possible with paleomagnetism, because there are errors in the measurements of declination (typically $\pm 10°$), and the rotation could have occurred at any time since the rock was magnetized. Some hope exists that GPS measurements can give better rotation rates, but there is the difficulty of knowing whether observed velocity gradients on the scale of fault-bounded blocks reflect elastic (and recoverable) strain or even short-term postseismic transients, rather than permanent motions.

Jackson et al. (1992) followed the lines of argument described previously in central Greece and found (a) that the zero-length-change directions in the velocity field corresponded roughly to the observed fault strikes, and (b) that elongated blocks of that orientation driven by flow on their bases should indeed rotate in the sense and approximate rate inferred from the paleomagnetic data. Here, apparently, was the beginning of a coherent picture, illustrated schematically in figure 2.3. The northeast-southwest right-lateral shear necessary for the motion of central Aegean and Turkey relative to Eurasia could have passed through Greece as a single strike-slip fault (figure 2.3a). But for reasons unknown, though probably related to the reactivation of previous early or middle Tertiary deformation fabrics, the active fault system changes from a northeast-southwest strike to an east-west strike in central Greece (figure 2.1). This fault system could still accommodate the northeast-southwest shear in a simple way if the slip vectors on the east-west

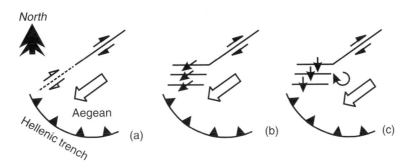

Figure 2.3 Sketches to illustrate the accommodation of southwest movement of the southern Aegean (white arrow) relative to Europe by faulting in central Greece. The northeast-southwest right-lateral strike-slip faulting in the northern Aegean (black line) could have continued across central Greece (a), but instead is seen to change to normal faulting with an east-west strike. Such normal faults could accommodate the overall motion with slip vectors directed southwest, parallel to those on the strike-slip faults, as is common in oceanic ridge-transform systems (b). But instead the normal faults have slip vectors directed south, and must rotate clockwise (c).

faults remained parallel to those on the strike-slip faults and were directed southwest, as in figure 2.3b (and all oceanic ridge-transform intersections). If, however, blocks oriented oblique to the flow are obliged to rotate clockwise in response to the forces on their bases, then the slip vectors on the faults must adjust accordingly to allow for this, as in figure. 2,3c.

One criticism with this analysis is that the blocks do not need to be driven by forces on their bases for this scheme to work, and Goldsworthy et al. (2002) produced a simple block model to illustrate a fault geometry and kinematics similar to that observed in figure 2.3c, but in which the blocks are driven by forces on their ends. A second, and more profound, problem is associated with how such fault configurations accommodate finite deformations and is discussed in the next section.

Another attempt to resolve this question of what drives block motions in the upper crust was undertaken by Bourne et al. (1998a,b), who examined velocity profiles (measured by GPS) across systems of parallel strike-slip faults in New Zealand and California. In both cases the velocity profiles measured perpendicular to the faults showed that the total Pacific–Australia or Pacific–North America plate motion was accommodated across the zone. Bourne et al. (1998a,b) then fitted a smooth function to those profiles and showed that, assuming this function represented the velocity in the underlying creeping lithosphere, each fault-bounded block (whose margins were defined by the known faults) moved with the average velocity of the flow beneath it. Furthermore, the velocities on the faults between the blocks estimated in this way agreed with the longer-term fault slip rates estimated from late Quaternary geological offsets or trenching. These results were initially interpreted to imply that the fault-bounded blocks were driven principally by traction on their bases (Bourne et al. 1998a). But, as Bourne et al. (1998b) pointed out, the observations are ambiguous, as the fault spacing is sufficiently close that the observed velocity profiles could also represent elastic accumulation of strain between the blocks, rather than the long-term velocity in the flowing lithosphere beneath. This later interpretation indeed seems more likely if, as the geological slip estimates imply, the total plate motion is accommodated completely by permanent and discontinuous movement on the faults, because then any short-term deformation between the faults can only be elastic and recoverable. In other words, the same observations would result if the blocks were driven by forces on their ends rather than underneath. Such ambiguity, despite excellent datasets, appears typical of this rather elusive question.

Finite Deformation

Serious difficulties arise when we try to understand the relations between faulting and large finite strains, in particular, in regions where the fault-bounded blocks are known to rotate about vertical axes. The difficulty arises because the relation between the instantaneous velocity field and the faults that can accommodate it is so tight (equation 2.1) that once the faults have rotated into a different orientation, they can no longer do the job; in that new orientation, they would accommodate a different velocity field. This objection applies regardless of whether the blocks are driven by forces on their edges or on their bases, and is simply a consequence of the blocks being rigid. Yet it is clear from the paleomagnetic data in Greece and the Western Transverse Ranges that large finite rotations do occur during the deformation. Thus, at least in places where rotations about vertical axes are important, even though we know both the instantaneous velocity field and how it is achieved by faulting, that is insufficient to see how the finite deformation is achieved. It is therefore unsafe to use the magnitude of finite rotations alone to conclude that the rotating blocks respond principally to forces on their bases (England and Wells 1991; Molnar and Gipson 1994), because this ignores the question of how those rotations are achieved by faulting as the deformation progresses; if the same faults are active throughout, then they cannot accommodate an unchanging velocity field as they rotate. In such places something must change with time; either new faults form or the velocity field changes.

In central Greece there is certainly evidence that the fault pattern has changed in an organized and rapid way. The northwestern Aegean and eastern seaboard of central Greece has a strong northwest-southeast structural fabric inherited from earlier Tertiary compressional or extensional deformation (Gautier et al. 1999). The region contains many large normal faults of this orientation that were active in the Quaternary and dominate the trend of the coastline. Yet, almost without exception, all the normal faulting earthquakes with well-constrained mechanisms show east-west striking nodal planes, cutting across this northwest-southeast trend (figure 2.1; Hatzfeld et al. 1999). This evidence, combined with geomorphological and structural evidence in the field, strongly suggests that activity moved from the northwest-southeast faults to the east-west faults within the Quaternary (Caputo and Pavlides 1993; Leeder and Jackson 1993; Hatzfeld et al. 1999; Jackson 1999; Goldsworthy and Jackson 2001; Goldsworthy et al. 2002). What is remarkable is how quickly, and how completely, this change has occurred.

There is therefore support for the notion that fault patterns can change with time, perhaps in response to rotation about vertical axes. Although there are possible indications that faults can adapt to some extent by changing the direction of slip in their fault planes (figure 2.4; Jackson and McKenzie 1999), each set of faults obviously accommodates a substantial finite offset, which is why they are so prominent in the landscape and geology, before becoming inactive when motion is taken up by the new set of faults (figure 2.5). It is therefore unrealistic to think that faults adapt continually to allow the velocity field to remain unchanged. If the faults account for most of the strain and the blocks they bound remain effectively rigid, then their continued activity while substantial offsets are achieved must have some effect on the velocity field.

On the other hand, it is also unrealistic to think that regional velocity fields are controlled by the local requirements of rigid rotating blocks. The shear through central Greece is ultimately related to the southwest motion of the central Aegean and Turkey, in response to either a push from the high ground in eastern Turkey or to a pull from the Hellenic Trench, or both (McKenzie 1972; Le Pichon 1982). This large-scale movement is unlikely to be altered by the rotation of rigid blocks in central Greece, except on a relatively local scale.

The likelihood is of some compromise between continued activity on major faults that perhaps adapt to rotation, to some extent, by changing their rake, and the creation of new faults when the initial set becomes severely misoriented for the requirements of the large-scale regional velocity field. This is a well-known structural phenomenon that is familiar in extensional regions, where blocks bounded by normal faults are known to rotate about horizontal axes on discrete and sequential generations of subparallel fault sets (Proffett 1977; Morton and Black 1975). These are geometric arguments, however, that have no bearing on whether the motions of the blocks are driven by forces on their bases or not.

Strength of the Continental Lithosphere

Our speculations about the relationship between faulting in the upper crust and the creep in the lower lithosphere are inevitably colored by our beliefs on lithosphere strength. For twenty years the popular view of continental strength profiles has consisted of a weak lower crust sandwiched between relatively strong layers in the upper crust and mantle. Maggi et al. (2000a,b) have challenged this view, which they believe to be incorrect. In a review of continental seismicity they found no evidence for significant earthquakes in the continental mantle, in contrast to earlier studies which thought that occasional, though rare, earthquakes beneath the Moho indicated an important strength contrast between the lower crust and uppermost mantle (Chen and Molnar 1983). Instead, Maggi et al. (2000a,b) found that continental earthquakes are contained within the continental crust: usually

within the upper crust alone, but in some places including the lower crust as well (figure 2.6). In addition they found that variations in the seismogenic thickness T_s correlate with variations in the effective elastic thickness T_e, both of them having similar values, although T_e is usually the smaller of the two. [The reader should be aware that the subject of estimating T_e from gravity and topography is contentious. The conclusions of Maggi et al. (2002a) are based on the analysis of McKenzie and Fairhead (1997) and McKenzie (2003), in which the controversy is discussed.] The simplest interpretation of these results is that the strength of the continental lithosphere is in its seismogenic layer, and that the underlying creeping lithosphere, including the mantle part, is relatively weak. The alternative is to argue that, because earthquakes indicate frictional instability rather than strength, the continental mantle could still be strong despite being aseismic. The problem then is that, because $T_e < T_s$, it is also necessary to argue that the seismically active layer has long-term weakness whereas the aseismic part has long-term strength, otherwise T_e would be greater than T_s. This possibility cannot logically be ruled out, but seems improbable and unnecessarily complicated compared with the simpler view that the long-term strength resides in the seismogenic layer. Maggi et al. (2000a) go on to discuss possible controls on the relative strength of the continental crust and mantle, concluding that some effect besides that of temperature is important, probably water content.

The implications of this revised, and simpler, view of continental strength profiles, if it is correct, are profound and are discussed further by Jackson (2000a,b). With no obvious evidence for significant strength in the continental mantle, little support exists for the view that its behavior dominates the deformation of the lithosphere and that the upper crust responds in a passive way. Indeed, Maggi et al. (2000a) suggest that the height of mountains and plateaus is correlated with the strength of their bounding forelands, which is contained within the seismogenic layer. If, at least in some places, the lower crust is stronger than the upper mantle, it is also necessary to reexamine the conditions under which the lower crust can flow to even out crustal thickness contrasts, as it has manifestly done in some areas, particularly in extensional metamorphic core complexes (Gans 1987). These issues are discussed further by McKenzie and Jackson (2002), who conclude that special circumstances are needed to make the lower crust flow on this scale, such as the intrusion of igneous melts or the addition of water-rich fluids.

◀───

Figure 2.4 Two photos of an active normal fault surface exposed in a quarry at Arkitsa in central Greece (approximately 38.8°N, 23.0°E). About 50 m is exposed in the direction of slip; note the person for scale in (a). The striations and corrugations on the fault surface are gently curved, changing in rake by ~1° between top and bottom of the exposure. As Jackson and McKenzie (1999) point out, this change is in the right sense, and approximately the right amount, if the fault was adjusting to clockwise rotation about a vertical axis by changing the slip vector in its plane, although this interpretation is not necessarily correct.

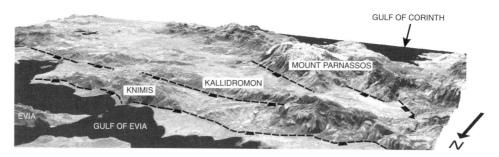

Figure 2.5 Block perspective view across the graben system of Locris in central Greece, centered at about 38.8°N 22.5°E and looking south to the Gulf of Corinth. Three subparallel normal fault systems bound the mountains of Knimis, Kallidromon, and Parnassos. The exposure in figure 2.4 is at the eastern (left) end of the Knimis fault. The geomorphology and stratigraphy suggest that the youngest and most active fault system is that following the coast and bounding Knimis. The other two are older, probably now inactive, and have strikes more northwest-southeast than the Knimis fault, which is approximately east-west. This is consistent with an evolution in which the older faults have been abandoned as a result of clockwise rotation about a vertical axis (see Goldsworthy and Jackson 2000).

Focal Depth Distributions

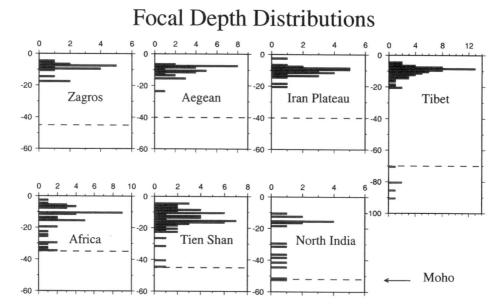

Figure 2.6 Histograms of earthquake centroid depths determined by body wave modeling (see Maggi et al. 2000a). Note that in the Zagros, Aegean, Iranian plateau, and Tibet seismicity is confined to the upper crust, whereas in parts of Africa, the Tien Shan, and north India the entire crustal thickness is active. The gray bars beneath Tibet are earthquakes that some have claimed to be in the mantle (see Maggi et al. 2000b), although the Moho beneath this particular part of southern Tibet (near 29°N, 90°E) is now thought to be at 80–85 km (Yuan et al. 1997), and the earthquakes may well be in the lowermost crust (Jackson 2002b). There is no evidence for significant seismicity in the continental mantle.

Conclusions

At the moment our views of continental tectonics are confused by not knowing what really controls the patterns of deformation we see at the surface. On the one hand, we observe that distributed faulting is usually organized on a regional scale into sets of subparallel faults that are responsible for large, discontinuous offsets and which bound relatively rigid blocks. This simple observation puts severe constraints on the average velocity fields that can be accommodated by such sets of faults, particularly if they rotate as they move. On the other hand, at length scales much larger than the lithosphere thickness, there has been some success in understanding how velocity fields, and even finite deformation in the form of compensated topography, are related to forces on the edges of the lithosphere that arise from plate motions or within the lithosphere that arise from crustal thickness contrasts (England and Houseman 1986; England and Molnar 1997b). The difficulty comes when trying to see how these two scales interact to produce features of interest to tectonic and structural geologists. An important factor is likely to be the relative strengths of the uppermost mantle, the seismogenic crust, and the large faults themselves. Current opinion may be changing toward a view that the upper mantle is relatively weak compared with the seismogenic crust, which, on the basis of seismogenic and elastic thickness variations, can be quite variable itself. Meanwhile, the strength of faults within the upper crust, perhaps the greatest uncertainty of all, remains a topic as controversial today as it was 30 years ago (Scholz 2000).

Acknowledgments

I thank Philip England, John Haines, Bill Holt, Dan McKenzie, Peter Molnar, and Chris Scholz, whose work and insights have strongly influenced my own over the years, though I alone am responsible for any misconceptions or errors presented here. I also thank R. Buck, G. Karner, D. Lizarralde, J. Stock, and B. Taylor for careful reviews. This work is Cambridge Earth Sciences contribution ES 6093.

References

Avouac, J.-P. and P. Tapponnier. 1993. Kinematic model of active deformation in central Asia. *Geophys. Res. Lett.* 20:895–898.

Beck, M. E. 1976. Discordant paleomagnetic pole positions as evidence for regional shear in the western Cordillera of North America. *Am. J. Sci.* 276:694–712.

Bourne, S., P. England, and B. Parsons. 1998a. The motion of crustal blocks driven by flow of the lower lithosphere: implications for slip rates of faults in the South Island of New Zealand and southern California. *Nature* 391:655–659.

Bourne, S. J., T. Arnadottir, J. Beavan, D. J. Darby, P. C. England, B. Parsons, R. I. Walcott, and P. R. Wood. 1998b. Crustal deformation of the Marlborough fault zone in the South Island of New Zealand: geodetic constraints over the interval 1982–1994. *J. Geophys. Res.* 103: 30147–30165.

Caputo, R. and S. Pavlides. 1993. Late Cainozoic geodynamic evolution of Thessaly and surroundings (central-northern Greece). *Tectonophysics* 223:339–362.

Chen, W-P. and P. Molnar. 1983. Focal depths of intra-continental and intraplate earthquakes and their implications for the thermal and mechanical properties of the lithosphere. *J. Geophys. Res.* 88:4183–4214.

Clarke, P., R. R. Davies, P. C. England, B. Parsons, H. Billiris, D. Paradissis, G. Veis, P. A. Cross, P. H. Denys, V. Ashkenazi, R. Bingley, H.-G. Kahle, M.-V. Muller, and P. Briole. 1998. Crustal strain in central Greece from repeated GPS measurements in the interval 1989–1997. *Geophys. J. Int.* 135:195–214.

England, P. C. and G. A. Houseman. 1986. Finite strain calculations of continental deformation: 2, Comparison with the India-Eurasia collision zone. *J. Geophys. Res.* 91:3664–3676.

England, P. and J. Jackson. 1989. Active deformation of the continents. *Annu. Rev. Earth Planet. Sci.* 17:197–226.

England, P. C. and P. Molnar. 1990. Right-lateral shear and rotation as the explanation for strike-slip faulting in eastern Tibet. *Nature* 344:140–142.

England, P. and P. Molnar. 1997a. The field of crustal velocity in Asia calculated from Quaternary rates of slip on faults. *Geophys. J. Int.* 130:551–582.

England, P. and P. Molnar. 1997b. Active deformation of Asia: from kinematics to dynamics. *Science* 278:647–650.

England, P. and R. Wells. 1991. Neogene rotations and quasicontinuous deformation of the Pacific Northwest continental margin. *Geology* 19:978–981.

Gans, P. B. 1987. An open-system, two-layer crustal stretching model for the Eastern Great Basin. *Tectonics* 6:1–12.

Gautier, P., J.-P. Brun, R. Moriceau, D. Sokoutis, J. Martinod, and L. Jolivet. 1999. Timing, kinematics and cause of Aegean extension: a scenario based on a comparison with simple analogue experiments. *Tectonophysics* 315:31–72.

Goldsworthy, M. and J. Jackson. 2001. Migration of activity within normal fault systems: Examples from the Quaternary of mainland Greece. *J. Struct. Geol.* 23:489–506.

Goldsworthy , M., J. Jackson, and J. Haines. 2002. The continuity of active fault systems in Greece. *Geophys. J. Int,* 148:596–618.

Hatzfeld, D., M. Ziazia, D. Kementzetzidou, P. Hatzidimitriou, D. Panagiotopoulos, K. Makropoulos, P. Papadimitriou, and A. Deschamps. 1999. Microseismicity and focal mechanisms at the western termination of the North Anatolian Fault and their implications for continental tectonics. *Geophys. J. Int.* 137:891–908.

Holt, W. E., J. F. Ni, T. C. Wallace, and A. J. Haines. 1991. The active tectonics of the eastern Himalayan syntaxis and surrounding regions. *J. Geophys. Res.* 96:14595–14632.

Holt, W. E. and A. J. Haines. 1993. Velocity field in deforming Asia from inversion of earthquake-released strains. *Tectonics* 12:1–20.

Hornafius, J. S., B. P. Luyendyk, R. R. Terres, and M. J. Kammerling. 1986. Timing and extent of Neogene tectonic rotation in the western Transverse Ranges, California. *Geol. Soc. Am. Bull.* 97:1476–1487.

Jackson, J. 1994. Active tectonics of the Aegean region. *Annu. Rev. Earth Planet. Sci.* 22: 239–271.

Jackson, J. 1999. Fault death: A perspective from actively deforming regions. *J. Struct. Geol.* 21:1003–1010.

Jackson, J. 2002a. Faulting, flow and the strength of the continental lithosphere. *Int. Geol. Rev.* 44:39–61, in press.

Jackson, J. 2002b. Strength of the continental lithosphere: time to abandon the jelly sandwich? *GSA (Geol. Soc. Am.) Today* 12:4–10.

Jackson, J. A. and D. P. McKenzie. 1984. Active tectonics of the Alpine-Himalayan belt between western Turkey and Pakistan, *Geophys. J. R. Astron. Soc.* 77:185–264.

Jackson, J. and D. McKenzie. 1999. A hectare of fresh striations on the Arkitsa Fault, central Greece. *J. Struct. Geol.* 21:1–6.

Jackson, J. and P. Molnar. 1990. Active faulting and block rotations in the western Transverse ranges, California. *J. Geophys. Res.* 95:22073–22087.

Jackson, J. A., A. Haines, and W. Holt. 1992. The horizontal velocity field in the deforming Aegean Sea region determined from the moment tensors of earthquakes. *J. Geophys. Res.* 97:17657–17684.

Kissel, C. and C. Laj. 1988. The Tertiary geodynamic evolution of the Aegean arc: A paleomagnetic reconstruction. *Tectonophysics* 146:183–201.

Lamb, S. H. 1987. A model for tectonic rotations about a vertical axis. *Earth Planet. Sci. Lett.* 84:75–86.

Leeder, M. R. and J. A. Jackson. 1993. The interaction between normal faulting and drainage in active extensional basins, with examples from the western United States and central Greece. *Basin Res.* 5:79–102.

Le Pichon, X. 1982. Land-locked oceanic basins and continental collision: The eastern Mediterranean as a case example. In K. Hsu, ed., *Mountain Building Processes*, pp. 201–211. London: Academic Press.

Luyendyk, B. P., M. J. Kammerling, and R. R. Terres. 1980. Geometric model for Neogene crustal rotations in southern California. *Geol. Soc. Am. Bull.* 91:211–217.

Maggi, A., J. A. Jackson, D. McKenzie, and K. Priestley. 2000a. Earthquake focal depths, effective elastic thickness, and the strength of the continental lithosphere. *Geology* 28:495–498.

Maggi, A., J. A. Jackson, K. Priestley, and C. Baker. 2000b. A re-assessment of focal depth distributions in southern Iran, the Tien Shan and northern India: do earthquakes really occur in the continental mantle? *Geophys. J. Int.* 143:629–661.

McClusky, S., S. Balassanian, A. Barka, C. Demir, S. Ergintav, I. Georgiev, O. Gurkan, M. Hamburger, K. Hurst, H. Kahle, K. Kastens, G. Keklidze, R. King, B. Lotzev, O. Lenk, et al. 2000. Global Positioning System constraints on plate kinematics and dynamics in the eastern Mediterranean and Caucasus. *J. Geophys. Res.* 105:5695–5719.

McKenzie, D. 1972. Active tectonics of the Mediterranean region. *Geophys. J. R. Astron. Soc.* 30:109–185.

McKenzie, D. 2003. Estimating T_e in the presence of internal loads. *J. Geophys. Res.*, in press.

McKenzie, D. and D. Fairhead. 1997. Estimates of effective elastic thickness of the continental lithosphere from Bouguer and free air gravity anomalies. *J. Geophys. Res.* 102:27523–27552.

McKenzie, D. P. and J. A. Jackson. 1983. The relationship between strain rates, crustal thickening, paleomagnetism, finite strain and fault movements within a deforming zone. *Earth Planet. Sci. Lett.* 65:182–202, and correction *ibid.,* 70:444 (1984).

McKenzie, D. and J. Jackson. 1986. A block model of distributed deformation by faulting. *J. Geol. Soc. (Lond.)* 143:249–253.

McKenzie, D. and J. Jackson. 2002. Conditions for flow in the continental crust. *Tectonics* 21(6):1055, doi: 10.1029/2002TC001394.

Molnar, P. and J. M. Gipson. 1994. Very long baseline interferometry and active rotations of crustal blocks in the Western Transverse Ranges, California. *Geol. Soc. Am. Bull.* 106:594–606.

Molnar, P. and P. Tapponnier. 1975. Cenozoic tectonics of Asia: Effects of a continental collision. *Science* 189:419–426.

Morton, W. H. and R. Black. 1975. Crustal attenuation in Afar. In A. Pilger and A. Rosler, eds., *Afar Depression of Ethiopia*, Inter-Union Commission on Geodynamics, Science Report, vol. 14, pp. 55–65. Stuttgart: Schweizerbart'sche Verlangsbuchhandlung.

Proffett, J. M. 1977. Cenozoic geology of the Yerington district, Nevada, and implications for the nature of basin and range faulting, *Geol. Soc. Am. Bull.* 88:247–266.

Scholz, C. H. 2000. Evidence for a strong San Andreas fault. *Geology* 28:163–166.

Taymaz, T., J. Jackson, and D. McKenzie. 1991. Active tectonics of the north and central Aegean Sea. *Geophys. J. Int.* 106:433–490.

Yuan, X., J. Ni., R. Kind, J. Mechie, and E. Sandvol. 1997. Lithospheric and upper mantle structure of southern Tibet from a seismological passive source experiment. *J. Geophys. Res.* 102: 27491–27500.

CHAPTER THREE

Mechanics of Low-Angle Normal Faults

Gary J. Axen

Introduction

Since their discovery in the Basin and Range province (Longwell 1945; Anderson 1971; Armstrong 1972; Crittenden et al. 1980; Wernicke 1981) and subsequent recognition worldwide, "detachment faults" have been the center of heated debate. Detachment faults (figure 3.1) are gently dipping, commonly domed, fault surfaces of large aerial extent along which a significant part (commonly 5–15 km) of the crustal column is missing due to large-magnitude slip (typically 10–50 km). Debate centers on whether or not these faults formed and/or slipped as "low-angle normal faults" (dip <30°), because standard fault mechanical theory does not allow such orientations (Anderson 1942) and because earthquakes on such faults are rare (Jackson 1987; Jackson and White 1989).

The footwalls of detachments with net slip >15–20 km commonly expose a thick (0.1–3 km) ductile shear zone that evolved into a frictional slip surface as the footwall was unroofed and cooled on its trip to the surface. This evolution commonly forms these sequentially overprinting rock types preserved in the upper footwall (Coney 1980; Wernicke 1981; Davis 1983): (1) foliated, lineated mylonites formed in dominantly simple shear by plastic or semibrittle, pressure-insensitive mechanisms; (2) a thinner, structurally higher zone of tectonic breccias (commonly chlorite-epidote rich) that formed in a cataclastic flow regime; and (3) a thin (0–3 m) zone of finer-grained "microbreccia" with a sharp, striated detachment fault surface at its top. These products are particularly common in quartzo-feldspathic footwalls; detachments with lower net slip or those developed on other footwall rock types may lack mylonites, thick breccias, and/or microbreccias.

Detachment faults in North America form a semicontinuous belt that runs from Mexico to Canada (Coney 1980; Axen et al. 1993), and they clearly underlie many tens of thousands of square kilometers. Most large-displacement detachments (with or without a ductile footwall) are exposed in structural domes 10–30 km in diameter that are reflected in the topography and form of the mountain range. In these, the footwalls are topographically high, the mylonitic foliation (if

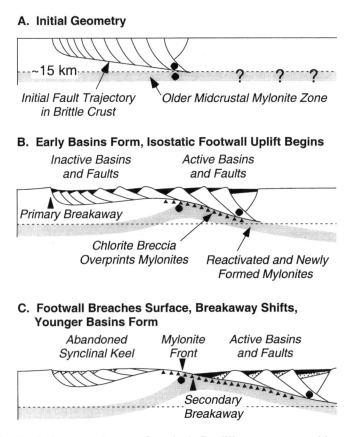

A. Initial Geometry

~15 km

*Initial Fault Trajectory
in Brittle Crust*

Older Midcrustal Mylonite Zone

B. Early Basins Form, Isostatic Footwall Uplift Begins

*Inactive Basins
and Faults*

*Active Basins
and Faults*

Primary Breakaway

*Chlorite Breccia
Overprints Mylonites*

*Reactivated and Newly
Formed Mylonites*

**C. Footwall Breaches Surface, Breakaway Shifts,
Younger Basins Form**

*Abandoned
Synclinal Keel*

*Mylonite
Front*

*Active Basins
and Faults*

*Secondary
Breakaway*

Figure 3.1 Evolutionary scheme of typical Cordilleran metamorphic core complex through normal shear on a ductile-brittle detachment system (e.g., Whipple Mountains, California). See figure 3.2 for aerial distribution of tectonic elements shown here in cross section. Low cut-off angle between syntectonic strata and the detachment would be expected above the secondary breakaway. After Spencer (1984).

any) and detachments themselves are domed, and the hanging walls are preserved as erosionally isolated remnants over the top of the domes and around the margins of the domes at the bases of the ranges. This doming is generally accepted as primarily due to isostatic rebound (figure 3.1) of the footwall resulting from buoyancy forces generated by removal of the hanging wall (Spencer 1984; Buck 1988; Wernicke and Axen 1988) and must be compensated by materials of crustal density (Block and Royden 1990; Wernicke 1990; Wdowinski and Axen 1992).

Most detachment domes probably reflect fold interference patterns. Isostatic rebound forms folds with axes perpendicular to the transport direction (figures 3.1 and 3.2) and parallel to the "breakaway" where the detachment originally surfaced. These folds commonly include crustal-scale antiformal welts that progressively widen as the hanging wall is withdrawn, a process known as a "rolling hinge" evolution (Spencer 1984; Buck 1988; Hamilton 1988; Wernicke and Axen 1988; Axen and Bartley 1997). Rolling hinge evolution causes rotation of moderate- and

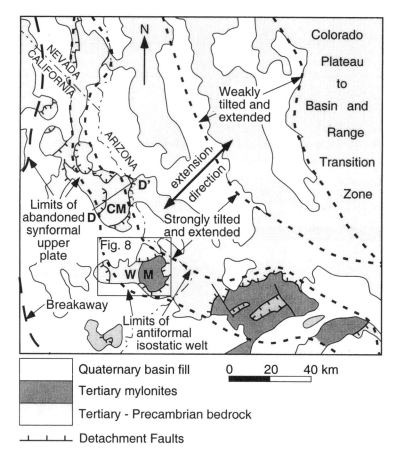

Figure 3.2 Map of lower Colorado River extensional corridor, showing detachment faults, surface exposures of related mid-Tertiary mylonite gneisses, and other tectonic elements common to highly extended terranes (see figure 3.1 and text). The major detachment faults of this region had upper plate-to-northeast motion and root to the northeast under the Colorado Plateau, causing a broad transition zone. CM, Chemehuevi Mountains; WM, Whipple Mountains; cross section DD' is shown in figure 3.7; box shows location of figure 3.8. Modified from Spencer and Reynolds (1989).

low-angle faults to gentler dips (Wernicke and Axen 1988; Axen 1993) and "back rotation" of gently dipping faults to opposite dips (Spencer 1984). The latter commonly causes abandonment of the detachment on the side of the dome near the breakaway, stranding hanging wall blocks there while continued slip on the opposite side of the dome forms a "secondary breakaway" in that location (figure 3.1; Dorsey and Becker 1995). In contrast, the folds with axes parallel to transport direction may be spaced 1–30 km apart, and folds of various wavelengths are commonly present within a single dome. These folds may have three causes: original corrugations in the detachment surface (Frost et al. 1982), extension-perpendicular shortening (Holm and Lux 1991; Mancktelow and Pavlis 1994), or differential sedimentary loading in the hanging wall (Axen and Bartley 1997). These are not mutually exclusive.

Alternative interpretations of detachment faults are (1) initially steep, rooted normal faults that rotated to gentle dips during and after their slip, either in a domino style (Proffett 1977; Davis 1983) or by a rolling hinge mechanism (Buck 1988; Wernicke and Axen 1988); (2) surfaces of landslide movement that are not rooted and do not accommodate lithospheric extension; or (3) the exhumed brittle-ductile transition (Miller et al. 1983). Exclusion of these three explanations is essential in compelling (versus consistent) cases of low-angle normal fault (LANF) slip or initiation. The second case is usually easy to exclude because most detachment faults clearly cut down into the crust for kilometers in their initial trajectories and none exhibit the large, thrust-sense "toe" required where enormous landslides would have overridden the Earth's surface. The third case is precluded by the fact that essentially all detachment faults are knife-sharp frictional sliding surfaces that were clearly brittle during their final slip.

A primary argument against the existence of LANFs is an apparent absence of LANF earthquakes in the seismic record (Jackson 1987; Jackson and White 1989; Thatcher and Hill 1991; Collettini and Sibson 2001). However, this absence of evidence, if true, is not compelling evidence of absence. For example, LANFs may slip aseismically (Jackson 1987), in which case they would be a special class of faults. Alternatively, LANF earthquakes simply may be infrequent with respect to the length of the instrumental seismic record (Wernicke 1995). Recent work suggests that historical LANF earthquakes have occurred but that they are difficult to recognize by standard first-motion or centroid moment-tensor analyses (see Abers 1991; Abers et al. 1997; Axen 1999 for reviews).

The second main argument against LANFs is theoretical; existing fault mechanical theory does not adequately explain either LANF slip (Sibson 1985) or initial formation of LANFs. Even scientists that accept their existence lack consensus about their mechanics. Significantly, LANFs share many mechanical problems with the San Andreas fault, for which mechanical consensus is also lacking despite the fact that it is certainly the most well studied fault on Earth. This suggests that some aspect(s) of accepted fault mechanical theory is (are) inadequate, a significant problem for Earth scientists.

This chapter considers only continental LANFs that undergo frictional slip. Similar structures have now been recognized at many slow-spreading midocean ridges, but direct observations of these faults are fewer and arguments parallel to those set out in the following text are much more difficult to make. In the next section a few prime LANF examples are reviewed first. For these, compelling arguments can be made that their dip at the time of formation and/or during frictional slip was <30°. Other compelling examples exist and many consistent examples exist, but these will not all be reviewed here. (For example, those with an early ductile shear-zone history generally pose inherent geometrical difficulties due to penetrative footwall strain, so usually do not provide compelling evidence for formation or frictional slip at low dips.) Second, the status of LANF earthquake seismology and paleoseismology is briefly reviewed. This section shows that seismically active LANFs exist. In the section on low-angle normal fault mechanics, LANF mechanics are discussed, using simple mechanical models to show that

LANF slip in the shallow crust is mechanically feasible, but slip in the strong midcrust is difficult, and that accepted mechanics are inadequate to explain primary LANFs. Finally, the implications of existing LANF mechanics for fault mechanics in general are considered briefly.

Evidence for Primary LANFs and for LANF Slip

Even one compelling example of a primary LANF or of LANF slip is sufficient to prove that they may form and slip at low dip, respectively. It is important to distinguish between evidence for a low-angle dip on a fault at the time it formed (a primary LANF) and evidence for slip while at low angle (LANF slip on a fault that may have formed at a steeper dip). The former is much more difficult to explain mechanically than the latter. Evidence for the initiation and slip of LANFs falls into two categories: geometric and seismologic/paleoseismologic.

Compelling geometric evidence typically takes the form of geologic or reflection seismic data. It must be shown that formation and/or slip of LANFs occurred at low dips and that subsequent passive rotation to low dip did not occur. Common arguments include the following. (1) Low cut-off angles ($<30°$) between detachment faults and upper-plate strata that were horizontal when the fault formed or slipped (figure 3.1). In most cases, geological observations at the outcrop to detailed map scale prove low cut-off angles for syn-detachment sedimentary strata only in the upper few kilometers of the crust (Davis and Lister 1988). Reflection seismic data commonly extend these observations to several kilometers depth and tens of kilometers of length across strike. (2) Geometric reconstructions (see later in this chapter) that show LANF trajectories with respect to paleohorizontal. These reconstructions build on numerous surface observations that, in some cases, constrain LANF trajectories over several kilometers of depth, tens of kilometers in the dip direction, and through the fault's history. Thus, reconstructions are generally the best method to unambiguously deduce the formation angle and geometric history of a fault. (3) Thermochronology of footwall rocks. Some dating methods ($^{40}Ar/^{39}Ar$, (U-Th)/He, fission track) yield the time at which a specific mineral cooled through a limited temperature range. The cooling rate may be constrained by a few specific methods (e.g., $^{40}Ar/^{39}Ar$ multidomain diffusion modeling of K-feldspar) or by obtaining a variety of time–temperature pairs from different minerals and/or methods from the same sample or from a restricted area. Thus, the thermal history of LANF footwalls can be constructed along profiles parallel to the fault dip, constraining fault dip over reasonably large areas. This technique has three main problems. First, an initial (at least) geothermal gradient must be assumed to obtain a dip angle. Second, much heat may be advected with the rising footwall, thus perturbing the geothermal gradient around the fault. Modeling this effect introduces additional parameters that tend to preclude unique determination of fault dip history (fault dip is commonly assumed), whereas not modeling this effect requires the dubious assumption that isotherms were horizontal during LANF slip (see Ketcham 1996). Third, complicated fault histories (e.g., rolling hinge evolution) often cannot be excluded with thermochronologic data (see Axen and Bartley 1997). Fourth, some commonly applied combinations of minerals and dating methods (particularly $^{40}Ar/^{39}Ar$ dating of biotite, muscovite, and hornblende) yield ages of cooling through temperatures >300°C, which are likely to reflect ductile rather than frictional regimes. (4) Paleomagnetism of footwall rocks. Paleomagnetism can provide constraints on the orientation of rocks at the time they cooled through their Curie temperature or acquired a secondary magnetism. However, angular confidence limits can be problematic ($\pm 5-10°$ at 95 percent confidence is common), it is difficult to know the time at which a rock obtained its magnetic signature, it must be shown that the rock-fault geometry has remained the same since that time, and the Curie temperature is commonly >300°C (see comment 3 on thermochronology).

Seismologic evidence for historical LANF earthquakes is increasingly common but is still sparse and remains disputed among seismologists. Paleoseismologic evidence of LANFs exists in a few cases, but in general, in the shallow crust only.

Because frictional slip on LANFs is of prime concern here, thermochronologic and paleomagnetic data are not discussed. Wernicke (1995) summarizes these approaches and some examples. Specific compelling examples of LANF initiation or frictional slip are discussed in the following text.

Southeastern Nevada and Southwestern Utah

The Mormon Mountains-Tule Springs Hills-Beaver Dam Mountains region of southeastern Nevada and southwestern Utah (figure 3.3) contains faults that can be compellingly demonstrated to have formed and slipped as LANFs. These faults overprint well mapped Mesozoic structures of the Sevier foreland thrust belt. This area is ideally suited for geometric analysis of fault orientations during formation and slip for four main reasons (Wernicke and Axen 1988; Axen et al. 1990). (1) The area is extremely well exposed and well mapped, so orientations and locations of faults and stratigraphic contacts are well constrained and accurate. (2) Most of the rocks in the footwalls of the detachment faults were subhorizontal prior to extension. The footwall rocks of all three detachment faults in the area are mainly made of a layer-cake Paleozoic-Mesozoic sequence deposited on old, cold crystalline basement of the North American craton (Wernicke et al. 1985; Axen et al. 1990). These rocks were in the footwall, or autochthon, of the Mesozoic Sevier thrust belt prior to extension, so were only very mildly affected by Mesozoic thrust-related deformation. The footwall of the structurally lowest detachment fault is the Colorado Plateau, a tectonically stable craton in which strata are still regionally subhorizontal except where locally deformed in monoclines. Very important postthrust but preextension deposits are preserved locally. There is little or no angular discordance where these rest depositionally on strata of the thrust

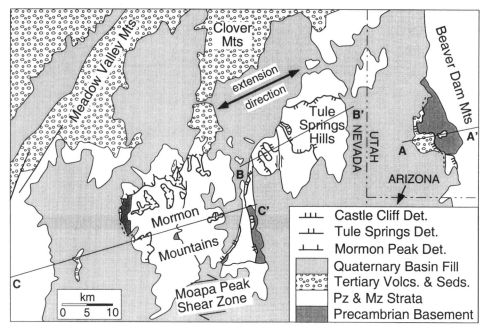

Figure 3.3 Simplified geologic map of the Mormon Mountains-Tule Springs Hills-Beaver Dam Mountains transect. The major detachments of this transect had upper plate-to-west-southwest motion and also root to the west-southwest; CCD, Castle Cliff; TSD, Tule Springs; and MPD, Mormon Peak detachments The Beaver Dam Mountains lie at the western edge of the Colorado Plateau in the footwall of the regional breakaway, causing an abrupt boundary between the Colorado Plateau and Basin and Range, with significant, isostatically driven, mid-Miocene uplift of the plateau edge in the Beaver Dam Mountains. Sections AA′, BB′, and CC′ are shown in figures 3.4, 3.5, and 3.6, respectively. Simplified from Axen et al. (1990).

autochthon (Wernicke and Axen 1988). Thus, it is certain that the strata of the thrust autochthon were regionally subhorizontal prior to mid-Miocene extension and that the present dips of these rocks are due to Basin and Range tectonics. (3) The thrust-belt structure is very well characterized, so that it provides additional constraints on reconstructions of detachment faults to their preslip condition. In key places, the postthrust, preextension strata rest on rocks of the thrust sheets, so constrain the preextensional orientations of thrust-related structures (Axen et al. 1990; Axen 1993). (4) Penetrative ductile strain is insignificant in the footwalls of these detachment faults, so does not complicate geometric restorations.

This sequence of thin, unmetamorphosed, well described, laterally continuous, and well mapped rock units that contain a well characterized group of laterally continuous Mesozoic thrust-belt structures is very important because it allows simple geometric constraints to be placed on the preextensional orientation of the detachment-fault footwalls and, hence, of the detachment faults themselves.

Between ~17 and ~10 Ma, the thin-skinned thrust belt of the Mormon Mountains-Tule Springs region was overprinted by three detachment faults on which the upper plates moved in a west-southwest direction. From west to east these are the Mormon Peak, Tule Springs, and Castle Cliff detachments. The three detachments probably formed broadly synchronously, but cross-cutting relations indicate that final slip migrated from west to east (Axen 1993).

Five independent geometric reconstructions show that the exposed parts of the Mormon Peak and Castle Cliff detachments were gently west dipping both when they formed and as they slipped to paleodepths of 5–7+ km (well into the strong, seismogenic crust), and that the Tule Springs detachment was very gently dipping in the upper crust (≤5 km) and moderately dipping at greater depth. Four of these reconstructions are based solely on fault and bed geometries within single range blocks, so do not rely on geological correlations between ranges nor on the types of structures that separate individual range blocks, where basin-fill obscures relations.

Castle Cliff Detachment

The eastern and youngest detachment, the Castle Cliff detachment (figures 3.3 and 3.4), forms the present western boundary of the geologically stable and little-deformed Colorado Plateau, which lies in its footwall (Hintze 1986). The detachment probably slipped at LANF dips but probably formed at a dip slightly greater than 30°.

The Castle Cliff detachment places various Paleozoic strata on sheared, crushed, altered, and/or silicified Precambrian crystalline basement (Hintze 1986). Carpenter and Carpenter (1994) interpreted these blocks as remnants of gravity slide sheets; this interpretation is discussed later in the chapter. Some upper-plate Paleozoic rocks are, in fact, landslide masses that rest depositionally on moderately east-tilted mid-Tertiary conglomerate (Hintze 1986). These landslide masses and the sediments they are enclosed within are cut by and tilted east above the Castle Cliff detachment (Hintze 1986; Anderson and Barnhard 1993b), showing that the detachment was active in mid-Tertiary or later time. Apatite fission-track data show that exhumation of the Beaver Dam Mountains basement began at ∼16 Ma (O'Sullivan et al. 1994; Stockli 1999).

In the detachment footwall, strata of the westernmost plateau are warped into a range-scale antiform that is cored by Proterozoic crystalline basement, the "Beaver Dam anticline." The western limb of this antiform is generally absent below the detachment and, where present, is composed of a highly attenuated, internally normal-faulted, and incomplete Paleozoic section (Hintze 1986) that was attenuated during detachment slip. In fact, most of the "west limb" probably never existed at all, because available data indicate that the fold formed due to upflexing of the plateau edge by isostatic rebound of the footwall as the hanging wall was withdrawn westward: mid-Tertiary strata that lie with low-angle disconformity on the Mesozoic plateau sequence are tilted as steeply as the older strata, showing that the fold formed in mid- or late-Tertiary time (Wernicke and Axen 1988), while Basin and Range extension was ongoing. (The fold has been interpreted as a "Laramide" structure of contractional origin [Hintze 1986; Carpenter

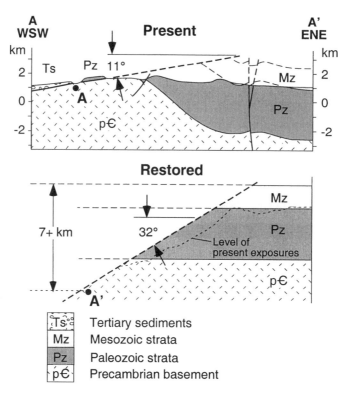

Figure 3.4 Present-day and restored cross section across the Beaver Dam Mountains at the edge of the Colorado Plateau (see figure 3.3 for location), drawn approximately parallel to transport direction on the Castle Cliff detachment, showing the effects of isostatic rebound of the detachment footwall. Restoration is by simple line-length balancing. Along the plateau margin in this area, Tertiary deposits rest disconformably or in very low angle unconformity on Mesozoic strata of the plateau, so the fold depicted is clearly a Tertiary feature and the Mesozoic-Paleozoic sequence was subhorizontal before extension began. After Wernicke and Axen (1988).

and Carpenter 1994], but these authors either did not know the age of tilted Tertiary strata or ignored it [Hintze 1986; Carpenter and Carpenter, 1994]). Wernicke (1985) and Wernicke and Axen (1988) argued that this uplift formed due to isostatic rebound of the unloading footwall of the Castle Cliff detachment as the hanging wall was withdrawn westward.

The Castle Cliff detachment presently dips ~11° west under the largest klippe at Sheep Horn Knoll (figures 3.3 and 3.4). Simple line-length-balanced palinspastic reconstruction of the tilted footwall strata back to horizontal returns the detachment there to an initial dip of 32°. The structurally lowest exposed footwall rocks restore to a paleodepth of at least 7 km. Most of this tilting probably occurred while the detachment was slipping and its footwall was rebounding, but some may be due to other causes. Under modern thin-skinned foreland-thrust belts, the basement-cover contact in the thrust-belt autochthon dips 3–8° toward the thrust belt (Wernicke et al. 1985). Stratigraphically higher autochthonous strata and strata closer to the thrust front dip less. Adding 3–8° to the 32° dip of the reconstruction at the onset of extension yields a maximum initial dip for the Castle Cliff detachment of ~35–40°, but this is probably too steep because the area was located at the Mesozoic thrust front where dip of basal autochthonous strata is expected to be less. Part of the eastward tilting of strata in the Beaver Dam Mountains may be due to a buried normal fault that is inferred from a gravity gradient west of exposed hanging wall blocks (Hintze 1986; Blank and Kucks 1989).

Tule Springs Detachment

The central, Tule Springs detachment (figures 3.3 and 3.5) is the only one of the three that displays clear control by preexisting anisotropy; it followed and reactivated a shallow footwall thrust decollement on the thin (<150 m) Jurassic Moenave Formation for >10 km in the direction of detachment transport (Axen 1993). The detachment and most of the footwall strata currently are subhorizontal (Axen 1993). Hanging wall normal faults have up to ~3 km of stratigraphic separation, strike generally north-south, are closely spaced, and accommodate ~10 km of extension of the upper plate, but do not offset stratigraphic contacts in the footwall, so it is clear that their cumulative displacements are concentrated along the re-

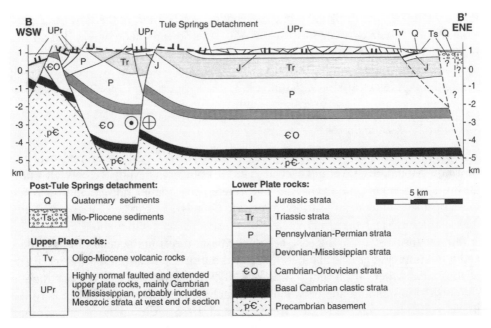

Figure 3.5 Cross section across the Tule Springs Hills, drawn approximately parallel to transport direction of the Tule Springs detachment. The detachment reactivated the Mesozoic Tule Springs thrust in a footwall- and hanging wall-flat part, so the upper plate comprises strongly extended Paleozoic strata and the footwall is little-deformed, subhorizontal Jurassic beds. At the west, the footwall is flexed up like the Colorado Plateau edge in the Beaver Dam Mountains, showing that Tule Springs detachment had a steeper western section. See figure 3.3 for section location. Simplified from Axen (1993).

activated thrust contact or within the immediately underlying Jurassic siltstones (Axen 1993).

Geologically reasonable reconstructions of the Tule Springs detachment favor an initial dip of 3–15° to depth of 2–5 km. If the basement-cover contact of the thrust-belt autochthon under the Tule Springs Hills dipped 3–8° west, then the thrust footwall flat, which also makes up the present footwall of the detachment, could not have dipped more than 3–8° when it was reactivated as a LANF. This is supported by regional reconstructions of the Castle Cliff detachment, which places the Tule Springs footwall adjacent to an identical section of subhorizontal Colorado Plateau strata (Axen et al. 1990; see also Anderson and Barnhard 1993b). An absolute maximum, but geologically untenable, initial detachment dip angle of 36° can be derived independently from rocks in the hanging wall (Axen 1993). This reconstruction (1) maximizes the angle that the sub-Tertiary unconformity cuts stratigraphically down eastward across hanging wall Paleozoic strata and maximizes the westward thickening of thrust-duplex slices along the thrust. However, this reconstruction is geologically flawed on several grounds and a dip as steep as 36° is rejected, although a dip as steep as 15° is acceptable in the hanging wall reconstruction (see Axen 1993 for full discussion). The low-angle part of the Tule Springs detachment, where it reactivated the thrust flat, was probably never deeper than ~5 km (maximum thickness of the upper-plate stratigraphy plus thin thrust duplexes), and may have been as shallow as 2–3 km at the onset of extension. West of the reactivated thrust flat, the Tule Springs detachment had an initially moderate dip, steepening westward in an antilistric geometry, cutting across the top of the thrust ramp and into the strata and basement of the thrust belt autochthon (Axen et al. 1990; Axen, 1993).

Mormon Peak Detachment

The Mormon Peak detachment (figures 3.3 and 3.6) is broadly domed over the Mormon Mountains (Wernicke et al. 1985), both as a result of footwall uplift during slip and because it was cut and rotated by smaller normal faults. When these postdetachment faults are restored, the detachment can be seen to cut gently down structural and stratigraphic section westward across the whole range: from within the thrust stack in the east, across the basal thrust in the center of the range, and into strata and crystalline basement of the thrust autochthon on the west side of the range, where it presently dips ~5–10°W. This palinspastic reconstruction shows that the initial angle between the detachment and the footwall strata of the thrust autochthon was ~17–20° (Wernicke et al. 1985). Adding the 3–8° permitted by dips of modern basement-cover contacts in thrust-belt footwalls indicates that the initial dip was likely in the range 20–28°. This paleodip angle is valid to at least 6 km depth.

This dip estimate agrees with an initial dip determined independently through reconstructions of the detachment fault's hanging wall against its footwall, the only reconstruction considered here that involves rocks of two different mountain ranges. The key to this restoration (figure 3.6) is recognition that a syncline in the

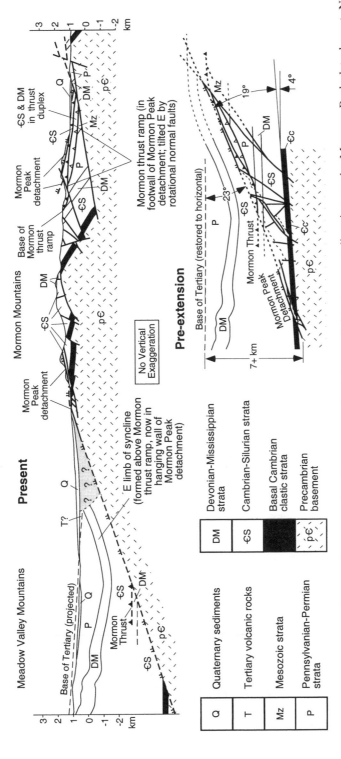

Figure 3.6 Present-day and restored cross section drawn approximately parallel to transport direction on the Mormon Peak detachment. Note that the reconstruction of the detachment footwall, and the Mormon Peak detachment to a 19° angle with the basal nonconformity is done entirely by restoring slip on small normal faults in the Mormon Mountains. This reconstruction is independent of, but confirmed by, restoration of the thrust-ramp syncline in the Meadow Valley Mountains against the thrust ramp in the Mormon Mountains, where very gently dipping Oligo-Miocene volcanic rocks constrain the dip of the east limb of the syncline prior to detachment slip. After Wernicke et al. (1985) and Axen et al. (1990).

Meadow Valley Mountains formed adjacent to a ramp in the Mormon thrust (Wernicke et al. 1988). Presently, the syncline is in the hanging wall of the Mormon Peak detachment, and the thrust ramp is exposed in the detachment footwall (Wernicke et al. 1988; Axen et al. 1990). Normal faulting is minor in the Meadow Valley Mountains, where the syncline is depositionally overlapped by subhorizontal mid-Tertiary volcanic strata. The syncline, both in the Meadow Valley Mountains and for more than 100 km to the south, is stratigraphically distinctive because it preserves the only Triassic strata in the frontal part of the thrust belt, and because those strata lie depositionally in low-angle unconformity over Permian rocks: directly on Permian clastic beds in the west and on the west-ernmost remnants of stratigraphically higher Permian Toroweap and Kaibab Formations in the east limb (Burchfiel et al. 1974; Wernicke et al. 1988). It is structurally distinctive as the only major, broad syncline in the hanging wall of the front-most major thrust sheet. In the little-extended Spring Mountains, its east limb lies adjacent to the Keystone thrust ramp (Burchfiel et al. 1974), which is correlative with the Mormon thrust (Wernicke et al. 1988).

In the Meadow Valley Mountains, the syncline was erosionally beveled before deposition of presently subhorizontal Oligocene-Miocene ignimbrite sheets that make an angle of ~30° with the east limb of the syncline (Axen and Wernicke, unpublished reconnaissance). The base of the thrust ramp and the thrust flat that adjoins it to the west are exposed in the eastern and central Mormon Mountains less than 1 km structurally beneath the Mormon Peak detachment (Wernicke et al. 1985). Thus, a reconstruction of the east limb of the syncline (now in the hanging wall of the detachment) against the thrust fault ramp (now preserved in the detachment footwall) yields the angular relationship between the detachment and the ignimbrite stack that was subhorizontal at the onset of extension (Axen et al. 1990). That angle is ~21°, consistent with the previous estimate (20–28°), and places the basement rocks of the western Mormon Mountains at a paleodepth of >7 km at the onset of extension. The angle between the base of the Tertiary sequence and the basement-cover contact is 4° in the reconstruction, within the modern range of 3–8°. Thus the reconstruction is compatible not only with details of the thrust belt locally, but also with the general architecture of foreland thrust belts worldwide.

Regional Reconstructions

The five independent reconstructions summarized previously are mutually compatible and can be combined into regionally consistent balanced cross sections (Axen et al. 1990). The reconstruction of the thrust ramp and its ramp syncline require ~25 km of displacement on the Mormon Peak detachment, which is definitely the best constrained displacement of the three detachments. Restorations of the other two detachments (Axen et al. 1990) and regional considerations (Wernicke et al. 1988; Bohannon et al. 1993) suggest that together they accommodated an additional ~30 km of extension.

Alternative Interpretations

These extension estimates and fault geometries were challenged by Anderson and Barnhard (1993a, 1993b) and by Carpenter and Carpenter (1994). Anderson and Barnhard (1993a, 1993b) argue that all three detachments are rooted tectonic faults but that they have antilistric (concave-down) geometries, steepening in the subsurface west of their present levels of exposure. They adopt most of the geometries of our cross sections (Wernicke et al. 1985; Wernicke and Axen 1988; Axen et al. 1990; Axen 1993) in the shallow crust (present levels of exposure), so apparently accept LANF slip to significant paleodepth. Anderson and Barnhard (1993b) postulate that the inferred steep parts of the detachments at depth accommodated significant sinistral slip, not dominantly dip slip as is evident for the exposed parts. Thus, their interpretations require a significant spatial change in fault kinematics, from low-angle normal dip slip in the exposed parts of the detachment systems to strongly oblique sinistral-normal slip where the faults enter the subsurface. We view this as an ad hoc explanation that ignores the fact that several sinistral-normal faults cut across the Tule Springs detachment (figures 3.3 and 3.5; Axen 1993) and across detachment-related corrugations in the Beaver Dam Mountains, Utah (Anderson and Barnhard 1993a, Plate 1) and, therefore, postdate detachment slip. In addition, seismic reflection data in the basin southwest of the Beaver Dam Mountains reveals listric, but not antilistric, fault geometries (Bohannon et al. 1993; Carpenter and Carpenter 1994). Also, their cross sections do not adequately treat the thrust structures. In particular, their depiction of the western Mormon Peak detachment (Anderson and Barnhard 1993b, Plate 1) implies that it does not carry thrust belt rocks in its hanging wall, which is required by both the local and regional geology. Although the possibility that the Castle Cliff and Mormon Peak detachments steepen at depth west of their present exposures cannot be excluded completely, the arguments mentioned previously are compelling that the Mormon Peak and Tule Springs detachments formed and slipped as LANFs in the upper 6–7 and 2–5 km of the crust, respectively.

Carpenter and Carpenter (1994) interpret the exposed detachments in the region as surficial landslides, on the basis of reinterpreted oil industry seismic reflection lines (mainly from the basins to the south), regional considerations, and the presence of conglomerate at exposures of the Mormon Peak detachment. However, none of their arguments are compelling.

The seismic reflection lines presented by Carpenter and Carpenter (1994) do not constrain the present or initial dips of detachment faults in the Mormon Mountains-Beaver Dam Mountains transect for these reasons. (1) Most of their seismic lines do not cross the ranges in question. (2) The one line that does so (their line 5–5A) runs at a high angle to the extension direction—it is a "strike" section, not a "dip" section, so does not constrain detachment geometry in the direction of transport (c.f., Carpenter et al. 1989; Axen and Wernicke 1989). (3) Their lines contain significant areas of poor data quality, and even where reflectors are strong, interpretations are very subjective (see later in the chapter). (4) Their lines are referenced for velocity structure to a single well that is ~25 km south of the transect and separated from it by the major, east-northeast-trending, dextral Moapa Peak shear zone (figure 3.3; Axen et al. 1990; see also Bohannon et al. 1993, figures 8 and 9).

The eastern two-thirds of their line 4–4A, the line with the best reflector sequences, runs nearly parallel to the exposed bedrock transect and to the regional extension direction but is separated from that transect by the Moapa Peak shear zone, an east-northeast-trending zone of oroflexure and south-down faulting that accommodates 6–18 km of dextral shear (Axen et al. 1990). Carpenter and Carpenter (1994) show high-angle and listric normal faults that accommodate only modest extension. In several places they interpret discordant reflector sets as unconformities with large onlap angles where a well aligned fault is clearly imaged at slightly shallower levels in the section. They infer that faulting controlled sedimentation from Oligocene time to the present. This contrasts with decades of surface studies that show that extension in the region began in earnest at or after ~17 Ma (Bohannon 1984; O'Sullivan et al. 1994; Stockli, 1999). In contrast, Bohannon et al. (1993) interpret similarly oriented reflection lines through the same area in terms of listric normal faults flattening at ~4–6 km depth, accommodating tens of kilometers of crustal extension, with the bulk of syntectonic sedimentation occurring between 13 and 10 Ma. Thus, the seismic reflection data are themselves subject to various interpretations (including large-slip LANFs).

Several aspects of the geology in the transect preclude a gravity-slide interpretation for the detachments discussed in the previous section. First, in the Tule Springs detachment, there is no high ground from which the >120 km^2 of preserved allochthon could have slid: the range is topographically low and is surrounded on all sides by lower basins. Second, there is no evidence of radial sliding off of the Mormon Mountain dome, which the Carpenters postulate as the elevated source region. In general, the hanging walls of all three detachments are cut by predominantly north-northwest- to north-striking faults, implying that landsliding would have been directed only to the east or west (landslides are commonly internally extended parallel to their transport direction). In addition, there are no large basins to the north or west into which large amounts of upper-plate material could have slid. The upper plate of the Mormon Peak detachment in the northern Mormon Mountains is structurally continuous to the north with a bedrock ridge of east-tilted late Oligocene(?)-Miocene volcanic rocks that connects north to the adjacent Clover Mountains (figure 3.3). If the upper plate of the detachment in the northern Mormon Mountains slid northward, then it also comprises the entire Clover Mountains. To the west-northwest the basin is no more than 2–5 km wide for a distance of ~10 km (figure 3.3), and the exposed strata in the immediate detachment hanging wall belong to the same formation and are oriented similarly to the strata across the basin to the west, which belong to the east limb of the ramp syncline discussed previously. Third, landslide surfaces almost universally steepen upward into a head scarp and side walls, none of which are apparent in the Mormon Mountains. To the contrary, the Mormon Peak detachment is a smoothly contourable surface that is gentler at higher elevations (Wernicke et al. 1985), like other Cordilleran detachments. Fourth, the Mormon Peak detachment, in general, is cut by north-northwest- to north-striking faults that tilt it eastward and that probably formed while the Mormon Mountains dome was forming above the structurally deeper Tule Springs detachment (Axen et al. 1990). Thus, the highland from which the upper plate supposedly slid had not yet fully formed when the slides were emplaced. Fifth, the landslide hypothesis for the Castle Cliff detachment does not explain (1) why the footwall is altered, crushed, sheared, and silicified for many meters below the supposed landsliding surface (these features are common along rooted normal faults), (2) why the Castle Cliff detachment is warped on a 2–3 km wavelength (as are many other Cordilleran detachments) (Anderson and Barnhard 1993b), nor (3) why the sediments of the adjacent basin are tilted east into the fault (which is expected in a detachment hanging wall).

The present author has not seen the sediments along the Mormon Peak detachment that Carpenter and Carpenter (1994) and Wernicke (1982) discuss. Wernicke (1982) inferred that these were deposited on the Mormon thrust sheet in Mesozoic or early Tertiary time and then were overridden by a higher thrust sheet, from which the upper plate of the Mormon Peak detachment was derived. However, it is now known that the upper plate of the Mormon Peak detachment is derived from the Mormon-Tule Springs thrust sheet, precluding this interpretation. Wernicke (oral communication, 2002) presently interprets these sediments as having been deposited after or near the end of detachment activity in cavern systems that followed the crushed carbonates adjacent to the detachment. This is in agreement with the presence of coarse-grained carbonate spar masses that the present author has seen along the Mormon Peak detachment in the western Mormon Mountains. It is possible that some of these "deposits" are, in fact, cataclastic rocks formed in situ along the detachment, but Wernicke (oral communication, 2002) says that there are layered beds with rounded clasts in some examples. Further study of these enigmatic deposits is warranted.

Chemehuevi and Whipple Mountains, California

The Chemehuevi Mountains of southeastern California expose three stacked LANFs that are broadly domed over the range (figures 3.2 and 3.7; Howard and John 1987). The footwall consists of predetachment intrusive rocks and their metamorphic country rocks but the detachments were all brittle faults—they did not evolve from ductile shear zones, so the large penetrative shear strains typical of metamorphic core complexes do not complicate the geometric analysis in the Chemehuevi Mountains. Of mechanical significance (see section on LANF mechanics), these detachments cut through relatively homogeneous, isotropic plutons for significant distances.

The detachments in the Chemehuevi Mountains are exposed for ~25 km in the direction of upper-plate transport (southwest to northeast). Translation of the upper plates is a minimum of 8 km, on the basis of local relationships but probably was in the range 20–40 km, on the basis of regional geology (John 1987; figure 3.7). Fault rocks along the detachments formed as a result of frictional processes,

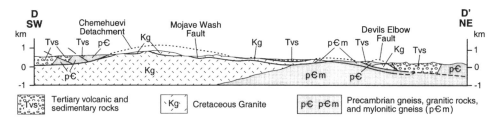

Figure 3.7 Cross section through the Chemehuevi Mountains drawn approximately parallel to the transport direction of three brittle detachment faults there. The Chemehuevi and Mojave Wash faults cut across an undeformed, brittle Cretaceous pluton and across foliation in Precambrian gneisses. Fault rock types limit the depths at which the fault could have formed in its eastern and western parts, showing that it could not have dipped more steeply than ~15°. From John (1989).

except for local protomylonites and rare, thin mylonite zones (John 1987). In general, the nature of the early fault rocks indicates that the eastern exposures of the detachment faults were at initially greater depths (~10 km) than the western exposures (~4–6 km), indicating an initial eastward dip in the range of 5–15° using reasonable geothermal gradients to constrain the depths at which the different fault rocks were formed (Howard and John 1987; John 1987). Under normal crustal conditions, typical fault slip rates, and typical temperatures for these fault rocks, steeper initial dips would require either that the faults record ductile flow in their eastern footwalls or that the geothermal gradient in the east was substantially lower than it was ~20 km farther west.

The Chemehuevi Mountains record LANF slip in the upper part of the crystalline middle crust, a few kilometers deeper than can be demonstrated in the southeastern Nevada region discussed previously. Pseudotachylyte along the detachment faults in the Chemehuevi Mountains (John 1987; and locally along the Whipple detachment, Wang 1997) shows that they were capable of seismogenic slip. (Pseudotachylyte is an ultrafine-grained fault rock, commonly composed of glass, devitrified glass, and/or ultracataclasite, that forms during seismogenic slip; Sibson 1975; Magloughlin and Spray 1992; Spray 1995.)

The Whipple Mountains (figures 3.2 and 3.8) lie south of the Chemehuevi Mountains and also expose a domed detachment fault that bounds a footwall metamorphic core complex made of middle crustal intrusive and metamorphic rocks (Davis 1988). The Whipple detachment slipped at least ~40 km on the basis of the horizontal offset of a distinctive dike swarm found in the footwall west of the dome and in the hanging wall east of the dome (Howard et al. 1982). This distribution precludes a landslide interpretation of the Whipple detachment unless the slide mass went entirely over the topographically high Whipple dome. The Whipple footwall has clearly rebounded isostatically ~15 km and it probably experienced a rolling hinge evolution (e.g., Spencer 1984; Dorsey and Becker 1995), but there is no compelling evidence that the fault ever dipped even moderately (~45°), much less steeply (~60°).

The last ~10 km of slip was clearly on a LANF plane. This is shown by two

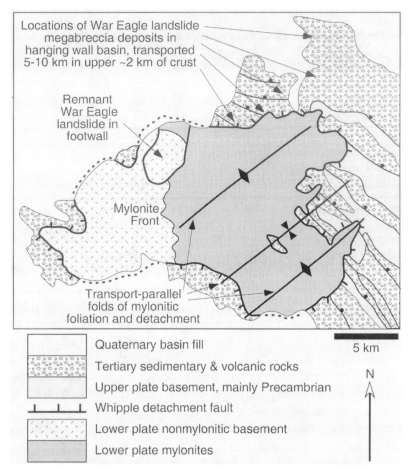

Locations of War Eagle landslide megabreccia deposits in hanging wall basin, transported 5-10 km in upper ~2 km of crust

Remnant War Eagle landslide in footwall

Mylonite Front

Transport-parallel folds of mylonitic foliation and detachment

Quaternary basin fill

Tertiary sedimentary & volcanic rocks

Upper plate basement, mainly Precambrian

Whipple detachment fault

Lower plate nonmylonitic basement

Lower plate mylonites

5 km

N

Figure 3.8 Simplified geologic map of the Whipple Mountains. Note that the War Eagle landslide (partly preserved in the footwall) formed monolithic megabreccia sheets in the upper-plate sedimentary basin. These sheets are now offset 5–10 km from their source but were never buried more deeply than ~2 km, requiring very low-angle slip on the Whipple fault in between. Three transport-parallel folds or corrugations in the fault surface and underlying mylonitic foliation are also shown. After Dorsey and Roberts (1996) and Davis (1988).

pieces of compelling evidence. First is the presence of subhorizontal sediments in fault-bounded, upper-plate blocks that are much wider than they are thick and that are truncated at their bases by the Whipple fault, precluding any significant rotation of the fault since deposition of those sediments (Davis and Lister 1988). Second is offset of the large War Eagle landslide by the Whipple fault (figure 3.8). The War Eagle landslide slid approximately north from the northwestern part of the domed footwall into the adjacent hanging wall basin (Dorsey and Becker 1995; Dorsey and Roberts 1996) before being offset ~10 km to the northeast by final detachment slip (Yin and Dunn 1992). Along that portion, the Whipple detachment is the only major structure to intervene between the landslide masses in its hanging wall and footwall. There, the detachment dips 12–28°N (Yin and Dunn 1992) due to doming that largely predated the landslide emplacement, so detachment slip was low-angle oblique dextral and normal. Very little structural relief exists along the detachment between the positions of the landslide in the lower and upper plates (only ~300 m across ~10 km), so the net slip vector is nearly horizontal (~2°). The base of the landslide in both the lower and upper plates is gently dipping, precluding any major differential rotations since landsliding and since it was offset by the Whipple detachment (Yin and Dunn 1992).

Subsurface Examples

Many seismic reflection profiles collected worldwide convincingly image listric or low-angle normal faults. Some tie range-bounding normal faults and/or their neotectonic scarps to LANFs imaged to several kilometers depth in the subsurface. These include the Santa Rita fault, Arizona (Johnson and Loy 1992) and a fault under Lamoille Valley, Nevada (Smith et al. 1989). Others display a listric geometry in which low-angle faults at depth dip as little as ~5°. Many of these are in oil producing regions and have been drilled as well (in fact, most of these examples are imaged on proprietary industry lines). Wernicke (1995) shows examples that are in the public domain that are particularly relevant because they cut down at low dips into crystalline basement rocks and underlie wide tracts of young (Tertiary-Quaternary), subhorizontal, upper-plate strata that preclude the LANFs having been rotated to low dips. He also discusses the main caveats involved in interpretation of such profiles (e.g., the effects of velocity structure on dip angle). These arguments will not be elaborated on here.

Earthquake Seismology

Compilations of normal-fault earthquake focal mechanisms show that most such events occur on planes steeper than 30° (Jackson 1987; Jackson and White 1989; Collettini and Sibson 2001). These compilations included a small number (15–25) of large-magnitude (M_s or m_b > 5.2) normal events for which the correct nodal plane could be picked unambiguously. Recently, a large (M_w = 6.8) LANF earthquake was documented in the Woodlark Basin (Abers 1991, 2001; Abers et al. 1997). Waveform modeling precludes a dip as steep as 30° (Abers 1991), and local geology verifies that the gentle nodal plane is the fault plane (Abers 2001). Several additional smaller earthquakes occurred nearby with nodal planes dipping 25–35° (Abers et al. 1997; Abers 2001). One of these earthquakes, with a nodal plane that dips 23°N, is particularly convincing in terms of its teleseismic location (Abers 2001) near a LANF that is well imaged by seismic reflection profiles and that dips 27 ± 3° from the seafloor to 6 km below the seafloor (Taylor et al. 1995, 1999]. The footwall of this fault was drilled on Moresby Seamount where a few meters of clayey gouge containing the weak minerals talc and serpentine were encountered (Shipboard Scientific Party 1999). This weak gouge formed hydrothermally and allows a standard mechanical explanation (see later in the chapter) for slip on the LANF to the depth at which the gouge formed. It is unclear that such gouge is present to the depth of the earthquake hypocenter, but the fact that frictional instability occurs there suggests that it may not be. Thus, it appears that at least two LANF earthquakes can be added to the compilation of normal events for which the appropriate nodal plane can be picked unambiguously (see Wernicke 1995).

Other probable seismogenic LANFs are reviewed by Axen (1999) and Abers (2001). Axen (1999) discussed several examples where LANF slip was apparently triggered by earlier earthquakes or early moment release in earthquakes with complex, long-duration ruptures. Ironically, some of the complex earthquakes with probable triggered LANF subevents were included in the compilations by Jackson (1987) and Jackson and White (1989) of high-angle normal-fault seismicity. Microseismicity outlines LANFs that also have experienced significant earthquakes and/or are well known from surface geology and seismic reflection in Greece (Reitbrock et al. 1996; Rigo et al. 1996; Bernard et al. 1997; Sorel 2000) and Italy (Boncio et al. 2000).

To date there are apparently no well documented, large LANF earthquakes on very gently dipping planes (e.g., ~20°), such as geology and seismic reflection studies indicate are reasonably common. However, this may be simply a case of the well monitored seismic record being too short or not having been adequately explored. The latter inference stems from the fact that none of the known or probable LANF events or subevents were immediately apparent from initial focal mechanism or centroid moment tensor studies; all required significant extra effort (e.g., careful waveform modeling) to be documented (see Axen 1999).

Neotectonic Studies

Neotectonic studies indicate Quaternary to historic activity on LANFs. Most notable is the LANF surface rupture that occurred during the 1954 Dixie Valley earthquake (Caskey et al. 1996). That event was triggered by the complex dextral-normal Fairview Peak earthquake four minutes earlier, precluding seismic waveform analyses due to the coda of the Fairview Peak event (Doser 1986; Caskey and Wesnousky 1997). However, detailed mapping indicates that a significant portion of the surface rupture involved slip on a LANF. Seismic profiling supports this low dip to 1.75 km depth, and gravity suggests that it continues to the basin bottom at 2.7 km (Abbott et al. 2001). Axen et al. (1999) show that the Miocene-Pleistocene Cañada David detachment fault in northeastern Baja California probably had a significant seismic event that caused ~10 m of LANF slip sometime in the past ~20,000 years. Examples from the Death Valley area require Pliocene-Pleistocene LANF slip but do not preclude a late Quaternary switch to steeper normal-fault slip on cross-cutting range-bounding faults (Burchfiel et al. 1987, 1995; Cichanski 2000).

Mechanics of Low-Angle Normal Faulting

Mechanical models of LANFs fall into two general categories. (1) Continuum models (Yin 1989; Spencer and Chase 1989; Melosh 1990; Govers and Wortel 1993; Westaway 1998, 1999) generally follow Hafner (1951), model crustal-scale layers, and investigate initial and boundary conditions under which LANFs might form in intact rock. These models mainly address the initial formation of LANFs, for which cohesion of intact, homogeneous, isotropic rock must be overcome—a difficult and apparently unsolved problem (see later in the chapter). (2) Empirical models (Reynolds and Lister 1987; Axen 1992; Axen and Selverstone 1994; Axen et al. 2001; Abers 2001) combine observational data with results from experimental rock mechanics to explain LANF mechanics. These generally emphasize conditions under which frictional slip is possible on preexisting LANFs, although Axen (1992) also argued for LANF initiation controlled by mechanical anisotropy.

Continuum Models of LANF Initiation

Continuum models of the initiation of LANFs generally search for geologically reasonable initial and boundary conditions that allow brittle LANF formation in intact, homogeneous, and isotropic rock. In particular, model results usually satisfy the empirical Coulomb criterion for new brittle faults $\tau = C_0 + \mu_i\sigma_{ni}$, where the coefficient of internal friction $\mu_i = \tan \phi_i$, ϕ_i is the angle of internal friction, τ is the shear traction, and σ_{ni} is the effective normal traction (equal to the normal traction minus the pore fluid pressure) on the plane of the incipient fault, and C_0 is rock cohesion (a material constant).

The coefficient of internal friction of rocks is determined from the slope of the Mohr-Coulomb failure envelope; at low confining pressure it can be approximated by the linear Coulomb failure criteria. (It is well known that internal friction decreases with increasing confining pressure [Handin 1969], approaching zero at extreme pressures.) Tabulated values of the coefficient of internal friction μ_i of igneous and strong sedimentary rocks range from $\sim$1 to 1.7 but are <1 for many sedimentary and metamorphic rocks (Handin 1966, Table 11–4; Jaeger and Cook 1976, Table 6.15.1). Jaeger and Cook (1976:153–155) argue that $0.5 < \mu_i < 1$ for most rocks, but they do not distinguish between crystalline and sedimentary rocks and display the data normalized by the unconfined compressive strength, making it difficult to compare with stress regimes at different crustal levels. Experimental results from Westerly granite (Lockner 1995, figure 6; Marone 1995, figure 1a) show $\mu_i \approx 1$ for effective normal stress $\sigma_{ni} < 450$ MPa. (Note that Marone fits a straight line of $\mu_i = 0.73$ to all the data, including tests at higher normal stress, but the fit is poor at low normal stress.) The angle between the maximum principal compressive stress σ_1 and the fracture plane is $\theta = 45° - \phi_i/2$, usually $\sim$15–35°. Laboratory values for cohesion of granitic rocks range from $\sim$11 to $\sim$23 MPa (Handin 1966).

According to Anderson (1942), one principal stress is subvertical near the free surface of the Earth. In extensional stress regimes the maximum principal stress is subvertical. Boundary conditions in continuum models of LANF initiation must cause an inclined maximum principal stress at depth if they are to satisfy the Coulomb criterion.

Bartley and Glazner (1985) accomplished this with a combination of elevated pore-fluid pressure and surface slope, which imparts a shear stress oriented downhill. They concluded that with a surface slope of 8° and pore fluid pressure at 80 percent of the vertical stress, a normal fault could form with dip as low as ~35°, but the regional topographic relief would be enormous and the hydrothermal systems to which they appealed for pore fluid pressure are not sufficiently long lived.

Yin (1989) and Melosh (1990) showed that basal shear, such as might be exerted on the brittle crust by an underlying ductile shear zone, could cause stress rotations favoring LANFs. Spencer and Chase (1989) used laterally varying vertical stress on the base of their model such as might result from lateral changes in crustal thickness relict from previous shortening.

Buck (1990) pointed out that, in addition to appropriate principal stress orientation, the magnitude of shear traction resolved on failure planes must meet or exceed a minimum level given by Coulomb theory. He noted that in Yin's (1989) models much or all of the elastic layer is in a state of deviatoric compression, so adjacent parts of the crust would tend to shorten, not extend, and that even where deviatoric tension obtains and the maximum principal stress is rotated sufficiently for slip on LANFs to theoretically occur, the steeper conjugate fault system has higher resolved shear stress and should fail preferentially. On this basis, Buck (1990) and Wills and Buck (1997) concluded that LANFs remain mechanically untenable.

More recently, Westaway (1998) showed that shear traction and laterally varying buoyancy forces on the base of the brittle crustal layer combined with a slightly reduced friction coefficient within the layer allow LANFs to slip. Westaway (1999) combined basal shear stress, laterally varying basal buoyancy forces, and laterally varying cohesion to model conditions in which LANFs may initiate. In the latter case, the models presented are also largely in a state of deviatoric compression and the model LANFs form only near the lower corner of the model, on the up-flow end of the channel under the elastic layer.

Thus, the continuum models of LANF initiation do not agree well with field observations of LANFs and their surroundings. For example, LANFs in the Basin and Range province clearly formed in a large region in which both high- and low-angle normal faulting was widespread. Thus, deviatoric tension, not compression, dominated regionally, unlike the results of most of the models. Also, the continuum models produce spatially gradual changes in stress orientation and magnitude, which suggest that curved, listric normal faults should prevail. Listric faults occur but many examples exist of steep normal faults truncated abruptly by, or curving rapidly into LANFs, suggesting rapid gradients in stress orientation or formation of misoriented faults. Other common structures (summarized later in the chapter) show that maximum principal stress was oriented at high angles to some LANFs while they were slipping, but these angles may have been different at the time of formation.

Observationally Based Models for LANF Slip

Several studies of structures that formed adjacent to slipping LANFs show that the maximum principal stress was at a high angle to the LANFs, consistent with accepted extensional stress orientations (Anderson 1942) but inconsistent with Coulomb failure criteria. Steep normal faults and conjugate sets of normal faults are common in the hanging walls (Davis and Lister 1988; Wernicke et al. 1985; Axen 1993; many others) and footwalls (Reynolds 1985; Reynolds and Lister 1987; Axen and Selverstone 1994) of LANFs where cross-cutting relations require that they were active contemporaneously. The Coulomb criterion shows that the maximum principal stress bisects the acute angle between conjugate faults or lies ~30° from individual faults, so the maximum principal stress was subvertical in such cases. Similarly, steep tensile fractures (Reynolds and Lister 1987; Axen et al. 1995) and syntectonic dikes (Livaccari et al. 1995) in the footwalls of LANFs show that the minimum principal stress was subhorizontal and that the plane containing the maximum and intermediate principal stresses was therefore subvertical; for normal faulting, this requires that the maximum principal stress was subvertical.

Postulating postslip rotation of these LANFs from steep dips (e.g., rolling hinge or domino style) does not solve the mechanical problem because the LANF's surroundings must be rotated also. In that scenario, not only is the shear traction small on the detachment, but Andersonian stress orientations are also violated.

Reynolds (1985) and Reynolds and Lister (1987) argued that steep conjugate normal faults in the footwall of the South Mountain detachment formed synchronous with LANF slip. They suggested that these faults and older subvertical tensile fractures and dikes require that the maximum principal stress was steep during slip on the South Mountain detachment and during the ductile shear that preceded it. They argued that footwall structures record the LANF evolution at the ductile-brittle transition, from ductile shearing at nearly lithostatic pore fluid pressure and low differential stress (when steep tensile fractures were forming) to frictional slip with hydrostatic pore fluid pressure and higher differential stress (when conjugate faults were forming). Axen et al. (2001) reached similar conclusions about Alpine LANFs in the ductile-brittle transition.

Such conditions may provide a mechanical explanation for LANF initiation and slip, provided that the stress tensor in the fault zone need not be identical with that of its surroundings (Axen 1992). Rice (1992) showed that if a few reasonable criteria are met, then the principal stresses and fluid-pressure levels inside and outside a mature fault zone need not be the same in either orientation nor magnitude: stress increase and reorientation may occur within the fault core due to pore-fluid-pressure increase in the fault core. Axen (1992) applied the Rice model to LANFs and also argued that many LANFs, especially those bounding metamorphic core complexes, could initiate largely because of mechanical and permeability anisotropies that develop due to ductile shearing: rocks are weaker and more permeable parallel to the foliation than perpendicular to it. A data-based mechanical analysis of Alpine detachment faults (Axen et al. 2001) supports this idea of local stress tensor changes in general, but not in particular details.

Three specific models have been proposed for frictional slip on LANFs. Axen and Selverstone (1994) used orientation data from mixed-mode extensional/shear fractures developed in the upper ~200 m of the footwall of the Whipple detachment, California to show that the average angle θ_d between the maximum principal stress and the Whipple detachment ranged from ~55° to 80°, far outside the range permitted by Coulomb theory. These data suggest a subvertical maximum principal stress if the Whipple detachment was gently inclined when the fractures formed. On the basis of these data, they constructed a Mohr diagram that shows how the Whipple detachment could have slipped at low dip in such a stress regime, provided that (1) cohesion was important in the immediate surroundings of the detachment and (2) failure in the cohesive zone could be described by a Coulomb-Griffith failure envelope. In the model, the LANF itself fails according to a cohesionless friction law $\tau_d = \mu_d \sigma_{nd}$, where τ_d, σ_{nd}, and μ_d are the shear traction, effective normal traction, and coefficient of sliding friction, respectively, on the detachment. Using reasonable values of the tensile strength of the fault surroundings and average orientations of subsidiary faults, they showed that LANF slip could have occurred for fluid pressure that was hydrostatic at 4 km depth and 70 percent of lithostatic at 10 km depth. Axen and Selverstone (1994) argued that steep hanging wall faults would slip under cohesionless friction at lower differential stress than along the detachment and its immediate surroundings. This model has the distinct advantage of being based on and compatible with a variety of field and experimental data; the only assumptions involved are that rock mechanical results apply to fracture formation and frictional slip. Abers (2001) formulated a similar model using cohesionless friction for LANFs with tensile failure of LANF surroundings as a limit, and assuming that the maximum principal effective stress is vertical and equal to the effective lithostatic load. These models are discussed further in the next section.

Recently, Hayman et al. (2003) applied the extensional critical Coulomb wedge model (Xiao et al. 1991) to Quaternary (post-180 ka) LANFs in Death Valley that dip 19° to 36°. The model requires that the basal slope (LANF dip), surface slope, and both pore-fluid pressure and sliding friction for faults within the wedge as well as for the LANF all be specified. They concluded that critical wedge mechanics can explain slip on the LANF there if friction on the LANF is reduced to 0.4 or if fluid pressure along the LANF is raised to 70 percent of the lithostatic load on the LANF. Like Axen and Selverstone (1994), they used orientation of upper plate faults to infer a subvertical maximum principal stress.

LANF Friction and Limits on Differential Stress in Extending Crust

The models of Axen and Selverstone (1994) and Abers (2001) explain LANF slip by allowing differential stress in the crust to exceed that required for slip on optimally oriented faults. This section explores the potential limits on differential stress in extending crust and then compares the results from different limits and with existing data on LANF dips and depths.

Frictional sliding is described by

$$\tau = \mu\sigma_n \qquad (3.1)$$

where μ is the coefficient of sliding friction on some preexisting surface, τ is shear traction, and σ_n is effective normal traction (figure 3.9). Laboratory rock friction is characterized by coefficients of sliding friction μ given by "Byerlee's law." Byerlee (1978) compiled experimentally determined coefficients of maximum sliding friction from several laboratories and for a variety of rocks, fitting the data as

$$\tau = 0.85\ \sigma_n, \qquad 5\ \text{MPa} \leq \sigma_n \leq 200\ \text{MPa, and} \qquad (3.2)$$
$$\tau = 0.6\ \sigma_n\ +\ 50\ \text{MPa},\ 200\ \text{MPa} \leq \sigma_n \leq 2000\ \text{MPa} \qquad (3.3)$$

Thus, for most common rocks the coefficient of sliding friction μ is 0.85 at low pressure and 0.6 at greater pressure, where cohesion of 50 MPa is also needed. In fact, equation (3.2) is the form appropriate for the brittle continental crust in extension (see later in the chapter). Regardless, it has become common to ignore both the low-pressure form (equation 3.2) and the cohesion term in the high-pressure form (equation 3.3) in favor of a single simpler expression:

$$\tau = 0.6\ \sigma_n, \qquad (3.4)$$

Byerlee Friction on Optimally Oriented Faults Limits Differential Stress

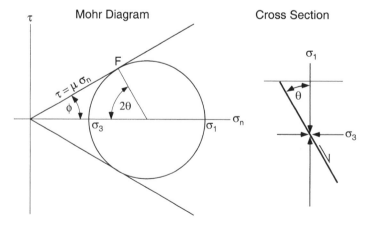

Figure 3.9 Mohr diagram showing a frictional failure envelope limiting differential stress in the crust. In this scenario, only optimally oriented faults, such as in the cross section or represented by point F, can slip.

which here is called the "minimum Byerlee limit." Examination of figures 5 and 7 of Byerlee (1978; see also figure 3 of Blanpied et al. 1998) shows that equation (3.4) provides a lower bound to most laboratory rock friction data (excluding certain weak minerals) but actually fits the bulk of the data poorly.

Byerlee's law has been recast as a limit on strength of the brittle lithosphere (Goetze and Evans 1979; Brace and Kohlstedt 1980), assuming that (1) one principal stress is vertical (Anderson 1942) and (2) optimally oriented weaknesses exist everywhere, for which $\theta_{opt} = 45° - \phi/2 = 0.5 \cot^{-1}\phi$. The shear and effective normal tractions τ and σ_n on such faults are given by

$$\tau = 0.5 (\sigma_1 - \sigma_3) \sin2\theta \text{ and} \tag{3.5}$$

$$\sigma_n = 0.5 (\sigma_1 + \sigma_3) - 0.5(\sigma_1 - \sigma_3) \cos2\theta \tag{3.6}$$

(Jaeger and Cook 1976:14; but note that I define θ differently). Here σ_1 and σ_3 are the maximum and minimum principal effective stresses, respectively, with compression positive. Slip on such faults may limit the differential stress $\sigma_D = \sigma_1 - \sigma_3$ in the crust (figure 3.9). However, the second assumption does not allow for slip on misaligned faults, which clearly exist (e.g., steep reverse faults, LANFs), proving the assumption invalid. Regardless, this formulation for crustal strength is common, although it is probably misleading in many cases and should be used with care.

The existence of misaligned faults requires consideration of cases in which slip on optimally oriented faults obeying Byerlee friction does not limit crustal strength. Viable alternative limits on differential stress in the crust are given by allowing the minimum principal effective stress σ_3 to approach either zero or the tensile strength T, or to limit differential stress with failure envelopes of the Griffith or Coulomb types. All allow σ_D to increase dramatically beyond a Byerlee limit and allow θ to be considerably greater (or less) than the optimal angle.

These cases are discussed in the next section in relationship to LANF slip and put in terms of quantities that can be measured in the field or estimated from experimental results. These are detachment dip (which can be measured and related to the angle θ); cohesion C, tensile strength T, and the coefficient and angle of internal friction, μ_i and ϕ_i, respectively, which have been measured for common rock types. Unless stated otherwise, I assume that the maximum principal effective stress equals the vertical effective stress with hydrostatic fluid pressure:

$$\sigma_1 = \sigma_v = \Delta\rho gz = (\rho_r - \rho_w)gz, \tag{3.7}$$

where ρ_r is density of typical crustal rock (2,700 kg/m^3), ρ_w is density of water (1,000 kg/m^3), g is gravitational acceleration (9.8 m/s^2), and z is depth (Anderson 1942). Hydrostatic pore pressure provides a limiting case and, although elevated fluid pressure exists in some cases, much evidence indicates that it is not likely during protracted periods (Manning and Ingebritson 1999; Townend and Zoback 2000).

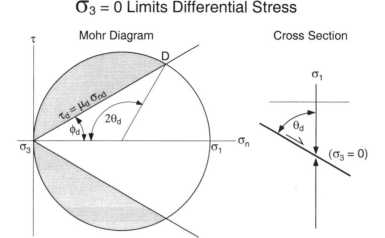

Figure 3.10 Mohr diagram for the case in which differential stress is limited by allowing the minimum principal effective stress to equal zero. Any fault with shear and effective normal tractions that fall in the shaded region may slip. D represents the detachment fault shown in the cross section.

Differential Stress Limited by $\sigma_3 = 0$

Figure 3.10 shows the limiting case in which the least principal effective stress is zero. This case is justified by assuming that tensile strength of rocks is so low that it can be neglected. Rocks will fail in tension if σ_3 reaches their tensile strength (see next section). Note that any fault with resolved shear and effective normal tractions τ and σ_n that fall in the shaded region can slip. We are concerned with detachment faults represented by point D, for which $\theta_d = \cot^{-1}\mu_d$. From equations (3.4) and (3.5) the shear and effective normal tractions on the detachment, τ_d and σ_{nd}, are

$$\tau_d = 0.5\ \sigma_1\ \sin2\theta_d \text{ and} \tag{3.8}$$

$$\sigma_{nd} = 0.5\ \sigma_1 - 0.5\ \sigma_1\ \cos2\theta_d \tag{3.9}$$

Substituting (7), (8), and the limiting value $\sigma_{nd} = 200$ MPa into equation (3.2) with $\Delta\rho = 1{,}700$ kg/m³ and solving for z shows that equation (3.2) is valid to ~20.6 km depth (in the absence of other limits on differential stress).

For $\mu = 0.85$ the shallowest normal fault dip possible is ~40°. Using the minimum Byerlee limit $\mu = 0.6$ allows detachments only as gentle as 31° to slip. More gently dipping detachments clearly exist, so upper-crustal slip on those is inconsistent with crustal strength limited by $\sigma_3 = 0$ and hydrostatic pore fluid pressure, both for standard Byerlee friction and with the minimum Byerlee limit.

Differential Stress Limited by Tensile Strength

Figure 3.11 shows the situation in which the differential stress is limited by the tensile strength T: $\sigma_3 = T$ (Sibson 1998). From triangles TDA and ODA we have

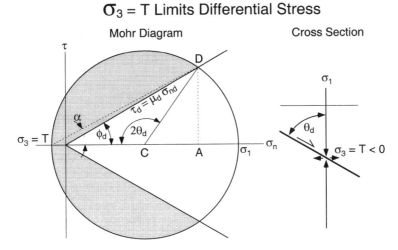

Figure 3.11 Mohr diagram for the case in which tensile strength limits differential stress. D and shaded region as in figure 3.10.

$\sigma_{nd} = \tau_d/\tan \phi_d = (\tau_d/\tan \alpha) + T$. Combining this with equation (3.5) using $\sigma_3 = T$ and $\cot \alpha = \tan \theta_d$ (from triangle TCD) we obtain

$$\frac{1}{\mu_d} = \tan \theta_d + \frac{2T}{(\sigma_1 - T) \sin 2\theta_d} \tag{3.10}$$

The second term on the right side reflects the difference from the case of $T = \sigma_3 = 0$. The magnitude of T for common crustal crystalline rocks (granite, diorite) ranges from ~1 to ~30 MPa with a mean of ~10 MPa (Lockner 1995). Following the procedure of the previous section, equation (3.2) is valid to ~20.2 km, using equation (3.5) instead of (3.8) and with $\sigma_3 = T = -10$ MPa. At crustal length scales and geologic timescales T may be lower, so this average value likely provides a maximum limit on differential stress.

Differential Stress Limited by Griffith Criterion

Intact rocks may fail along mixed-mode, extensional, and shear fractures when $\sigma_3 < 0$, but such failure is difficult to characterize with a simple analytical expression (Lockner 1995), probably because of the important effects of inhomogeneous microcrack distributions in rocks. For lack of a better alternative, the Griffith criterion is used (figure 3.12; Jaeger and Cook 1976:101–102; Axen and Selverstone 1994; Sibson 1998) as the limit on differential stress:

$$\tau_g^2 = 4|T|(\sigma_{ng} + |T|) \qquad \text{for } \sigma_1 + 3\sigma_3 < 0. \tag{3.11}$$

where τ_g and σ_{ng} are the shear and effective normal tractions, respectively, on a mixed-mode fracture formed at some point G on the Griffith failure envelope. This yields a parabola on a Mohr diagram, with slope at G given by

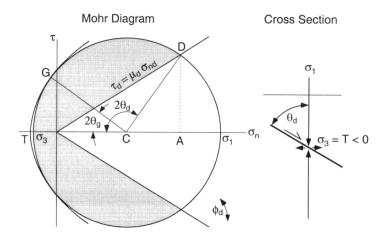

Figure 3.12 Mohr diagram for the case in which Griffith failure limits differential stress. D and shaded region as in figure 3.10.

$$\mu_g = \frac{d\tau_g}{d\sigma_{ng}} = \sqrt{\frac{T}{\sigma_{ng} + T}} = \tan \phi_g = \tan(90 - 2\theta_g) = \cot 2\theta_g \quad (3.12)$$

(Axen and Selverstone 1984), where μ_g is a "Griffith coefficient of internal friction" related to the angle ϕ_g (figure 3.12). This is valid up to about $\sigma_{ng} = 0$ (Brace 1960), where equation (3.11) predicts that the tensile strength T is exactly half the cohesive strength C. Above $\sigma_{ng} = 0$, cracks close and a modified Griffith criterion that includes friction across cracks is appropriate (McClintock and Walsh 1962). This criterion is equivalent to the standard Mohr-Coulomb failure criterion (Brace 1960).

The effective normal traction at G (figure 3.12) is $\sigma_{ng} = \sigma_1 - r - r \cos 2\theta_g$, where r is the radius of the circle, equal to half the differential stress σ_D. Substituting this into equation (3.12) and rearranging gives the differential stress

$$\sigma_D = 2r = \frac{\sigma_1 - T(\tan^2 2\theta_g - 1)}{\cos^2 \theta_g}. \quad (3.13)$$

Similarly, the coefficient of friction μ_d on a detachment fault at point D with tractions σ_{nd} and τ_d is given by substituting $\sigma_{nd} = \sigma_1 - r - r \cos 2\phi_d$ and $\tau_d = r \sin 2\phi_d$ (equation 3.5 rewritten in terms of r into equation 1), and using equation (13) for r:

$$\mu_d = \frac{\sin 2\theta_d}{\dfrac{\sigma_1}{r} - 2 \cos^2 \theta_d} = \frac{\sin 2\theta_d}{\dfrac{\sigma_1(2 \cos^2 \theta_g)}{\sigma_1 - T (\tan^2 2\theta_g - 1)} - 2 \cos^2 \theta_d} \quad (3.14)$$

The maximum differential stress (corresponding to the maximum depth) occurs at $\sigma_{ng} = 0$ where (from equations 3.11 and 3.12) $\tau_g = \pm 2T$ and $\phi_g = 2\theta_g = 45°$, so $r = 2|T|/\sin 45 = 2.8\ |T|$. Thus, from equations (3.7) and (3.13) $z_{max} = 4|T|\cos^2 22.5°/\Delta\rho g \sin 45°$. Using $T = -10$ MPa (see previous section) and $\Delta\rho = 1,700$ kg/m³, the Griffith limiting case applies at depths less than $z_{max} = 2.9$ km if pore pressure is hydrostatic. At the maximum depth, equation (3.14) simplifies to $\mu_d = (\sin 2\theta_d)/(2 \cos^2 22.5° - 2 \cos^2 \theta_d)$, which for $\mu_d = 0.85$ yields $\theta_d = 56°$ and a minimum detachment dip of 34°; for $\mu_d = 0.6$, $\theta_d = 64°$ and the minimum detachment dip is 26°. These are very similar to the values given by equation (3.10) for 2.9 km depth and $T = -10$ MPa. Thus, the Griffith and tensile-strength limits are roughly equivalent with respect to LANF mechanics, and therefore, I restrict discussion in the next section to the simpler tensile strength limit.

Differential Stress Limited by Coulomb Failure

Figure 3.13 shows the case for which a linear Coulomb criterion limits the differential stress (Sibson 1998):

$$\tau_i = \mu_i \sigma_{ni} + C, \tag{3.15}$$

where τ_i and σ_{ni} are the resolved shear and effective normal tractions, respectively, on the newly formed fracture, μ_i is the coefficient of internal friction (the tangent of the angle of internal friction ϕ_i), and C is cohesion. As mentioned previously, we also have $\tau_i = r \sin 2\theta_i$ and $\sigma_{ni} = \sigma_1 - r - r \cos 2\theta_i$. Using these with $\phi_i =$

Coulomb Failure Limits Differential Stress

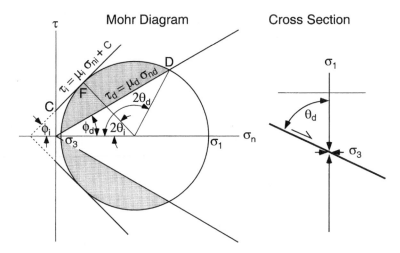

Figure 3.13 Mohr diagram for the case in which Coulomb failure limits differential stress. D and shaded region as in figure 3.10.

$90 - 2\theta_i$ in equation (3.15) and some trigonometric identities yields the differential stress:

$$\sigma_D = 2r = 2(\mu_i\sigma_1 + C)\tan\theta_i \qquad (3.16)$$

The coefficient of friction μ_d on the detachment is calculated as mentioned previously: this expression for r is combined with the left equality of equation (14) along with $\tau_d = r\sin 2\theta_d$:

$$\mu_d = \frac{\sin 2\theta_d}{\dfrac{\sigma_1}{(\mu_i\sigma_1 + C)\tan\theta_i} - 2\cos^2\theta_d}. \qquad (3.17)$$

Mohr-Coulomb failure permits lower differential stress in extending crust at greater depth than tensile strength does (figure 3.14): as depth and σ_1 increase, σ_3 ultimately must become greater than T and the entire circle shifts to the right. The minimum depth (for hydrostatic pore pressure) is obtained using equations (3.7) and (3.16) in $\sigma_3 = T = \sigma_1 - 2r$:

$$Z_{min} = \frac{T + 2C\tan\theta_i}{\Delta\rho g(1 - 2\mu_i\tan\theta_i)}, \qquad (3.18)$$

which describes Mohr circles that both are tangent to the failure envelope and pass through T. At depths shallower than this minimum (smaller σ_1), circles that reach the failure envelope have $\sigma_3 < T$, which is impossible. Using $T = -10$ MPa, $C = 20$ MPa, and $\mu_i = 1$ (see section on continuum models of LANF initiation) in equation (3.18) yields $z_{min} = 2.3$ km. It also is useful to consider parameter values that conspire to restrict the part of the crust in which differential stress is limited by cohesion. Using $C = 20$ MPa and $\mu_i = 1$ with $T = -5$ MPa and 0 MPa yields z_{min} of 4.0 km and 5.8 km, respectively.

Discussion

Abundant compelling evidence exists that several detachment faults and listric normal faults formed and/or slipped at low dips (3–30°) at 0–10 km depths. Clearly, the question of whether or not LANFs exist is answered in the affirmative, and focus should shift to understanding LANF mechanics and their implications for general fault and crustal mechanics.

LANFs provide a unique setting in which to study the products of "weak" faults because they directly and rapidly expose footwall fault rocks that were formed by interactions of various processes, allowing inexpensive, direct observation. In contrast, strike-slip faults typically leave all but the shallowest fault rocks at depth, and thrust faults literally bury the evidence. To study directly the

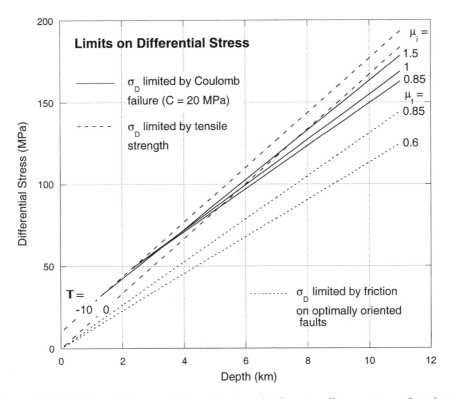

Figure 3.14 Various limits on differential stress in the extending crust as a function of depth, for different values of cohesion, internal friction, and tensile strength. Friction on optimally oriented faults is shown only for reference because large angles between LANFs and the maximum principal stress direction preclude it from being the limiting factor. Note that differential stress is limited by tensile strength in the shallow extending crust and by cohesive failure at depth.

products of strike-slip and thrust faulting, either expensive in situ studies or subsequent exhumation are required, the latter probably occurring after significant geologic time has passed during which potentially significant modifications of fault rocks may occur.

Crustal Strength and LANF Friction

Significant controversy exists about the strength of the crust, strength of crustal faults, and the nature of mechanical limits on these values. Thus, severely misoriented LANFs, being among the most difficult faults to explain mechanically, provide constraints on crustal and fault strength if standard models of rock friction and failure apply. Controversy centers on whether or not specific faults are weak relative to their surroundings, relative to other nearby faults, and relative to laboratory values of friction.

The differential stress in brittle extending crust may be controlled by Coulomb failure, tensile strength, and/or Byerlee's law. In turn, the differential stress controls the degree to which frictional slip may occur on misoriented faults because faults within weak surroundings cannot be as misoriented as those in stronger surroundings. Figure 3.14 compares the maximum differential stress allowed in extending crust by Coulomb failure, tensile strength, and Byerlee's law for optimally oriented faults, using accepted ranges of internal friction (0.85 to 1.5), sliding friction (0.6 to 0.85), and tensile strength (0 to −10 MPa). Because the crust is not everywhere cut by optimally oriented faults, it is unlikely that Byerlee's law limits differential stress in the crust, although frictional slip on misaligned faults may do so if enough misaligned faults exist to preclude failure of intact rock. If only optimally aligned faults are considered, then the maximum angle θ permitted between σ_1 and the weakest faults (e.g., ones with vanishingly small friction) is 45°. This is clearly untrue for LANFs where studies have constrained this angle (Reynolds and Lister 1987; Axen and Selverstone 1994).

Thus, where LANFs slip, Coulomb failure apparently limits differential stress in the deeper part of the brittle crust, and tensile (or Griffith) failure does so in the shallower part (figure 3.14; Sibson 1998). Coulomb failure permits lower differential stress in the deeper brittle crust than allowed with tensile failure alone as a limit, so makes slip more difficult on misaligned faults there (figure 3.15).

LANFs in the deeper brittle crust do not appear to slip under normal Byerlee friction and hydrostatic fluid pressure. Figure 3.15 shows two groups of curves that reflect detachment friction of 0.85 (low-pressure Byerlee's law) and 0.6 (minimum Byerlee limit). Within each group, the minimum normal fault dip changes from being limited by tensile strength at shallower depths (dashed lines) to being limited by cohesion at greater depths (solid lines). The crossover point depends on the specific tensile strength, cohesion, and internal friction used. The minimum allowed dip decreases for higher internal friction, higher cohesion, and more negative tensile strength. Detachment friction of 0.85 does not allow slip on normal faults dipping less than 30° deeper than ~2 km; and for $\mu_d = 0.6$, slip on normal faults dipping less than 31° is not allowed if tensile strength is negligible, nor below ~5 km for $\mu_i = 1$ and $C = 20$ MPa. In contrast, LANF slip is allowed in the upper 1–2 km of the crust if $T < 0$.

The fact that surficial landslides and LANFs exist is strong evidence against the use of $T = 0$ as a limit on crustal strength (a common assumption). Landslides and LANFs in the upper few kilometers of the crust indicate that it is inappropriate to assume that $T = 0$ limits differential stress at shallow crustal levels. Because Coulomb failure, not tensile strength, apparently limits differential stress at deeper levels (figure 3.14), it appears inappropriate to assume that $T = 0$ is a valid limit on differential stress at any crustal level.

It thus seems clear that LANFs require "abnormal" conditions to slip. The key question is what causes these conditions and how abnormal are they? The apparent lack of abnormally weak minerals along some LANFs suggests that a likely cause may be elevated fluid pressure (Axen 1992; Axen and Selverstone 1994). However, figure 3.16 shows that even extremely elevated fluid pressure ($\lambda > 0.9$, where $\lambda = P_f/\rho gz = P_f/\sigma_1$) does not adequately explain slip on the Whipple, Tule

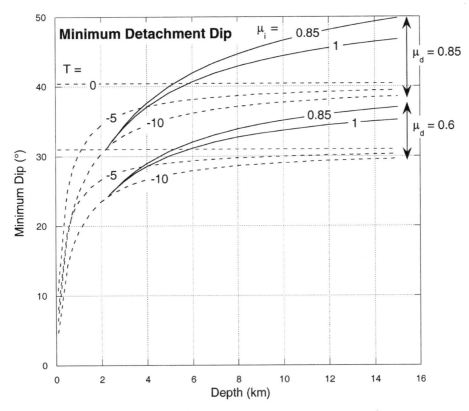

Figure 3.15 Minimum permitted normal fault dip as a function of depth for two values of internal friction, cohesion of 20 MPa, and three tensile strength values. Note (from figure 3.14) that minimum dip is determined by tensile failure in the shallow crust, where LANFs are permitted, and by Coulomb failure at depth, where they are not. Tensile failure limits are shown to depth only for reference.

Springs, or Chemehuevi detachments. The effect of high pore fluid pressure is to increase the depth to which LANF slip of a given dip is possible (figure 3.16), but high pore pressure does not change the geometry of the Mohr's diagrams discussed previously and becomes less effective at shallow depth.

A better alternative is that the coefficient of friction on detachments is lower than laboratory values. Figure 3.17 shows the maximum friction permissible for slip on detachments of specific dips, assuming hydrostatic pore pressure and vertical maximum principal stress. Dip and depth ranges are also shown for LANFs discussed previously, for which sliding friction apparently was in the range $0.2 < \mu_d < 0.5$ at depths from 1 to 10 km, given hydrostatic fluid pressure and the parameter values shown in figure 3.17 (recall that these values may allow differential stress that is too high and detachment dips that are too low, exacerbating the problem). Reduction of friction over minimum laboratory values explains slip at depth on the Chemehuevi and Tule Springs detachments better than elevated pore pressure does but requires very low friction for the last few kilometers of

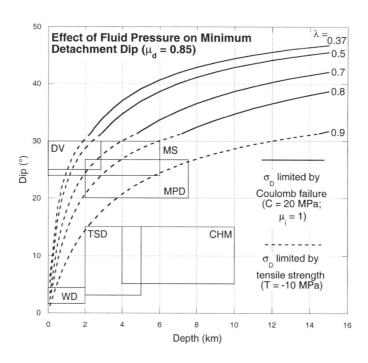

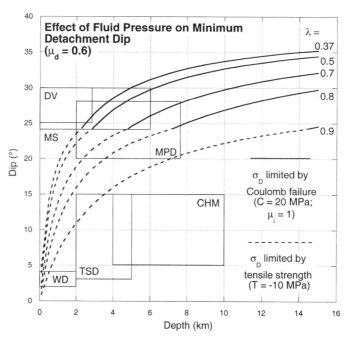

slip on the Whipple fault. The detachment at Moresby Seamount with talc along it and the siltstones subjacent to the Tule Springs detachment both indicate that weak minerals may help some LANFs slip, but others, such as the Mormon Peak, Castle Cliff, Whipple, and Chemehuevi detachments, do not have such minerals.

Thus, accepted laboratory values of sliding friction are probably not applicable to all natural faults, lending credence to low-friction models of other faults (Lachenbruch and Sass 1980; Mount and Suppe 1987; Zoback et al. 1987; Bird and Kong 1994). In particular, it seems likely that fault strength decreases for large slip. Most measures of rock strength decrease with increasing sample size (Lockner 1995), although this is not expected of friction (see Scholz 1990:46–48). Existing in situ constraints on rock friction suggest that laboratory values of friction apply to fractures and faults sampled by borehole measurements (centimeters to many meters; e.g., Townend and Zoback 2000; many others) and by small earthquakes (tens to hundreds of meters; e.g., Zoback and Harjes 1997). Townend and Zoback (2000, figure 3) showed that borehole stress measurements are consistent with sliding friction of 0.85 on hydraulically conductive fractures in the upper few kilometers of both stable, nonorogenic crust and crust adjacent to the San Andreas fault at Cajon Pass. The vast majority of such fractures had ratios of resolved shear to effective normal tractions of 0.6–0.85. Field observations (Sibson 1994) also support application of laboratory friction values to faults of moderate displacements (a few kilometers). Laboratory and in situ measurements span a few orders of magnitude in length scale, but they are nevertheless one to two orders of magnitude smaller in fault size and several orders of magnitude smaller in slip than detachments; field examples discussed by Sibson (1994) have one to two orders of magnitude less slip than most detachments. Because these data from smaller faults favor Byerlee's law and because strong evidence presented previously favors low frictional strength on large-displacement LANFs, it seems likely that fault strength decreases with slip. Future experimental studies focused on larger slip magnitudes may provide important insights.

An alternative possibility is that the standard mechanical analyses presented previously are not, in fact, applicable to large-slip LANFs or other large-slip faults. In the arguments mentioned previously, friction is essentially averaged over long geologic timescales and large distances, ignoring dynamic friction and short-term

Figure 3.16   (a) Effect of suprahydrostatic fluid pressure ($\lambda = P_f/\rho g z = P_f/\sigma_1$) on minimum normal fault dip, for Byerlee's law (coefficient of friction of 0.85) and for differential stress limited by tensile strength (10 MPa) in the shallow crust and by Coulomb failure (20 MPa cohesion and internal friction of 1.0) at depth. Boxes show geologic constraints on formation and slip angles for detachments discussed in text: MS, Moresby Seamount; MPD, Mormon Peak detachment; TSD, Tule Springs detachment; CHM, Chemehuevi detachment; WD, Whipple detachment. (b) Same as in (a) but for the minimum Byerlee limit (coefficient of friction of 0.6). Note that even high fluid overpressure does not explain some geologic examples.

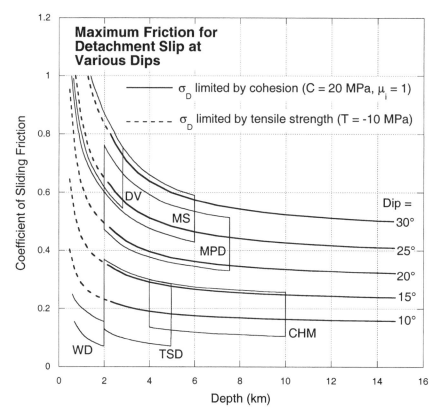

Figure 3.17 Maximum permissible detachment friction versus depth for fixed values of detachment dip, using same limits on differential stress as in figure 3.16. Depth-dip data are displayed for same examples as in figure 3.16. Note that detachment friction between 0.4 and 0.5 allows slip on some LANFs at depth, but that friction must fall to very low values (0.2–0.3) to explain others.

changes, both of which may be important. For example, if dynamic friction can drop precipitously from static friction values and if onset of dynamic slip events is controlled by chemical or stress corrosion of locked asperities on faults, then the analysis discussed previously, and much of the published rock-mechanics literature, would be irrelevant.

Spatially Heterogeneous Stress States around LANFs

The Mohr's constructions discussed in the previous section implicitly assume that LANFs are surrounded by intact rock. This may be reasonable for the footwalls, where large, through-going faults are uncommon and where brittle LANF slip was commonly preceded by mylonitization and chloritic brecciation, so that footwall rocks were, in effect, reconstituted immediately before discrete LANFs evolved (see Axen and Selverstone 1994). However, the upper plates of LANFs are typically cut by many normal faults, some of which formed early in the LANF history and have kilometers of slip. Thus, a strong case can be made that the mechanical states, and therefore strengths, of the upper and lower plates are different through much of LANF evolution. In turn, this implies that the stress states differ across detachments too.

It also has been suggested that the stress states within mature fault zones should be different from that of their surroundings. Mandl et al. (1977) showed experimentally that the orientations of maximum and minimum principal stresses in granular shear zones evolve to 45° angles to the shear zone boundary; Byerlee and Savage (1992) provided a simple mechanical explanation for this behavior. Byerlee (1990) and Rice (1992) suggested more radical changes in the stress tensor within mature shear zones based on fluid-pressure gradients and stress-continuity relations; Axen (1992) applied the Rice model to LANFs. Axen et al. (2001) argued that the stress field changed rapidly both in space and time during the ductile-to-brittle evolution of Alpine detachment faults. Most of these models share the concept that shear and effective normal tractions are continuous across the fault-zone boundaries.

The observationally based models (Axen 1992; Axen and Selverstone 1994; Abers 2001; Axen et al. 2001) attempt to explain LANF slip and evolution in terms of standard rock mechanical theory and spatial mechanical heterogeneity. Many structural orientation data suggest that LANFs slip while the maximum principal stress is at a high angle ($55° \leq \theta \leq 80°$) to the fault plane, in accord with Anderson's (1942) theory, even near the faults (within meters or tens of meters) (Reynolds and Lister 1987; Axen and Selverstone 1994). Axen and Selverstone (1994) and Abers (2001) explain this by embedding cohesionless, frictional LANFs in stronger cohesive surroundings. Because such models are fundamentally data based, and despite differences among them, they may point us in the correct direction. Of course, temporal mechanical variations may be as important as spatial mechanical heterogeneity (e.g., due to the seismic cycle or to varying fluid pressure and chemistry, fluidization of gouge, formation of frictional melt, etc.).

Initial Formation of LANFs

The most outstanding problem in LANF mechanics is that no satisfactory model exists that explains the initial formation of LANFs in intact, isotropic rock, or where they cut across obvious anisotropy, although both clearly have happened. For example, the Chemehuevi Mountains detachments (Howard and John 1987; John 1987) cut across wide expanses of relatively isotropic plutonic rocks. Similarly, the Mormon Peak detachment initially transected subhorizontal autochthonous strata at low angles without following the mechanical anisotropy present, and cut gently into the upper 2–3 km (minimum) of underlying crystalline basement (Wernicke et al. 1985; Wernicke and Axen 1988). Continuum models (discussed in a previous section) allow LANF initiation only locally and typically require widespread compression of adjacent crust simultaneously, at odds with numerous observations of widespread extension on steep normal fault systems during detachment faulting.

In many cases, the mechanical and permeability anisotropy provided by pre-existing mylonitic foliation that formed in the midcrust probably played a role in localizing LANFs (Axen 1992). In the Tule Springs Hills the mechanical contrasts between strong dolomites of the upper plate of a thrust and the weak siltstones of the thrust footwall probably controlled the very gently dipping LANF trajectory (Axen 1993). Control of initial trajectories by preexisting anisotropy may be common, but the exceptions to this rule provide strong evidence that current fault mechanical theories have fundamental flaws or overlook important effects.

It is accepted that many LANFs evolved from distributed ductile shear zones into discrete brittle faults over time spans of a few million years (Wernicke 1981; many others). What is not emphasized is that this occurs through a complex interaction of structures and deformation regimes that operate on widely different time and length scales. Deep, hot ductile shear zones that are kilometers thick may flow at steady long-term strain rates controlled by the microscopic to grain-scale characteristics of the weakest common mineral phase (e.g., quartz). However, strain rate probably varies widely near the ductile-brittle transition where the effects of the seismic cycle are felt. Cataclastic flow may occur in thick (order 10–100 m) zones but be controlled by fracture mechanics at a much smaller scale. In some settings, ductile shear zones may become embrittled by geologically instantaneous formation of cracks and fractures that range from centimeters to a few kilometers in length (Axen et al. 2001). Earthquake ruptures may propagate downward into ductile zones, take seconds to form, be kilometers in length, and form or follow narrow (centimeters) zones of fault gouge that may flow cataclastically at both long-term and seismogenic strain rates and that undergo chemical alteration on the timescales of pore-fluid passage (hours to years).

The character of these spatial and temporal heterogeneities and the nature of their interactions are probably key to a better understanding of LANF mechanics as well as to fault mechanics and mechanics of the brittle crust in general. If the existing models that emphasize such temporal and spatial variations are correct (Byerlee 1990; Rice 1992; Axen 1992, Axen and Selverstone 1994; Axen et al. 2001), then crustal strength may be controlled by complicated interactions among a variety of processes within narrow zones (order 10–1000 m thick) surrounding evolving faults. The challenge ahead is to obtain laboratory, field, and in situ data pertinent to such interactions and to integrate them into models that iterate among processes operating on a variety of time and length scales.

Conclusions

Abundant evidence exists for formation of and for slip on LANFs. Regardless, existing mechanical models do not fully explain these enigmatic faults. Existing continuum models for the formation of LANFs do not match observations that most LANFs form in areas undergoing regional extension and that the maximum principal stress makes a large angle with the LANFs even within tens of meters of the faults. The most widely applicable model for formation of LANFs is based on field observations but relies on mechanical and permeability anisotropies that support stress-field heterogeneity around LANFs. However, this model does not explain LANFs that formed in isotropic rock or cut across preexisting rock fabrics. The most successful models for frictional slip on LANFs require that the faults are embedded in stronger surroundings in which differential stress can become elevated. The assumption that tensile strength is negligible leads to no situations in which LANFs may form, so is rejected as a limit on differential stress in extending brittle crust. Including tensile strength in simple models allows LANF slip in the shallow crust, where it limits differential strength. Deeper in the extending brittle crust, Coulomb failure limits differential stress, but LANF slip is not allowed under hydrostatic fluid pressure and Byerlee friction values. Elevated pore-fluid pressure does not allow LANF slip in the deeper brittle crust unless extreme ($\lambda > 0.9$) and unlikely, fluid pressure is reached and maintained. However, low coefficients of sliding friction (relative to laboratory values) allow LANF slip throughout the brittle crust. Low friction may result from large-magnitude slip. These observations lend credence to low-friction models of other large faults. The difficulty of explaining LANF formation and slip suggest that our understanding of crustal mechanics is lacking some key ingredient(s), possibly related to spatial, temporal, and/or chemical changes that are not addressed in standard rock mechanical theory.

References

Abbott, R. E., J. N. Louie, S. J. Caskey, and S. Pullammanappallil. 2001. Geophysical confirmation of low-angle normal slip on the historically active Dixie Valley Fault, Nevada. *J. Geophys. Res.* 106:4169–4181.

Abers, G. A. 1991. Possible seismogenic shallow-dipping normal faults in the Woodlark-D'Entrecasteaux extensional province, Papua New Guinea. *Geology* 19:1205–1210.

Abers, G. A. 2001. Evidence for seismogenic normal faults at shallow dips in continental rifts. In: R.C.L. Wilson, B. Taylor, and N. Froitzheim, eds., *Nonvolcanic Rifted Margins*. Geological Society of London Special Publication 187, pp. 305–318.

Abers, G. A., C. Z. Mutter, and J. Fang. 1997. Shallow dips of normal faults during rapid extension: Earthquakes in the Woodlark-D'Entrecasteaux rift system, Papua New Guinea. *J. Geophys. Res.* 102:15301–15317.

Anderson, E. M. 1942. *The Dynamics of Faulting and Dyke Formation with Application to Britain.* Edinburgh: Oliver and Boyd.

Anderson, R. E. 1971. Thin-skin distension in Tertiary rocks of southwestern Nevada. *Geol. Soc. Am. Bull.* 82:43–58.

Anderson, R. E. and T. P. Barnhard. 1993a. Aspects of three-dimensional strain at the margin of the extensional orogen, Virgin River depression area, Nevada, Utah, and Arizona. *Geol. Soc. Am. Bull.* 105:1019–1052.

Anderson, R. E. and T. P. Barnhard. 1993b. Heterogeneous Neogene strain and its bearing on horizontal extension and horizontal and vertical contraction at the margin of the extensional orogen, Mormon Mountains area, Nevada and Utah. *U.S. Geol. Surv. Bull.* 2011:1–43.

Armstrong, R. L. 1972. Low-angle (denudation faults), hinterland of the Sevier orogenic belt, eastern Nevada and western Utah. *Geol. Soc. Am. Bull.* 83:1729–1754.

Axen, G. J. 1992. Pore pressure, stress increase, and fault weakening in low-angle normal faulting. *J. Geophys. Res.* 97(B6):8979–8991.

Axen, G. J. 1993. Ramp-flat detachment faulting and low-angle normal reactivation of the Tule Springs thrust, southern Nevada. *Geol. Soc. Am. Bull.* 105:1076–1090.

Axen, G. J. 1999. Low-angle normal fault earthquakes and triggering. *Geophys. Res. Lett.* 26:3693–3696.

Axen, G. J. and J. M. Bartley. 1997. Field test of rolling hinges: Existence, mechanical types, and implications for extensional tectonics. *J. Geophys. Res.* 102:20515–20537.

Axen, G. J., J. M. Bartley, and J. Selverstone. 1995. Structural expression of a rolling hinge in the footwall of the Brenner Line normal fault, eastern Alps. *Tectonics* 14:1380–1392.

Axen, G. J., J. M. Fletcher, E. Cowgill, M. Murphy, P. Kapp, I. MacMillan, E. Ramos-Velázquez, and J. Aranda-Gomez. 1999. Range-front fault scarps of the Sierra El Mayor, Baja California: Formed above an active low-angle normal fault? *Geology* 27:247–250.

Axen, G. J. and J. Selverstone. 1994. Stress-state and fluid-pressure level along the Whipple detachment fault, California. *Geology* 22:835–838.

Axen, G. J., J. Selverstone, and T. Wawrzyniec. 2001. High-temperature embrittlement of extensional Alpine mylonite zones in the midcrustal ductile-brittle transition. *J. Geophys. Res.* 106:4337–4348.

Axen, G. J., W. J. Taylor, and J. M. Bartley. 1993. Space-time patterns and tectonic controls of Tertiary extension and magmatism in the Great Basin of the western United States. *Geol. Soc. Am. Bull.* 105(1):56–76.

Axen, G. J., B. Wernicke, M. F. Skelly, and W. J. Taylor. 1990. Mesozoic and Cenozoic tectonics of the Sevier thrust belt in the Virgin River Valley area, southern Nevada. In B. Wernicke, ed., *Basin and Range Extensional Tectonics at the Latitude of Las Vegas, Nevada*, pp. 123–154. Boulder, CO: Geological Society of America.

Axen, G. J. and B. P. Wernicke. 1989. On the role of isostasy in the evolution of normal fault systems: Reply. *Geology* 17:775–776.

Bartley, J. M. and A. F. Glazner. 1985. Hydrothermal systems and Tertiary low-angle normal faulting in the Southwestern United States. *Geology* 13:562–564.

Bernard, P., P. Briole, B Meyer, H. Lyon-Caen, J.-M. Gomez, C. Tiberi, C. Berge, R. Cattin, D. Hatzfeld, C. Lachet, B. Lebrun, A. Deschamps, F. Courboulex, C. Larroque, A. Rigo, et al. 1997. The Ms = 6.2, June 15, 1995 Aigion earthquake (Greece): Evidence for low angle normal faulting in the Corinth rift. *J. Seism.* 1:131–150.

Bird, P. and X. Kong. 1994. Computer simulations of California tectonics confirm very low strength of major faults. *Geol. Soc. Am. Bull.* 106:159–174.

Blank, H. R. and R. P. Kucks. 1989. Preliminary aeromagnetic, gravity and generalized geological maps of the U.S.G.S. Basin and Range-Colorado Plateau transition zone study area in southwestern Utah, southeastern Nevada, and northwestern Arizona. *U.S. Geol. Surv. Open File Rept. 89–432*, 1–16 (three maps, 1:100,000 scale).

Blanpied, M. L., C. J. Marone, D. A. Lockner, J. D. Byerlee, and D. P. King. 1998. Quantitative measure of the variation in fault rheology due to fluid-rock interactions. *J. Geophys. Res.* 103:9691–9712.

Block, L. and L. H. Royden. 1990. Core complex geometries and regional scale flow in the lower crust. *Tectonics* 9:557–567.

Bohannon, R. G. 1984. Nonmarine sedimentary rocks of Tertiary age in the Lake Mead region, southeastern Nevada and northwestern Arizona. *U.S. Geol. Survey Profess. Pap. 1259*, 1–72.

Bohannon, R. G., J. A. Grow, J. J. Miller, and R.H.J. Black. 1993. Seismic stratigraphy and tectonic development of Virgin River depression and associated basins, southeastern Nevada and northwestern Arizona. *Geol. Soc. Am. Bull.* 105:501–520.

Boncio, P., F. Brozzetti, and G. Lavecchia. 2000. Architecture and seismotectonics of a regional low-angle normal fault zone in central Italy. *Tectonics* 19:1038–1055.

Brace, W. F. 1960. An extension of Griffith theory to fracture of rocks. *J. Geophys. Res.* 65:3477–80.Brace, W. F. and D. L. Kohlstedt. 1980. Limits on lithospheric stress imposed by laboratory experiments. *J. Geophys. Res.* 85:6248–6252.

Buck, W. R. 1988. Flexural rotation of normal faults. *Tectonics* 7:959–973.

Buck, W. R. 1990. Comment on "Origin of regional rooted low-angle normal faults: A mechanical model and its implications." *Tectonics* 9:545–546.

Burchfiel, B. C., R. J. Fleck, D. T. Secor, R. R. Vincelette, and G. A. Davis. 1974. Geology of the Spring Mountains, Nevada. *Geol. Soc. Am. Bull.* 85:1013–1022.

Burchfiel, B. C., K. V. Hodges, and L. H. Royden. 1987. Geology of Panamint Valley-Saline Valley pull-apart system, California, Palinspastic evidence for low-angle geometry of a Neogene range-bounding fault. *J. Geophys. Res.* 92:10422–10426.

Burchfiel, B. C., P. Molnar, P. Zhang, Q. Deng, W. Zhang, and Y. Wang. 1995. Example of a supradetachment basin within a pull-apart tectonic setting: Mormon Point, Death Valley, California. *Basin Res.* 7:199–214.

Byerlee, J. 1978. Friction of rocks. *Pure Appl. Geophys.* 116:615–626.

Byerlee, J. 1990. Friction, overpressure, and fault-normal compression. *Geophys. Res. Lett.* 17:2109–2112.

Byerlee, J. and J. C. Savage. 1992. Coulomb plasticity within the fault zone. *Geophys. Res. Lett.* 19:2341–2344.

Carpenter, D. G., J. A. Carpenter, U. A. Franz, and S. J. Reber. 1989. On the role of isostasy in the evolution of normal faults: Comment. *Geology* 17:774–775.

Carpenter, J. A. and D. G. Carpenter. 1994. Analysis of basin-range and fold-thrust structure, and reinterpretation of the Mormon Peak detachment and similar features as gravity slide systems: Southern Nevada, southwest Utah, and northwest Arizona. In: S. W. Dobbs and W. J. Taylor, eds., *Structural and Stratigraphic Investigations and Petroleum Potential of Nevada, with Special Emphasis South of the Railroad Valley Producing Trend*, pp. 15–52. Nevada Petroleum Society: Reno, Nev.

Caskey, S. J. and S. G. Wesnousky. 1997. Static stress changes and earthquake triggering during the 1954 Fairview Peak and Dixie Valley earthquakes, Central Nevada. *Bull. Seism. Soc. Am.* 87:521–527.

Caskey, S. J., S. G. Wesnousky, P. Zhang, and D. B. Slemmons. 1996. Surface faulting of the 1954 Fairview Peak (M_s 7.2) and Dixie Valley (M_s 6.8) earthquakes, Central Nevada. *Bull. Seism. Soc. Am.* 86:761–787.

Cichanski, M. 2000. Low-angle, range-flank faults in the Panamint, Inyo, and Slate ranges, California: Implications for recent tectonics of the Death Valley region. *Geol. Soc. Am. Bull.* 112:871–883.

Collettini, C. and R. H. Sibson. 2001. Normal faults, normal friction? *Geology* 29:927–930.

Coney, P. J. 1980. Cordilleran metamorphic core complexes: An overview. In M. D. Crittenden Jr., P. J. Coney, and G. H. Davis, eds., *Cordilleran Metamorphic Core Complexes*, pp. 7–31. Boulder, CO: Geological Society of America.

Crittenden, M. D., P. J. Coney, and G. H. Davis, eds. 1980. *Cordilleran Metamorphic Core Complexes*, Memoir 153. Boulder, CO.: Geological Society of America.

Davis, G. A. 1988. Rapid upward transfer of mid-crustal mylonitic gneisses in the footwall of a Miocene detachment fault, Whipple Mountains, southeastern California. *Geol. Rundsch.* 77:191–209.

Davis, G. A. and G. S. Lister. 1988. Detachment faulting in continental extension: Perspectives from the southwestern U.S. Cordillera. In: S. P. Clark Jr., B. C. Burchfiel and J. Suppe, eds., *Processes in Continental Lithospheric Deformation,* pp. 133–159. Boulder, CO: Geological Society of America.

Davis, G. H. 1983. Shear-zone model for the origin of metamorphic core complexes. *Geology* 11:342–347.

Dorsey, R. J. and U. Becker. 1995. Evolution of a large Miocene growth structure in the upper plate of the Whipple detachment fault, northeastern Whipple Mountains, California. *Basin Res.* 7:151–163.

Dorsey, R. J. and P. Roberts. 1996. Evolution of the Miocene north Whipple basin in the Aubrey Hills, western Arizona, upper plate of the Whipple detachment fault. In K. K. Beratan, ed., *Reconstructing the History of Basin and Range Extension Using Sedimentology and Stratigraphy*. Special Paper 303, pp. 127–146. Boulder, CO: Geological Society of America.

Doser, D. I. 1986. Earthquake processes in the Rainbow Mountain-Fairview Peak-Dixie Valley, Nevada, region 1954–1959. *J. Geophys. Res.* 91:12572–12586.

Frost, E. G., T. E. Cameron, and D. L. Martin. 1982. Comparison of Mesozoic tectonics with mid-Tertiary detachment faulting in the Colorado River area, California, Arizona, and Nevada. In J. D. Cooper, ed., Geologic Excursions in the California Desert, pp. 113–159. Anaheim, CA.: Geological Society of American, Cordilleran Section.

Goetze, C. and B. Evans. 1979. Stress and temperature in the bending lithosphere as constrained by experimental rock mechanics. *Geophys. J. R. Astron. Soc.* 59:463–478.

Govers, R. and M.J.R. Wortel. 1993. Initiation of asymmetric extension in continental lithosphere. *Tectonophysics* 223:75–96.

Hafner, W. 1951. Stress distributions and faulting. *Geol. Soc. Am. Bull.* 62:373–398.

Hamilton, W. B. 1988. Detachment faulting in the Death Valley region, California and Nevada. *U.S. Geol. Surv. Bull. 1790*, 51–85.

Handin, J. 1966. Strength and ductility. In S. P. Clark Jr., ed., *Handbook of Physical Constants*. Memoir, pp. 223–289. New York: Geological Society of America.

Handin, J. 1969. On the Coulomb-Mohr failure criterion. *J. Geophys. Res.* 74:5343–5348.

Hayman, N. W., J. R. Knott, D. S. Cowan, E. Nemser, and S. Sarna-Wojcicki. 2003. Quaternary low-angle slip on detachment faults in Death Valley, California. *Geology* 31:343–346.

Hintze, L. F. 1986. Stratigraphy and structure of the Beaver Dam Mountains, Utah. In D. T. Griffin and W. R. Phillips, eds., *Thrusting and Extensional Structures and Mineralization in the Beaver Dam Mountains, Southwestern Utah*. Special Publication, pp. 1–36. Salt Lake City, UT: Utah Geological Association.

Holm, D. K. and D. R. Lux. 1991. The Copper Canyon Formation: A record of unroofing and Tertiary folding of the Death Valley turtleback surfaces. *Geol. Soc. Am. Abstr. Progr.* 23:35.

Howard, K. A. and B. E. John. 1987. Crustal extension along a rooted system of imbricate low-angle faults: Colorado River extensional corridor, California and Arizona. In M. P. Coward, J. F. Dewey, and P. L. Hancock, eds., *Continental Extensional Tectonics*, pp. 299–311. London: Blackwell Scientific Publications.

Howard, K. E., J. W. Goodge, and B. E. John. 1982. Detached crystalline rocks of the Mojave, Buck, and Bill Williams Mountains, western Arizona. In E. G. Frost and D. L. Martin, eds., *Mesozoic-Cenozoic Tectonic Evolution of the Colorado River Region, California, Arizona, and Nevada*, pp. 377–392. San Diego: Cordilleran Publishers.

Jackson, J. A. 1987. Active normal faulting and crustal extension. In M. P. Coward, J. F. Dewey, and P. L. Hancock, eds., *Continental Extensional Tectonics*. Geological Society Special Publication, pp. 3–18. London: Blackwell Scientific Publications.

Jackson, J. A. and N. J. White. 1989. Normal faulting in the upper continental crust: observations from regions of active extension. *J. Struct. Geol.* 11:15–36.

Jaeger, J. C. and N. G. W. Cook. 1976. *Fundamentals of Rock Mechanics*. London: Chapman and Hall.

John, B. E. 1987. Geometry and evolution of a mid-crustal extensional fault system: Chemehuevi Mountains, southeastern California. In M. P. Coward, J. F. Dewey, and P. L. Hancock, eds., *Continental Extensional Tectonics*. Geological Society Special Publication, pp. 313–336. London: Blackwell Scientific Publications.

Johnson, R. A. and K. L. Loy. 1992. Seismic reflection evidence for seismogenic low-angle faulting in southeastern Arizona. *Geology* 20:597–600.

Ketcham, R. A. 1996. Thermal models of core-complex evolution in Arizona and New Guinea: Implications for ancient cooling paths and present day heat flow. *Tectonics* 15:933–951.

Lachenbruch, A. H. and J. Sass. 1980. Heat flow and energetics of the San Andreas fault zone. *J. Geophys. Res.* 85:6185–6222.

Livaccari, R. F., J. W. Geissman, and S. J. Reynolds. 1995. Large-magnitude extensional deformation in the South Mountains metamorphic core complex, Arizona: Evaluation with paleomagnetism. *Geol. Soc. Am. Bull.* 107(8):877–894.

Lockner, D. A. 1995. Rock Failure. In: T. J. Ahrens, ed., *Rock Physics and Phase Relations: A Handbook of Physical Constants*. AGU Reference Shelf 3, pp. 127–147. Washington, DC: American Geophysical Union.

Longwell, C. R. 1945. Low-angle normal faults in the Basin and Range province. *Am. Geophys. Union Trans.* 26:107–118.

Magloughlin, J. F. and J. G. Spray. 1992. Frictional melting processes and products in geological materials: Introduction and discussion. *Tectonophysics* 204:197–206.

Mancktelow, N. S. and T. L. Pavlis. 1994. Fold-fault relationships in low-angle detachment fault systems. *Tectonics* 13:668–685.

Mandl, G., L.N.J. de Jong, and A. Maltha. 1977. Shear zones in granular material. *Rock Mech.* 9:95–144.

Manning, C. E. and S. E. Ingebritsen. 1999. Permeability of the continental crust: Implications of geothermal data and metamorphic systems. *Rev. Geophys.* 37:127–150.

Marone, C. J. 1995. Fault zone strength and failure criteria. *Geophys. Res. Lett.* 22:723–726.

McClintock, F. A. and J. B. Walsh. 1962. Friction on Griffith cracks in rocks under pressure. In *Proceedings of the Fourth U.S. National Congress of Applied Mechanics*, pp. 1015–1021. New York: American Society of Mechanical Engineers.

Melosh, H. J. 1990. Mechanical basis for low-angle normal faulting in the Basin and Range province. *Nature* 343:331–335.

Miller, E. L., P. B. Gans, and J. Garing. 1983. The Snake Range decollement: An exhumed mid-Tertiary ductile-brittle transition. *Tectonics* 2:239–263.

Mount, V. S. and J. Suppe. 1987. State of stress near the San Andreas fault: Implications for wrench tectonics. *Geology* 15:1143–1146.

O'Sullivan, P., D. G. Carpenter, and J. A. Carpenter. 1994. Cooling history of the Beaver Dam Mountains, Utah: Determined by apatite fission track analysis. In S. W. Dobbs and W. J. Taylor, eds., *Structural and Stratigraphic Investigations and Petroleum Potential of Nevada, with Special Emphasis South of the Railroad Valley Producing Trend*, pp. 53–64. Reno, NV: Nevada Petroleum Society.

Proffett, J. M., Jr. 1977. Cenozoic geology of the Yerington District, Nevada, and implications for nature and origin of Basin and Range faulting. *Geol. Soc. Am. Bull.* 88:247–266.

Reynolds, S. J. 1985. Geology of the South Mountains, central Arizona. *Ariz. Bureau Geol. Mineral Tech. Bull.* 195:1–61.

Reynolds, S. J. and G. S. Lister. 1987. Structural aspects of fluid-rock interaction in detachment zones. *Geology* 15:362–366.

Rice, J. R. 1992. Fault stress states, pore pressure distributions, and the weakness of the San Andreas fault. In B. Evans and T.-F. Wong, eds., *Fault Mechanics and Transport Properties of Rocks: A Festschrift in Honor of W. F. Brace*, pp. 475–504. New York: Academic Press.

Rietbrock, A., C. Tiberi, F. Scherbaum, and H. Lyon-Caen. 1996. Seismic slip on a low-angle normal fault in the Gulf of Corinth: Evidence from high-resolution cluster analysis of microearthquakes. *Geophys. Res. Lett.* 23: 1817–1820.

Rigo, A., H. Lyon-Caen, R. Armijo, A. Deschamps, D. Hatzfeld, K. Makropoulos, P. Papadimitriou, and I. Kassaras. 1996. A microseismic study in the western part of the Gulf of Corinth (Greece): implications for large scale normal faulting mechanisms. *Geophys. J. Int.* 126:663–688.

Scholz, C. H. 1990. *The Mechanics of Earthquakes and Faulting*. New York: Cambridge University Press.

Shipboard Scientific Party. 1999. Leg 180 summary. In B. Taylor, P. Huchon, A. Klaus, et al., eds., *Proceedings of the Ocean Drilling Program, Initial Reports 180*. Ocean Drilling Program. Available at www-odp.tamu.edu/publications/180_IR/front.htm.

Sibson, R. 1975. Generation of pseudotachylyte by ancient seismic faulting. *Geophys. J. R. Astron. Soc.* 43:775–794.

Sibson, R. H. 1985. A note on fault reactivation. *J. Struct. Geol.* 7:751–754.

Sibson, R. H. 1994. An assessment of field evidence for "Byerlee" friction. *Pure Appl. Geophys.* 142:645–662.

Sibson, R. H. 1998. Brittle failure mode plots for compressional and extensional tectonic regimes. *J. Struct. Geol.* 20:655–660.

Smith, R. B., W. C. Nagy, K. A. Julander, J. J. Viveiros, C. A. Barker, and D. G. Gants. 1989. Geophysical and tectonic framework of the eastern Basin and Range–Colorado Plateau–Rocky Mountain transition. In L. C. Pakiser and W. D. Mooney, eds., *Geophysical Framework of the Continental United States*. Memoir, pp. 205–233. Boulder, CO: Geological Society of America.

Sorel, D. 2000. A Pleistocene and still-active detachment fault and the origin of the Corinth-Patras rift, Greece. *Geology* 28:83–86.

Spencer, J. E. 1984. The role of tectonic denudation in the warping and uplift of low-angle normal faults. *Geology* 12:95–98.

Spencer, J. E. and C. G. Chase. 1989. Role of crustal flexure in initiation of low-angle normal faults and implications for structural evolution of the Basin and Range province. *J. Geophys. Res.* 94:1765–1775.

Spencer, J. E. and S. J. Reynolds. 1989. Tertiary structure, stratigraphy, and tectonics of the Buckskin Mountains. In J. E. Spencer and S. J. Reynolds, eds., *Geology and Mineral Resources of the Buckskin and Rawhide Mountains, West-Central Arizona. Bull. Ariz. Geol. Surv.* 103–167.

Spray, J. 1995. Pseudotachylyte controversy: Fact or friction? *Geology* 23(12):1119–1122.

Stockli, D. F. 1999. *Regional Timing and Spatial Distribution of Miocene Extension in the Northern Basin and Range Province*, 239 pp. Stanford, CA: Stanford University.

Taylor, B., A. M. Goodliffe, and F. Martinez. 1999. How continents break up: Insights from Papua New Guinea. *J. Geophys. Res.* 104:7497–7512.

Taylor, B., A. M. Goodliffe, F. Martinez, and R. Hey. 1995. Continental rifting and initial sea-floor spreading in the Woodlark basin. *Nature* 374:534–537.

Thatcher, W. and D. P. Hill. 1991. Fault orientations in extensional and conjugate strike-slip environments and their implications. *Geology* 19:1116–1120.

Townend, J. and M. D. Zoback. 2000. How faulting keeps crust strong. *Geology* 28:399–402.

Wang, C. 1997. A microstructural study on pseudotachylytes and microbreccias from the Whipple low-angle normal fault: Products of seismogenic slip. *Geol. Soc. Am. Abstr. Progr.* 72–73.

Wdowinski, S. and G. J. Axen. 1992. Isostatic rebound due to tectonic denudation: A viscous flow model of a layered lithosphere. *Tectonics* 11:303–315.

Wernicke, B. 1981. Low-angle normal faults in the Basin and Range Province: Nappe tectonics in an extending orogen. *Nature* 291:645–648.

Wernicke, B. 1990. The fluid crustal layer and its implications for continental dynamics. In M. H. Slisbury and D. M. Fountain, eds., *Exposed Cross Sections of the Continental Crust*, pp. 509–544. Dordrecht, The Netherlands: Kluwer Academic Publishers.

Wernicke, B. 1995. Low-angle normal faults and seismicity: A review. *J. Geophys. Res.* 100:20159–20174.

Wernicke, B. and G. J. Axen. 1988. On the role of isostasy in the evolution of normal fault systems. *Geology* 16:848–851.

Wernicke, B., G. J. Axen, and J. K. Snow. 1988. Basin and Range extensional tectonics at the latitude of Las Vegas, Nevada. *Geol. Soc. Am. Bull.* 100:1738–1757.

Wernicke, B., J. D. Walker, and M. S. Beaufait. 1985. Structural discordance between Neogene detachments and frontal Sevier thrusts, central Mormon Mountains, southern Nevada. *Tectonics* 4:213–246.

Wernicke, B. P. 1982. *Processes of Extensional Tectonics*. Boston: Massachusetts Institute of Technology.

Westaway, R. 1998. Dependence of active normal fault dips on lower-crustal flow regimes. *J. Geol. Soc. London*. 155: 233–253.

Westaway, R. 1999. The mechanical feasibility of low-angle normal faulting. *Tectonophysics* 308: 407–443.

Wills, S. and R. Buck. 1997. Stress-field rotation and rooted detachment faults: A Coulomb failure analysis. *J. Geophys. Res.* 102:20503–20514.

Xiao, H. B., F. A. Dahlen, and J. Suppe. 1991. Mechanics of extensional wedges. *J. Geophys. Res.* 96:10, 301–310, 318.

Yin, A. 1989. Origin of regional rooted low-angle normal faults: a mechanical model and its implications. *Tectonics* 8:469–482.

Yin, A. and J. F. Dunn. 1992. Structural and stratigraphic development of the Whipple-Chemehuevi detachment system, southeastern California: Implications for the geometrical evolution of domal and basinal low-angle normal faults. *Geol. Soc. Am. Bull.* 104:659–674.

Zoback, M. D. and H.-P. Harjes. 1997. Injection-induced earthquakes and crustal stress at 9 km depth at the KTB deep drilling site, Germany. *J. Geophys. Res.* 102:18477–18491.

Zoback, M. D., M. L. Zoback, V. S. Mount, J. Suppe, J. P. Eaton, J. H. Healy, D. Oppenheimer, P. Reasenberg, L. Jones, C. B. Raleigh, I. G. Wong, O. Scotti, and C. Wentworth. 1987. New evidence on the state of stress of the San Andreas fault system. *Science* 238:1105–1111.

CHAPTER FOUR

Depth-Dependent Lithospheric Stretching at Rifted Continental Margins

Mark Davis and Nick Kusznir

Introduction

While the uniform stretching model (McKenzie 1978) and its derivatives have been applied with considerable success to the formation of intracontinental rift basins, the mechanism for the formation of rifted continental margins is at best controversial. Rifted margins have traditionally been assumed to form by extreme extension and thinning of continental lithosphere (Le Pichon and Sibuet 1981), ultimately leading to the initiation of seafloor spreading at high stretching factor. Recent work on the northwest Australian rifted continental margin (Driscoll and Karner 1998; Baxter et al. 1999) and the Norwegian rifted continental margin (Roberts et al. 1997) suggests that the stretching of the continental lithosphere adjacent to the continent-ocean transition zone is highly depth dependent and increases with depth. The observation of depth-dependent lithosphere stretching at rifted margins is at variance with observations for intracontinental rift basins (White 1990; Marsden et al. 1990; Roberts et al. 1993), where stretching estimates derived from upper-crustal faulting balances that predicted for the whole lithosphere from postrift subsidence analysis, as predicted by McKenzie (1978). The stretching of the upper crust at rifted margins (Driscoll and Karner 1998; Baxter et al. 1999; Roberts et al. 1997) is observed to be significantly less than that of the whole crust or lithosphere, so that in the context of a simple-shear lithosphere extension detachment model (Wernicke 1985; Lister et al. 1991), all rifted continental margins appear to correspond to "upper plate." This observation has been named the "upper-plate paradox" by Driscoll and Karner (1998). Other recent work suggests that there is heterogeneous stretching on some rifted continental margins so extreme that it leads to the exhumation of mantle rocks. Wide regions of exhumed continental mantle of (up to ~100 km in width) have been observed at nonvolcanic rifted continental margins lying between the rotated fault blocks of highly extended continental crust and unequivocal oceanic crust (Pickup et al. 1996; Discovery 215 Working Group 1998; Whitmarsh et al. 2001). Observations of mantle exhumation are supported by wide-angle seismology, direct sampling, geochemical analysis, magnetic anomalies analysis, and geological mapping of orogenically exhumed rifted margins.

Existing models of the formation of rifted continental margin fail to explain the observations of depth-dependent stretching and continental mantle exhumation at rifted continental margins. Given the fundamental importance of rifted continental margins in the plate-tectonic cycle and the advent of deep-water hydrocarbon exploration, it is important that the processes responsible for the formation of rifted continental margins are better understood. The objectives of this chapter are to describe new evidence for depth-dependent stretching at rifted continental margins from the analysis of data for the Goban Spur, Galicia Bank, Vøring, and South China Sea continental margins. In addition the mechanism of depth-dependent stretching and mantle exhumation is explored using simple analytical and finite-element fluid-flow models of the effect of early seafloor spreading on the adjacent continental margin lithosphere. A possible resolution of the "upper-plate paradox" identified by Driscoll and Karner (1998) is also examined.

Methodology for Determining Stretching Factors at Rifted Continental Margins

Three independent methods and data sets (figure 4.1) have been used to determine lithospheric stretching factors data for the Goban Spur, Galicia Bank, Vøring, and South China Sea continental margins. Upper-crustal extension has been calculated by summing the heaves measured on seismically imaged faults. Whole-crustal stretching is derived from crustal-thinning profiles constrained by seismic refraction, seismic reflection, or gravity inversion. "Whole-lithosphere" stretching has been determined from postbreakup thermal subsidence using flexural backstripping, decompaction, and reverse thermal subsidence modeling.

Upper-Crustal Stretching and Thinning

Extension in the brittle upper crust is derived from offsets on basement faults imaged on two-dimensional seismic reflection sections. Faults are assumed to be planar,

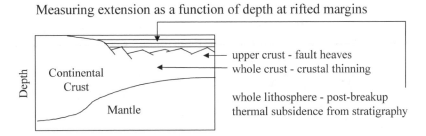

Figure 4.1 Schematic diagram illustrating the three independent methods used to independently estimate extension and stretching at depth within the lithosphere.

Table 4.1 Table of Parameters

Symbol	Quantity	Value	Units
β	stretching factor	—	
β_a	apparent stretching factor	—	
β_r	real stretching factor	—	
ε	thinning factor	—	
ε_{uc}	upper crustal thinning factor	—	
ε_{lc}	lower crustal thinning factor	—	
ε_{ml}	mantle lithosphere thinning factor	—	
f	1 + fault underestimation factor	—	
α	bedding rotation		°
θ	fault plane angle		°
l	horizontal fault offset		m
E	extension		m
v	vertical offset		m
W	distributed pure shear width	100	km
x	horizontal co-ordinate		km
x_0	fault location co-ordinate		km
t_0	initial crustal thickness	32	km

although extension estimates are not significantly altered if this assumption is invalid because the fault angle changes little over the measured vertical offset (see later discussion). Extension on a single fault (E_i) is given by simple trigonometry:

$$E_i = \frac{v_i}{\tan \theta_i} - l_i(1 - \cos \alpha_i) \qquad (4.1)$$

where E is extension, v is the vertical offset between footwall and hangingwall fault cutoffs, θ is fault dip, and l is the horizontal separation (see table 4.1 for a full definition of terms). Measured vertical offsets are used to generate fault heaves (through assumed dips on the fault plane), because they are measured more reliably than horizontal fault offsets. True dip section profiles are used, although since only the dip-slip component of faults contributes to crustal thinning, faults with slip directions that are not coincident with the two-dimensional depth profiles do not influence our analysis. We choose to ignore the block rotation component because several studies have demonstrated that faults are more likely to deform by vertical shear than as rigid "dominoes" (Westaway and Kusznir 1993). By making this assumption, we may slightly *overestimate* fault-related extension. To capture the error and uncertainty inherent in deriving stretching factors from faults, we vary the present-day fault dip θ from 25 to 70° for each fault.

It is convenient to define the 'thinning' factor, which is related to the stretching factor by equation (4.2).

$$\varepsilon = 1 - \frac{1}{\beta(x)} \qquad (4.2)$$

To compare upper-crustal stretching with similar estimates for the whole crust and "whole lithosphere," we express fault-derived extension (measured discretely from faults) as a laterally varying profile of stretching factor β. In the reference frame of the extended continental lithosphere, extension is derived from β by

$$E = \int \varepsilon dx = \int \left(1 - \frac{1}{\beta(x)}\right) dx \qquad (4.3)$$

Extension from a single fault is mapped to a profile of stretching factor using a continuous function such that the distributed extension is identical to that measured discretely on faults. The form and width of the function that maps fault-derived strain to a continuous function is arbitrary (White and McKenzie [1988] use a Gaussian function) but not critical, because neither assumption alters the total extension. We choose to use a cosine-squared function, which is continuous and has continuous first derivatives:

$$\beta_i = 1 + \beta_0 \cos^2 \left[\frac{\pi(x - x_0)}{W}\right] \qquad (4.4)$$

In the stretched reference frame, extension is given by equation (4.5), where the value of β_0 is determined numerically:

$$E = \int \left[1 - \frac{1}{\beta}\right] dx = \int \left[1 - \frac{1}{\beta_0 \cos^2 \left[\frac{\pi(x - x_0)}{W}\right]}\right] dx \qquad (4.5)$$

The total stretching factor (β_{tot}) associated with all faults is the product of stretching-factor profiles associated with each fault. Values of W in the range 75 to 150 km have been used (Roberts et al. 1993).

Whole Crustal Stretching and Thinning

Crustal thickness variation along rifted continental margins is assumed to be the consequence of crustal extension and thinning and is used to infer the lateral distribution of strain. Crustal thickness is derived from wide-angle seismic studies; where these are unavailable, the "reflection Moho" or data from gravity studies are used. Assuming a constant initial crustal thickness (t_0), the crustal-derived stretching factor is

$$\beta = \frac{t_0}{t_c} \qquad (4.6)$$

In our analysis we generally assume that the crustal thickness variations are due to stretching during the event under study. Where this assumption is clearly invalid (e.g., along the Norwegian margin that has a long tectonic history of lithosphere extension), instead of controversially attempting to partition crustal-thickness variations to separate events, we consider these data to be unreliable and hence do not use them in our analysis. The assumed initial crustal thickness is constrained where possible with local data, and we capture the associated likely uncertainty in the calculation of error bars in stretching and thinning factors (see later discussion).

Whole Lithosphere Stretching and Thinning

Reequilibration of lithosphere temperature anomalies generated during lithosphere stretching gives rise to postrift "thermal" subsidence (McKenzie 1978). The post-breakup subsidence history of a rifted margin recorded in stratigraphic data contains information that allows the magnitude of lithosphere stretching to be determined. Two-dimensional flexural backstripping, decompaction, and reverse postbreakup thermal subsidence modeling has been applied to two-dimensional depth converted stratigraphic cross sections derived from seismic reflection data to determine "whole lithosphere" stretching factors. The methodology is described in Roberts et al. (1998). The flexural backstripping, decompaction, and reverse postbreakup thermal subsidence modeling takes a present-day section and produces a series of palaeo cross sections whose palaeobathymetries are dependent on the magnitude of the lithosphere-stretching factor (McKenzie 1978) used in the reverse postbreakup thermal modeling. Calibration of the palaeo cross sections by observed palaeobathymetry indicators allows the magnitude and lateral variation of lithosphere-stretching factor to be determined. Throughout this chapter we compute equivalent "whole lithosphere" stretching factors using the assumption of uniform stretching. Where depth-dependent stretching occurs, the "true" lithospheric mantle-stretching factor may be greater than the equivalent "whole lithosphere" stretching factor, as enhanced stretching in the lithospheric mantle is required to offset suppressed crustal extension.

Stretching Observations on the Goban Spur Rifted Margin (UK Margin)

The Goban Spur rifted continental margin lies off southwest England (figure 4.2a). The margin lies in deep water and was formed by lithospheric extension as separation occurred between Europe and North America from the early Cretaceous to mid-Albian (Masson et al. 1985; de Graciansky and Poag 1985). In this analysis the structure of the margin is assumed to be two-dimensional, which is probably a reasonable approximation because major lateral changes in margin morphology occur on length scales greater than 100 km (Horsefield et al. 1993). Figure 4.2b–d illustrates line drawings of the interpretations of seismic reflection profile data at Goban

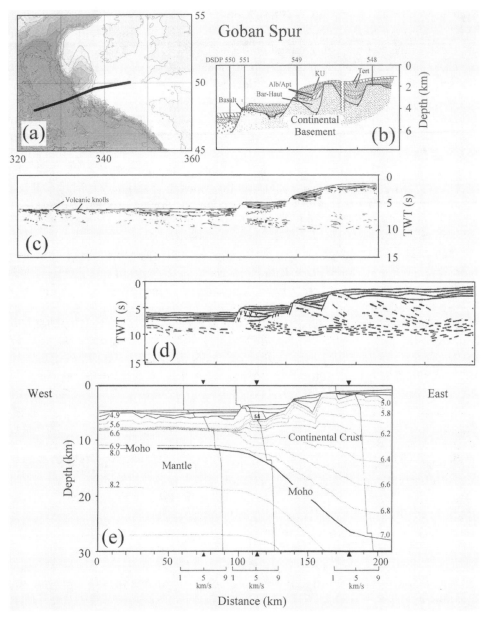

Figure 4.2 (a) Bathymetry at the Goban Spur rifted continental margin showing location of the profile used in this study. (b) Illustration of the interpreted depth section at Goban Spur, interpreted from seismic line CM10, and reproduced after (de Graciansky et al. 1985). (c and d) Line drawings of the outer section of BIRPS WAM line (Klemperer and Hobbs 1989; Peddy et al. 1989; Louvel et al. 1997). (d) Line drawing of BIRPS deep-seismic profile WAM. (e) Crustal structure and velocity structure at Goban Spur from seismic refraction data (Horsefield et al. 1993). The conventional "continent-ocean boundary" is usually interpreted to be coincident with DSDP Site 551. The reflection Moho in the interpreted oceanic domain is associated with very few identifiable reflectors. Black circles indicate two volcanic knolls.

Spur (de Graciansky and Poag 1985; Klemperer and Hobbs 1991; Louvel et al. 1997). The outer margin consists of a series of rotated basement-fault blocks that are clearly imaged on seismic reflection because of the relative lack of sedimentary cover. Rifting at the Goban Spur margin has had a long history. Permo-Triassic and Jurassic basins lie to the east of the continent-ocean boundary (Cook 1987), and the age of the structures generally young to the west. A Jurassic rift system manifested by northeast-southwest trending faults has been overprinted in the west, within ~100 km of the "continent-ocean boundary," by Cretaceous northwest-southeast trending faults. The age of the latest rift event that leads to continental breakup is constrained by DSDP wells to be in the Albian-Aptian interval (105–112 Ma; Joppen and White 1990), and the oldest seafloor magnetic anomaly is M34 (Albian-Santonian; Scrutton 1984). The seismic reflection line drawing shown in figure 4.2b taken from de Graciansky and Poag (1985) shows pre-, syn-, and postbreakup stratigraphy and clear breakup age upper-crustal faulting.

Deep normal incidence seismic reflection profiles from the WAM survey at Goban Spur (Peddy et al. 1989; Klemperer and Hobbs 1991; Louvel et al. 1997) reveal that the upper crust is relatively transparent and the lower crust is reflective (figure 4.2c and d), and as such, Goban Spur is typical of many deep-seismic reflection sections from around the British Isles (Cheadle et al. 1987). While the origins of the reflectivity are debated (Mooney and Meissner 1992), the base of the lower-crustal reflective zone is approximately coincident with the seismic refraction Moho and is therefore usually interpreted to represent the "seismic reflection" Moho. The results of refraction seismology work at Goban Spur (figure 4.2e) by Horsefield et al. (1993) confirms that the base of the lower-crustal reflective zone corresponds to the Moho for Goban Spur. Peddy et al. (1989) note that the reflective lower crust appears to thin in proportion to the whole crust. They suggest that lower-crustal reflectivity is a prerift feature and that extension of the lithosphere is consistent with uniform pure-shear (McKenzie 1978).

The conventional continent-ocean boundary is usually interpreted to be the volcanic knoll sampled by DSDP Site 551 (figure 4.2b–d). The "oceanic crust" to the southwest of the continent-ocean boundary is characterized by an irregular surface with little evidence of a reflection Moho. Goban Spur is often cited as a classic "nonvolcanic" margin (Watts and Fairhead 1997). Wide-angle seismic studies (Horsefield et al. 1993; figure 4.2e) provide no evidence for the high-velocity lower crust ($V_p = 7.2$–7.4 km s^{-1}) which would be diagnostic of basaltic underplating commonly found at "volcanic" rifted continental margins. Furthermore, Horsefield et al. (1993) and Masson et al. (1985) support the notion that (a) typical oceanic crust lies westward of a narrow (~15 km) continent-ocean boundary and that (b) there is *no evidence for depth-dependent stretching* (Horsefield et al. 1993). More recent work (Minshull et al. 1998), involving both reinterpretation of existing data and new data, challenges these notions. We suggest that the intracrustal velocity structure is not sufficiently unique to enable determination of intracrustal stretching factors; indeed, much of the internal structure may be an artifact of the modeling technique. Seismic reflection reveals (figure 4.2c) that the conventional oceanic crust at Goban Spur is characterized by an irregular surface and shows little evidence of a reflection Moho. It may be that unequivocal oceanic crust lies immediately to the southwest of the two prominent ridges located close to the western end of the WAM profile (Klemperer and Hobbs 1991; figure 4.2c), with transitional material to the landward side. The original seismic data (not reproduced here) reveals a strong reflector at ~1.3 s two-way time below the top of the (presumed) oceanic crust. This may suggest that this location is the "true" continent-ocean boundary and that the poor fitting of the amplitudes of P_n arrivals from the wide-angle data set (Horsefield et al. 1993) at Goban Spur does not provide strong support for a well defined and abrupt continent-ocean boundary.

Fault Extension Stretching Factors

The analysis of upper-crustal faulting was made from an interpretation largely derived from the deep-seismic reflection profile WAM (Klemperer and Hobbs 1991; figure 4.2c and d) and published sections of the higher-resolution data sets (CM-10 from de Graciansky and Poag 1985; figure 4.2b), with stratigraphic calibration and depth-conversion information from DSDP reports (Masson et al. 1985). Note that the sections interpreted from the WAM deep-seismic line (figure 4.2c and d) and the higher-resolution line CM-10 (figure 4.2b) are similar and image few upper-crustal faults. Total extension estimates for the outer ~100 km (the region corresponding to the wide-angle survey shown in figure 4.2e) is 12 ± 5 km, or $\beta = 1.13 \pm 0.05$. The error estimate assumes the uncertainty in estimating fault dip is ±10°. Present-day fault dips range from 35° to 70°, but the "average" observed dip is ~55°. The lateral profile of stretching and thinning factors derived from the upper crust is illustrated in figure 4.3a and b.

Crustal Thinning Stretching Factors

The crustal structure used to determine crustal thinning stretching factors is summarized in figure 4.3c and is derived from wide-angle refraction results taken from Horsefield et al. (1993), and shown in figure 4.2e, and depth-converted normal incidence deep-seismic data (Klemperer and Hobbs 1991). There is generally good agreement (<2 km misfit) between the reflection and refraction estimates of the seismic Moho. The crustal stretching (β) and thinning (ε) factors are calculated assuming an initial homogenous prerift crustal thickness of 32 ± 2 km (Bott and Watts 1970).

Figure 4.3a and b shows the profiles of crustal thinning-derived stretching and thinning factors. Errors in stretching factor (β) have been calculated assuming uncertainties of ±1/2, ±2 km, and ±2 km for top basement, Moho (Horsefield et al. 1993), and prerift (homogenous) crustal thickness, respectively. While errors in the thinning factor (ε) are symmetric, errors in the associated stretching factor

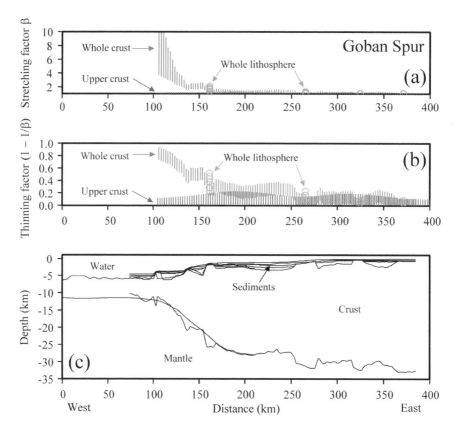

Figure 4.3 (a and b) Stretching and thinning factors for upper crust, whole crust, and "whole-lithosphere" for the Goban Spur rifted margin. Circles represent estimates of "whole-lithosphere" stretching factors with effective elastic thicknesses (T_e) of 0, 3, 5, 10 km. Whilst stretching estimates are similar toward the continental interior, a significant discrepancy is evident within ~75 km of the continent-ocean boundary. (c) Crustal sections derived from seismic data summarized in figure 4.2. Vertical lines of stretching and thinning factor derived from upper-crustal faulting and crustal thinning represent error bars of ±1.5 standard deviation.

(β) are not. The crustal thinning-derived extension estimate is 42 ± 8 km, or in terms of mean stretching factor, $\beta = 1.7 \pm 0.25$.

"Whole-Lithosphere" Stretching Factors

Flexural backstripping, decompaction, and reverse postbreakup thermal subsidence modeling has been carried out at Goban Spur to determine "whole-lithosphere" stretching factors. The present day two-dimensional stratigraphic cross sections used as input to the flexural backstripping, decompaction and reverse postbreakup thermal subsidence modeling were derived from the WAM seismic reflection profile (Klemperer and Hobbs 1991) and published sections of the higher-resolution CM-

10 data set (de Graciansky and Poag 1985), with stratigraphic calibration and depth-conversion information from DSDP reports (Masson et al. 1985). The depth-converted stratigraphy is summarized in figure 4.3c. As seen on seismic reflection data for Goban Spur (de Graciansky and Poag 1985; Klemperer and Hobbs 1991; Louvel et al. 1997) and summarized in figure 4. 2b–d, several of the fault blocks have bevelled tops that have been eroded at or near sea level. These eroded fault blocks are believed to be reliable palaeobathymetric markers and are therefore used to constrain our reverse postbreakup thermal subsidence modeling. Compaction parameters from Sclater and Christie (1980) have been used, and the pre-, syn-, and postbreakup lithologies are taken to be sandy-shale and chalk, respectively. An instantaneous rift age of 105 Ma is assumed and has been used to define the post-breakup thermal subsidence. To simulate the thermal load from adjacent oceanic crust, the stretching-factor profile has been ramped to $\beta = \infty$ at the conventional continent-ocean boundary (at $\sim$100 km on figure 4.3c).

The estimates of lithosphere stretching and thinning factor derived from flex-ural backstripping, decompaction, and reverse thermal subsidence modeling are shown in figure 4.3a and b. The lateral coupling of the adjacent oceanic thermal load into the rifted continental margin region is a strong function of the assumed effective elastic thickness (T_e) used in the flexural backstripping, decompaction, and reverse thermal subsidence modeling. As a consequence, the modeling has been performed with values of $T_e = 0, 3, 5, 10, 25$ km to determine sensitivity to T_e. No eustatic sea level curves are applied, because their timing and magnitude are debated (Hall and White 1994). By ignoring possible eustatic sea level change (which may have been up to $\sim$300 m higher 70 Ma BP; Pitman 1978), the β factor is potentially underestimated by $\Delta\beta \sim$0.2.

Stretching Factor Variation with Depth
on the Goban Spur Rifted Margin

Figure 4.3a and b shows the lateral profiles of stretching factor along the Goban Spur margin where fault and crustal thinning derived estimates are shown with their appropriate error bars, and backstripping-derived estimates of " whole-lithosphere" extension are marked with circles at locations with good palaeoba-thymetric control. The clusters of β estimates from flexural backstripping span the likely bounds of lithospheric effective elastic thickness (T_e) as described in a previous section. While the spectral signature of the sediment is dominated by long wavelengths, the short wavelength of the thermal load associated with the adjacent oceanic crust generates a strong dependence of the stretching factor to the lithosphere effective elastic thickness. As figure 4.3b illustrates, the "whole-lithosphere" stretching factor is significantly greater (i.e., without the error bars) than upper-crustal stretching factors for all values of effective elastic thickness (T_e). The figure demonstrates that there appears to be no significant increase in the upper-crustal stretching factor toward the continent-ocean boundary, and sug-gests that depth-dependent stretching occurs within $\sim$75 km of new oceanic crust.

Stretching Observations on the Galicia Bank Rifted Margin (Iberian Margin)

The rifted continental margin off the west coast of Iberia (figure 4.4a) is a "nonvolcanic" margin and formed during the continental separation between Iberia and Newfoundland that led to the opening of the North Atlantic ocean. The margin developed during a Berriasian/latest Aptian rifting episode (140–114 Ma; Boillot et al. 1989b), and is conjugate to the southeast Flemish Cap margin. While the margin has small free-air gravity anomalies (Sandwell and Smith 1997) and is approximately in isostatic equilibrium, the free-air edge-effect couplet typical of the continent-ocean boundary in many regions appears to be absent or suppressed. Oceanward of the narrow shelf lies the Galicia Interior Basin, which is probably a continuation of the Lusitania Basin of onshore Portugal. The Galicia Interior Basin suffered several episodes of rifting from Triassic to early Cretaceous and proceeded to break up to the west some 25 Ma later (Perez-Gussinye et al. 1998).

The outer margin comprises a series of rotated fault blocks striking north, dipping both to the east and west in the east and mainly to the west in the west (figure 4.4b; Whitmarsh et al. 1996). Fault blocks in the extreme west, near to the postulated continent-ocean boundary, are remarkable in that they have little or no postbreakup sedimentary cover (figures 4.4b and 4.5a). Normal incidence reflection data at the Galicia margin indicates the presence of a strong midcrustal (socalled S-) reflector which is continuous and almost linear over the outer 50 km of the margin (figure 4.5b; Reston 1996). While for many years the origin of the

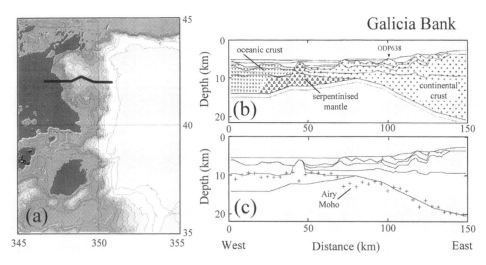

Figure 4.4 (a) Location map showing bathymetry offshore Portugal and the location of the Galicia Bank profile used in this study. (b) Crustal model derived from wide-angle data after Whitmarsh et al. (1996). (c) Wide-angle data set with estimates of the depth to Moho (asterisks) from an Airy-isostatic model. The small misfit between independent models provides confirmation that the margin is presently at or close to Airy isostatic equilibrium, a point that is discussed later.

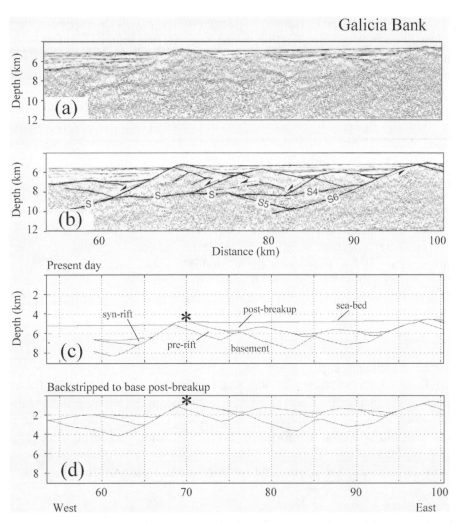

Figure 4.5 (a and b) Depth-migrated seismic reflection section and interpretation for Galicia Bank. (c) Present-day depth cross section; asterisk denotes fault block crest possibly beveled by erosion at or near sea level. (d) Depth cross section restored to base breakup using flexural backstripping and reverse postbreakup thermal subsidence with a β factor of infinity.

S-reflector was highly controversial, waveform modeling (Reston 1996; Whitmarsh et al. 1996) strongly suggests that the S-reflector represents a major detachment fault with a top-to-the-west sense of shear, a notion that is supported by similar "D" reflectors on the southern Iberian Abyssal Plain margin (Pickup et al. 1996). The Iberian margin shows a major detachment feature as predicted by a Wernicke-type simple-shear model (Wernicke 1985; Lister et al. 1991).

Figure 4.4b summarizes the wide-angle velocity model for the northern Galicia margin (Whitmarsh et al. 1996). The Moho in the west is overlain by a lens of high-velocity material, for which there are two explanations, namely: (a) serpentinized peridotite, or; (b) igneous underplating. P_mP arrivals were not observed at the west end of the refraction line, which suggests that velocity contrast is small. Since there is very little evidence of syn-rift melt (Whitmarsh et al. 2001) as is typical of "volcanic" margins (White and McKenzie 1989), the serpentinzation model is preferred. Figure 4.4c shows (with asterisks) a Moho predicted by Airy isostasy superimposed on the crustal structure derived from wide-angle seismology shown in figure 4.4b (Whitmarsh et al. 1996). The agreement between the seismic Moho (figure 4.4b) and Airy isostatic Moho (figure 4.4c) suggests the margin is in near local isostatic equilibrium.

A composite regional two-dimensional reflection profile has been generated by compiling four published seismic reflection sections. Seismic lines GAP-106 and GAP-014 splice to the regional wide-angle model, which in turn is coincident with reflection line segment GP-101. Depth conversion of the eastern portion of the composite reflection seismic was performed using velocities described in Murillas et al. (1990). This composite section is not reproduced to conserve space, although a simplified line drawing is shown in figure 4.6c.

Fault-Derived Stretching Factors

The determination of upper-crustal extension was made from the analysis of upper-crustal faulting observed on the composite section described previously. The analysis used 35 seismically observable faults using the methodology described under Methodology for Determining Stretching Factors at Rifted Continental Margins. Total recorded extension on this profile is 80 ± 12 km giving an equivalent mean stretching factor $\beta = 1.3 \pm 0.1$. The lateral profile of stretching and thinning factors derived from the upper crust is illustrated in figure 4.6a and b.

Crustal Thinning Stretching Factors

To extend our knowledge of depth to Moho eastward of that observed seismically on the margin, Moho depth has been determined from inversion of gravity anomalies, constrained by sediment thickness and bathymetry data. Where overlap with seismically determined Moho depth occurs, the Moho derived from Bouguer gravity inversion and wide-angle refraction are consistent and have misfits typically <1 km (figure 4.6c). Crustal thickness has been determined using the methodology outlined under Whole Crustal Stretching and Thinning from the crustal structure-thickness profile shown in figure 4.6c. An initial crustal thickness of 30 km was used consistent with the crustal-thickness estimates from unstretched crustal thickness under Iberia (Cordoba et al. 1987) and the Airy Moho depth (figure 4.4c). The resulting profile of stretching and thinning factors derived from crustal thinning is shown in figure 4.6a and b.

Errors in stretching factor (β) have been calculated assuming uncertainties of ± 1, ± 2, and ± 2 km for top basement, Moho, and prerift (homogenous) crustal

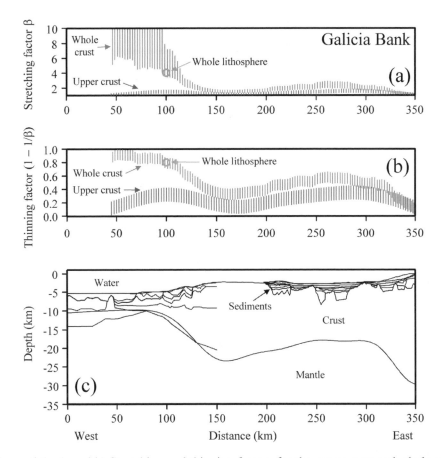

Figure 4.6 (a and b) Stretching and thinning factors for the upper crust and whole crust at the Galicia Bank rifted continental margin. The open circle at coordinate ~100 km is an estimate of whole-lithosphere stretching from Airy backstripping. (c) Crustal section derived from seismic and gravity.

thickness, respectively. Crustal-derived extension at Galicia is 203 ± 37 km giving an equivalent mean stretching factor $\beta = 2.9 + 1.5/-0.7$. Note that the errors are significantly asymmetric at large stretching factors.

Stretching Factor Variation with Depth on the Galicia Bank Rifted Margin

Stretching and thinning factors for the upper crust and whole crust are compared in figure 4.6a and b and are significantly different within ~130 km from the end of the composite profile, or ~75 km from the conventional continent-ocean boundary. This difference has been noted before (Boillot et al. 1989a, 1989b; Chenet et al. 1982; Sibuet 1992).

"Whole-Lithosphere" Stretching Factors and Postbreakup Subsidence

Profile GP101 (figure 4.5a and b; Reston 1996), coincident with the outer part of our composite line (figure 4.6c), is remarkable in that the postrift sedimentary cover is thin (mostly <1 km), and most unusually, the fault block at ~70 km has no postrift cover, presumably because the sediment supply was low and the spill point has never been reached. A simplified depth section of line GP101 is shown in figure 4.5c. The possibility exists (but it is by no means certain) that the beveled fault block identified with an asterisk at 70 km (figure 5c) represents a palaeo-bathymetric marker at or near sea level at the end of breakup. The results of restoring this section to base postbreakup (end breakup) using flexural backstripping, decompaction, and reverse thermal subsidence modeling with $\beta = \infty$ and $T_e = 3$ km is shown in figure 4.6d. Even with the largest stretching factor possible driving the postbreakup thermal subsidence ($\beta = \infty$), the beveled fault blocks fail to restore to sea level independently of the T_e used for the flexural response of the lithosphere.

Whilst high stretching factors are not unexpected because the section lies in water ~5 km deep, the postrift thermal subsidence predicted by McKenzie (1978) only predicts ~4 km after ~150 Ma for a stretching factor $\beta = \infty$. Such anomalous subsidence is difficult to explain using conventional models and cannot be accounted for by eustatic sea level variation. Using an instantaneous rift age of 141 Ma (Whitmarsh et al. 1996) does little to change the palaeobathymetry predicted at the end of the syn rift, which is >750 m for the older rift age for $T_e = 3$ km and $\beta = \infty$. While it is uncertain whether the fault blocks were eroded at sea level, from what is known about the end syn-rift palaeobathymetries, such deep water is unlikely. Similarly, and as noted by Whitmarsh et al. (1996) and Louden et al. (1991), the stretching factor estimated from Airy backstripping of ODP sites 639, 640, and 641 (situated in shallower water to the east of GP101 and figure 4.5) is larger than that determined from faulting and fits either an instantaneous model $\beta \sim 4$ or a finite-duration rift event of similar total magnitude. A "whole-lithosphere" stretching factor $\beta \sim 4$ is consistent with the crustal-thinning estimate but is significantly greater than the fault-derived estimate.

Speculation on the Cause of Anomalous Subsidence on the Galicia Bank Margin

The observation that the northern Iberian rifted continental margin is in near local isostatic equilibrium and the initial crustal thickness was 30 ± 1 km suggests that the prerift margin should have been sufficiently buoyant to be at sea level before rifting. As noted in the previous section, the ~5 km of water-loaded subsidence since breakup is greater than the thermal subsidence predicted by the uniform stretching model (McKenzie 1978), even for very large stretching factors (for $\beta = \infty$, only 4 km of subsidence is predicted since breakup). If the fault block was at sea level at breakup then ~1 km of fault-block water-loaded subsidence of the present-day bathymetry may be derived from syn-breakup subsidence that occurred during sediment-starved conditions. Under traditional models of rifting this observation is most unusual because syn-rift footwall uplift on large faults is well documented (Stein and Barrientos 1985) and usually causes foot-wall erosion. The failure of the section to backstrip to sea level may therefore suggest that extension of the western Iberian lithosphere occurred in two phases driven by two dissimilar processes. A modest initial stretching phase (β ~1.3, say, consistent with the upper-crustal fault-derived extension estimate) may have generated the observed fault geometries. Subsequently, depth-dependent stretching of the lithosphere may have occurred within the ductile lower crust and lithospheric mantle and generated further syn-breakup subsidence by lower crustal thinning without additional foot-wall uplift. Such a two-phase stretching model may provide a general explanation for the observation of apparent depth-dependent stretching of the lithosphere at rifted continental margins and is discussed subsequently in more detail.

Stretching Observations on the Vøring Rifted Margin (Mid Norway Margin)

The Norwegian rifted continental margin (figure 4.7a) is the culmination of a complex history of extensional tectonics since the Devonian Caledonian orogeny (Gage and Dore 1986). The most recent extensional events were the late Jurassic-early Cretaceous intracontinental rift event and the Palaeocene rift event that immediately preceded the late Palaeocene-early Eocene breakup during isochron 24r at ~55 Ma (Eldholm et al. 1995). Given the complex stretching history along the Norwegian margin, in this chapter estimates of stretching factors are only calculated for the Palaeocene (and late Upper Cretaceous) stretching events immediately leading up to continental breakup and rifted margin formation at ~55 Ma. The central and outer parts of the Norwegian margin have large accumulations of sediment (figure 4.7b). While the stratigraphy in the east is constrained by published hydrocarbon exploration wells and shallow boreholes, the deep-water stratigraphy determined by recent commercial drilling remains largely confidential. Consequently the stratigraphy used in this study is largely derived from the pre-drilling interpretations provided by Statoil (and published in Dore et al. 1997). The deepest visible reflector is believed to be the base Cretaceous (Blystad et al. 1995) with maximum depth of ~12 km (~7 s two-way travel time) below the sea surface. With the exception of the vicinity of the conventional continent-ocean boundary, where extrusive volcanics limit seismic penetration, extensional structures of Palaeocene age are normally well imaged by normal-incidence seismic reflection profiles.

The Norwegian margin is characterized by large amounts of igneous volcanic activity and is classified as a "volcanic" margin to distinguish it from other "non-volcanic" margins such as Goban Spur and west Iberia (White and McKenzie

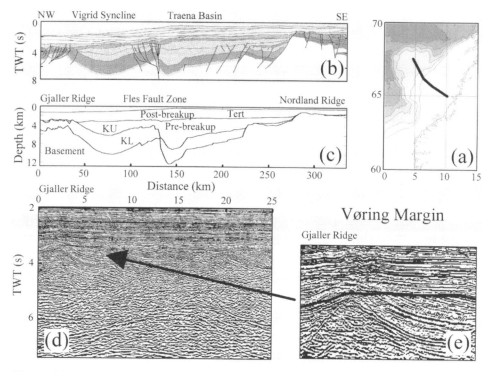

Figure 4.7 (a) Location map showing bathymetry offshore mid-Norway and the location of the Vøring Margin profile used in this study. (b and c) Depth and seismic reflection cross sections (e and f) for the Vøring Margin profile used in this study. Depth cross section and seismic data taken from Dore et al. (1997).

1989). The igneous activity at "volcanic" margins is evidenced by seaward dipping reflectors (SDRs), sill intrusions, high velocity and high density bodies with high MgO content, and extrusive lavas (White and McKenzie 1989). Close to the conventional continent-ocean boundary lies a significant "marginal high"—a feature common to many "volcanic" margins—which was formed during early Cenozoic rifting and breakup with emplacement of large volumes of igneous rocks, which were extruded at the surface and intruded as sills and underplating bodies (Mjelde et al. 1997a, 1997b). In this vicinity seaward-dipping reflectors are imaged on seismic reflection sections that were established by drilling (DSDP Leg 81 and ODP Legs 104, 152, and 163) to be T-MORB tholeiitic basalt lava flows extruded under subaerial or shallow marine conditions (Planke and Eldholm 1994). Pb-isotope studies of DSDP Leg 81 suggest that the extrusive lavas that form the seaward dipping reflectors are contaminated with a crustal signature (Morton and Taylor 1987) and therefore lie landward of the continent-ocean boundary, as proposed by Hinz (1981), in contrast to the subaerial seafloor spreading model of Mutter et al. (1982).

The Vøring basin (figure 4.7a) lies off the west coast of Norway at 66–68 °N. Figure 4.7b and c shows a section, taken from Line D of Dore et al. (1997), across the Vøring basin running from the Nordland Ridge, over the Traena basin, the Vigrid Syncline, and Gjaller Ridge toward the Vøring Marginal High (not shown), which is close to the continent-ocean boundary. Age and lithology data derived from regional interpretations by Statoil tie into commercial well 6508/5–1 to the east of the Nordland Ridge. Time-depth functions for depth conversion of seismic reflection data (figure 7c) are derived from wells from the Magnus/Southern Møre Basin area.

The Gjaller Ridge (figure 7b–e) is a prominent northeast-southwest trending high comprising mainly westward-dipping, deeply eroded, rotated, mainly pre-Tertiary fault blocks (Dore and Lundin 1996; Blystad et al. 1995). This ridge has a well developed unconformity at the Base Tertiary (65 Ma) and deeply eroded Cretaceous fault blocks that result from end-Cretaceous extension and contemporaneous footwall uplift (figure 4.7d and e). The Gjaller Ridge, and similarly the Fles Fault Zone, are believed to have been at (or near) sea level in the Palaeocene so providing a palaeobathymetric constraint for flexural backstripping. In addition, the Vøring Marginal High to the west (not illustrated) is interpreted to be a regional (near) sea level marker at Top Palaeocene times (57 Ma) since DSDP and ODP drilling found basaltic lavas interbedded with shallow-marine sediments (Planke and Eldholm 1994).

Fault-Derived Stretching Factors

The analysis of upper-crustal faulting to determine Palaeocene (and late Upper Cretaceous) upper-crustal extension associated with continental breakup was made from the section shown in figure 4.7b and c and taken from Line D of Dore et al. (1997). The seismic data on which this interpretation by Dore et al. (1997) is based is of good quality and allows the identification of small extensional faults (figure 7d and e). Fault-derived stretching and thinning factors are shown in figure 4.8a and b. Upper-crustal stretching factors are small, peaking at $\beta \sim 1.1$ around the Gjaller Ridge. Total extension derived from Palaeocene faulting is 10 ± 4 km. The mean stretching factor from the Fles Fault Zone to the "continent-ocean boundary" is $\beta = 1.07 \pm 0.03$.

Crustal Thinning-Derived Stretching Factors

Crustal thinning-derived stretching factors for Palaeocene breakup are not presented since it is difficult to determine the relative thinning contributions from the earlier rift events in the Triassic, Jurassic, and Cretaceous. Furthermore, crustal thickening from Palaeocene igneous underplating adds additional controversy. For these reasons we do not present estimates of whole-crustal extension.

"Whole-Lithosphere" Stretching Factors

Palaeocene and late Upper Cretaceous lithosphere stretching has been derived from the flexural backstripping, decompaction, and reverse postbreakup subsi-

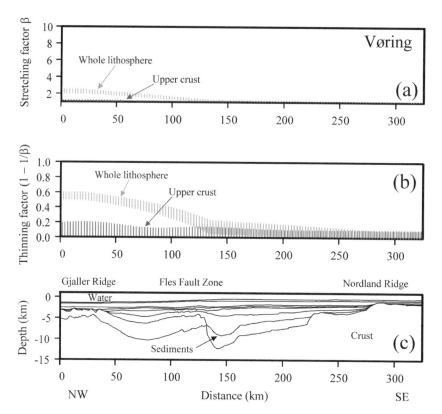

Figure 4.8 (a and b) Palaeocene stretching and thinning factors determined for upper crust and whole lithosphere for the Vøring rifted continental margin. (c) Upper-crustal section for the Vøring margin profile based on seismic data.

dence modeling of the stratigraphic section shown in figure 4.7c, which is derived from Dore et al. (1997). The assumption that the Base Tertiary unconformity at the Gjaller Ridge and Fles Fault Zones was at (or near) sea level in the Palaeocene has been used as the palaeobathymetric constraint for the flexural backstripping. While a range of values of T_e (0, 5, 10, and 25 km) were used in the flexural backstripping, these results are relatively insensitive to the value of T_e used, because the wavelength associated with loading is long. The resulting β-stretching-factor profile for the Palaeocene lithosphere-stretching event is shown in figure 4.8a and b assuming $T_e = 3$ km. The Palaeocene lithosphere-stretching factor, assuming no earlier late Jurassic-early Cretaceous rifting and no dynamic uplift from the Iceland plume, shows a value of 2.5 in the west of the profile decreasing toward 1 in the east.

Along the Norwegian margin, the determination of lithosphere-extension estimates derived from postbreakup subsidence history is complicated for the following reasons:

(a) Earlier late Jurassic-early Cretaceous and Triassic intracontinental rifting

The Norwegian margin has suffered a protracted rift history with large and distinct intracontinental rifting events occurring in the late Jurassic-early Cretaceous and Triassic prior to the formation of the rifted margin in the early Tertiary. Estimates of the magnitude of Palaeocene lithosphere stretching associated with continental breakup determined by flexural backstripping, decompaction, and reverse postbreakup subsidence modeling are sensitive to residual thermal subsidence from these earlier rift events. We therefore present results assuming likely bounding ranges on the timing and magnitude of earlier rift events.

Sensitivity of the determined magnitude of Palaeocene lithosphere stretching to earlier rift events is summarized in figure 4.9a and b. The effects of a large early Cretaceous rift event at 128 Ma (representing the end of late Jurassic-early Cretaceous rifting) with β stretching factors of 1.5 and 2 are examined. Inclusion of an earlier late Jurassic-early Cretaceous rift event reduces the magnitude of Palaeocene lithosphere stretching from $\beta = 2.5$ to 2.2 in the west of the profile for the preferred upper bound of $\beta = 1.5$ for the earlier late Jurassic-early Cretaceous rift event (Roberts et al. 1997).

(b) Igneous underplating

Igneous underplating generates permanent uplift providing that the density of the new material is less than the density of the asthenospheric mantle (Brodie and

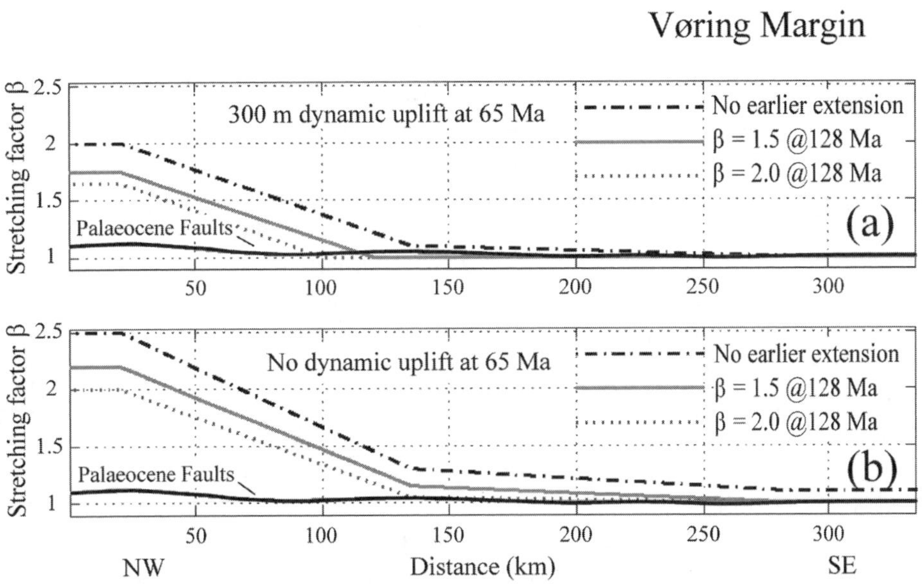

Vøring Margin

Figure 4.9 (a and b) Palaeocene stretching factors for whole-lithosphere extension of the Vøring rifted continental margin, determined by flexural backstripping and reverse postbreakup modeling, showing sensitivity to Palaeocene transient uplift from the Iceland plume and earlier Upper Jurassic-Lower Cretaceous rifting. Upper-crustal extension is significantly lower than "whole-lithosphere" extension under all reasonable circumstances of earlier rift event and transient plume uplift.

White 1995). The uplift magnitude may be estimated from the thickness of igneous underplating (inferred from wide-angle seismology) if the underplating density is known. In our analysis, we deliberately make the simplifying assumption that subsidence is assumed to result purely from cooling of the lithosphere. Therefore igneous underplating that predates the age of the palaeobathymetric marker probably makes little difference to the estimate of extension, whereas igneous underplating that postdates the palaeobathymetric marker buffers the predicted subsidence (by generating relative uplift) and reduces the estimated stretching factor. White and Lovell (1997) suggest that much of the emplacement of basaltic underplating in the North Sea occurs between 63 and 54 Ma. If, as it seems likely, significant underplating has occurred post 65 Ma (or in some cases post 57 Ma), then our "whole-lithosphere" stretching factors and extension discrepancies are underestimates. Simple calculations assuming Airy isostasy imply that a 5-km-thick underplated body distributed at any depth within the lithosphere generates 400–800 m of initial permanent uplift. Subsidence curves suggest that for an apparent stretching factor $\beta \sim 2$, the true stretching factor after correction for (say) ~500 m of water-loaded uplift might increase by $\Delta\beta < 0.5$.

(c) Transient uplift from the Iceland mantle plume

Many basins in the northeast Atlantic experienced regional Palaeocene uplift and accelerated Eocene subsidence. Although several authors proposed that the accelerated Eocene subsidence is generated by a Tertiary stretching event (Joy 1992; Hall and White 1994), a consensus has emerged that the Eocene subsidence anomaly is generated by the decay of Palaeocene regional uplift event associated with the initiation of the Iceland plume (Bertram and Milton 1989; Nadin and Kusznir 1995). Regional plume-related dynamic uplift was probably close to zero at 65 Ma and reached a peak around ~57 Ma (Nadin et al. 1997). Transient regional uplift due to a mantle plume at the time that the palaeobathymetric markers are to be restored must be considered when estimating the lithosphere-stretching factor using flexural backstripping and reverse postbreakup modeling. Nadin et al. (1997) presented evidence to show that the regional dynamic uplift associated with the initiation of the Iceland plume had a magnitude of the order of 300–500 m in the northern North Sea increasing to 900 m in the Faroe-Shetland Basin, and commenced in the mid-Palaeocene and rapidly decayed in the early Eocene. White and Lovell (1997) suggest that much of the emplacement of basaltic underplating in the North Sea occurred between 63 and 54 Ma, and is presumably synchronous with transient plume-related uplift.

Sensitivity of the determined magnitude of Palaeocene lithosphere to transient Palaeocene dynamic uplift associated with the initiation of the Iceland plume is shown in figure 4.9a and b. A magnitude of 300 m for transient plume uplift is assumed by comparing the distances of the North Sea observations (Nadin et al. 1997) and the Vøring Basin from the location of the initiating plume along the early Tertiary Faroes-Iceland-East Greenland aseismic ridge. The effect of including 300 m of Palaeocene dynamic plume uplift in the analysis is to decrease the

magnitude of the Palaeocene lithosphere stretching to $\beta = 2$ for the model with no earlier extension and to $\beta = 1.75$ when the earlier Upper Jurassic-Lower Cretaceous rift is included. The preferred estimate of Palaeocene lithosphere stretching by flexural backstripping assumes no transient plume uplift (figures 4.8a and b, 4.9b), because there is no evidence to suggest that the North Atlantic was dynamically supported by the early Iceland plume at 65 Ma.

Stretching Factor Variation with Depth on the Vøring Margin

For the Vøring margin, the preferred subsidence-derived estimate of the Palaeocene stretching event is $\beta = 1.6 \pm 0.2$; the stretching factor from seismically observable faulting in the Palaeocene section is $\beta = 1.07 \pm 0.03$. The profiles of Palaeocene stretching and thinning factors for the upper crust and for the whole lithosphere for the Vøring margin are shown in figure 4.8a and b. The corresponding stratigraphic cross section is shown in figure 4.8c. The preferred lithosphere-stretching factor profile assumes no dynamic plume uplift at 65 Ma and an earlier Upper Jurassic-Lower Cretaceous rift of magnitude $\beta = 1.5$. Stretching and thinning factors (figure 4.8a and b) for the lithosphere are significantly greater than those for the upper crust on the oceanward (west) side of the profile. These results are similar to and consistent with those of Roberts et al. (1997) who performed a similar analysis. Crustal thinning-derived extension estimates are not included because the partitioning of the observed present-day stretching factor between earlier rift events and/or igneous underplating is uncertain. Depth-dependent stretching appears to be present within ~100 km of the continent-ocean boundary for all reasonable assumptions of the magnitude of dynamic topography present at 65 Ma and the magnitude of the early Cretaceous rift event. The possibility of significant extension generated by dyke intrusion is discussed later in this chapter.

Stretching Observations on the South China Sea Rifted Margin (Pearl River Mouth Basin)

The South China Sea (figure 4.10a) is floored by oceanic crust, is surrounded by stretched continental crust, and has a Tertiary extensional origin. In the east, the oceanic crust of the South China Sea is subducting eastward and obliquely at the Manila Trench beneath the Luzon Arc (Pautot and Rangin 1989). This study focuses on the section of the northern continental margin of the South China Sea adjacent to the Pearl River Mouth Basin (also called the Zhujiangkou Basin) offshore of South China. The Pearl River Mouth Basin is 300 km wide, 800 km long, and trends northeast-southwest. The basin comprises a series of fault-bounded grabens and is filled with up to ~10 km of sediments (Lee and Lawver 1994). Regionally, the Pearl River Mouth Basin forms part of a chain of Tertiary basins in the Pacific Northwest that extend from Vietnam in the southwest to the Kuril

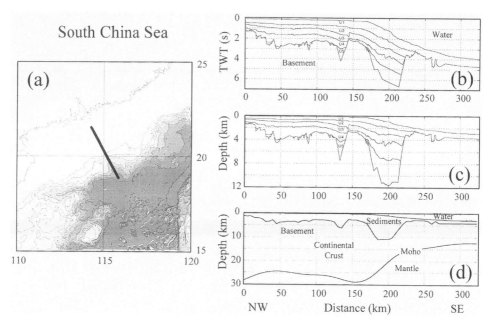

Figure 4.10 (a) Location map showing bathymetry of the South China Sea and the location of the Pearl River Basin profile used in this study. (b and c) Time- and depth-converted upper-crustal sections for the Pearl River Basin. Base U1 corresponds to base late Miocene; U2 and U3 Mid Miocene; U4 mid Palaeogene; U5 top basement (Turonian). (d) Crustal section based on seismic data and gravity data.

Basin, eastern Russia in the northeast. The spreading history of oceanic crust in the South China Sea is well known. East-west trending magnetic lineations correlate with magnetic anomalies 11–5D (32–17 Ma; Taylor and Hayes 1980, 1983; Briais et al. 1993) in the northwest subbasin and are in agreement with ages determined from a comparison of the heat-flow anomalies and plate-model predictions (Parsons and Sclater 1977; Lee and Lawver 1994). The South China Sea ocean basin is somewhat V-shaped (thinner in the southeast), where magnetic anomalies indicate that seafloor spreading began later (24–15 Ma; Briais et al. 1993). Seafloor spreading terminated with the emplacement of basaltic seamounts and other igneous bodies with ages 10–15 Ma.

Lithospheric extension in the Pearl River Mouth Basin occurred during the Palaeocene-Oligocene, although there is considerable uncertainty over the rifting history (Ru and Pigott 1986; Su et al. 1989). Ru and Pigott (1986) suggested that the Pearl River Mouth Basin has undergone at least three distinct tectonically pulsed rifting phases (late Cretaceous-Palaeocene, late Eocene-early Oligocene, and middle Miocene) with two intervening stages of seafloor spreading, whereas Su et al. (1989) suggest that the subsidence is consistent with a prolonged stretching event ($\beta \sim 1.8$) from the late Cretaceous to late Oligocene with a minor event in the Miocene ($\beta \sim 1.1$). However, in recent times age-dating of the stratigraphy has been debated with new interpretations favoring systematic shifts to younger ages (Edwards 1992). In contrast, the spreading history of oceanic crust in the South China Sea is well constrained by magnetic anomalies.

The data used in this study are largely provided by BP and are summarized in figure 4.10b–d. Sedimentation rates in the Pearl River Mouth Basin appear to have maintained pace with subsidence, and the palaeobathymetries at all times are small. Su et al. (1989) report that the foraminifera, pollen, and spore assemblages from ∼40 offshore wells indicate that the maximum palaeowater depths were <200 m and are significantly less than this from the post middle Miocene. Since few wells penetrate the syn rift, interval velocities for depth conversion (figure 4.10c) are taken from a BP-Amoco report (Moorcraft and Roberts 1991). Top basement in figure 4.10b–d corresponds to intra Upper Cretaceous (Turonian) age.

Upper-Crustal Stretching Factors

Upper-crustal extension was estimated by summing heaves on seismically resolvable faults offsetting top basement (figure 4.10b and c). A total of 33 seismically mappable faults were used to derive the upper-crustal stretching and thinning factor profiles shown in figure 4.11a and b. Fault-derived extension is 55 ± 10 km, which corresponds to a mean stretching factor $\beta = 1.3 \pm 0.05$. This result is consistent with Westaway (1994), but the stretching factor is significantly less than in the analysis of an interpretation of similar data by Su et al. (1989).

Crustal Thinning Stretching Factors

The depth to Moho is derived from an inversion of Bouguer gravity anomaly data that simultaneously solves for the residual thermal contribution from recent lithosphere stretching (Davis 1999) and is calibrated by wide-angle seismic estimates of Moho depth obtained by a nearby expanding spread profile (ESP) experiment (Nissen et al. 1995). The resulting crustal cross section is shown in figures 4.10d and 4.11c. Stretching and thinning factors determined from crustal basement thickness are shown in figure 4.11a and b. The total extension determined from crustal thinning from unstretched continental crust to the "continent-ocean boundary" is 140 ± 25 km (mean $\beta \sim 2.4 \pm 0.2$). and is consistent with the extension determined from the pseudo two-dimensional profile derived from ESP data (Nissen et al. 1995).

"Whole-Lithosphere" Stretching Factors

Given the substantial uncertainty in the ages of the mapped stratigraphic horizons and in the age of rifting, "whole-lithosphere" stretching factors from the Pearl River Mouth Basin are not reliable and are not presented.

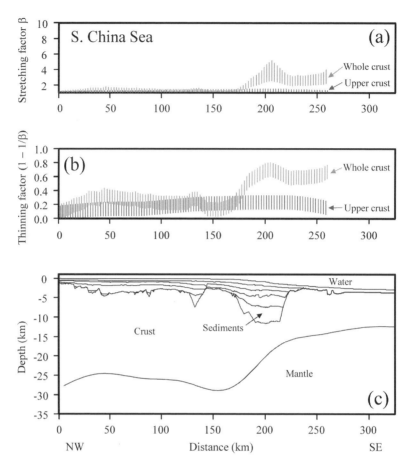

Figure 4.11 (a and b) Stretching and thinning factors determined for upper crust and whole crust for the South China Sea rifted continental margin (Pearl River Mouth Basin). (c) Crustal section for the Pearl River Basin margin based on seismic and gravity data.

Stretching Factor Variation with Depth on the South China Sea Margin

Stretching and thinning estimates derived from upper-crustal faulting and crustal thinning are compared in figure 4.11a and b for the Pearl River Mouth Basin segment of the South China Sea margin. Thinning estimates for the upper crust and whole crust are consistent in the northern part of the profile adjacent to mainland China, but as the continent-ocean boundary is approached southward whole crustal thinning greatly exceeds that for the upper crust and depth-dependent stretching becomes pronounced.

The interpretation of depth-dependent stretching in the Pearl River Mouth Basin is consistent with Westaway (1994) but contrary to that of Su et al. (1989). Su et al. (1989) derived stretching factors from faults, crustal thinning, and postrift thermal subsidence and suggested that there is no evidence for a significant extension discrepancy and that the main stretching event occurred during the late

Cretaceous-late Oligocene (ending at ~35 Ma) with β ~1.8. Although the consistency of the analysis of Su et al. (1989) seems impressive, Westaway (1994) used the same data to argue that extension factors demonstrate depth dependency with upper-crustal stretching factor β ~1.3. An advantage of this study of the Pearl River Mouth Basin is that the primary seismic data were used and depth conversion was performed with a well constrained velocity-depth function. Significantly more faults have been identified in the primary data. (Thirty-three faults are used in this study versus ten in Su et al. [1989]).

Observations of Depth-Dependent Stretching at Rifted Continental Margins: Reality or Artifact?

Independent methods and data sets have been used to determine upper-crustal, whole-crustal, and whole-lithospheric stretching factors for the Goban Spur, Galicia Bank, Vøring, and South China Sea continental margins; and the resulting profiles of stretching and thinning factor are summarized in figures 4.3, 4.6, 4.9, and 4.11. The total extension for each profile may be determined by integrating the thinning factor with respect to horizontal distance (equation 4.3). Total extension for each of the Goban Spur, Galicia Bank, Vøring, and South China Sea continental margins profiles are shown in figure 4.12 for upper crust and either whole crust or whole lithosphere. Upper-crustal and whole-lithosphere extension

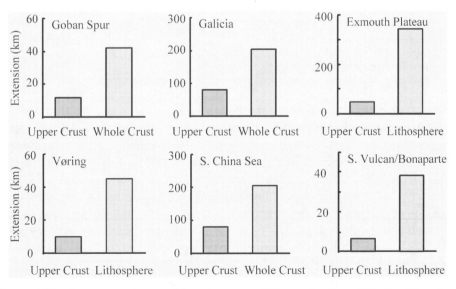

Figure 4.12 Summary of extension estimates at different depths within the lithosphere for Goban Spur, Galicia, Vøring, and the South China Sea (this study) and for the Exmouth Plateau (Driscoll and Karner 1998) and South Vulcan (Bonaparte) Basin (Baxter et al. 1999). In the Vøring margin, the estimates refer to the Palaeocene rifting event only.

estimates have also been determined for the northeast Australian rifted continental margin for the Exmouth Plateau, from stretching profiles published in Driscoll and Karner (1998), and the South Vulcan (Bonaparte) Basin from stretching profiles published in Baxter et al. (1999). In the Exmouth Plateau margin, the horizontal integration of thinning factor has been conducted from the continental edge of the transition zone at $x = -75$ km (Driscoll and Karner 1998) toward the continent to be consistent with the other extension estimates of figure 4.12.

In all of the rifted continental-margin examples shown in figure 4.12, the summed extension observed in the upper crust is substantially less than that observed for the whole crust or whole lithosphere. All cases studied show depth-dependent stretching of the continental lithosphere, where extension appears to increase with depth, over a width of 50–200 km adjacent to the continent-ocean transition. Whole-crustal and "whole-lithosphere" extension is typically a factor of ~3 larger than upper-crustal extension. Possible explanations, the majority of which we reject, are discussed in the next section.

Random Errors

Errors have been formally calculated for the thinning and stretching factors from upper-crustal faulting and whole-crustal thinning. As illustrated in figures 4.3, 4.6, 4.9, and 4.11, computed errors for thinning and stretching factors are significantly different (>1.5 standard deviations) within ~100 km of the "continent-ocean boundary" and converge toward the continent. The discrepancy between upper-crustal and whole-crust or whole-lithosphere stretching is therefore interpreted as being statistically significant.

Subseismic Resolution Faulting

Some of the extension by faulting of the upper crust is associated with small faults for which displacements are beneath the resolution of seismic reflection imaging. As a consequence, some upper-crustal extension by faulting will not be detectable using seismic reflection imaging techniques. Fault-scaling relationships enable quantification of subseismic extension and suggest that, for seismic reflection data, predicted missing extension amounts to ~35% of the observed extension (Walsh et al. 1991). Indeed, several authors have demonstrated that when subseismic extension is accounted for in intracontinental rift basins, independent estimates of extension based on thermal subsidence and faulting become comparable (Roberts et al. 1993; Marrett and Allmendinger 1992). Figure 4.13 illustrates the corrected (real) stretching factor (β_r) as a function of the observed (apparent) stretching factor (β_a) where f is defined as one plus the missing extension fraction.

$$\beta_r = \frac{\beta_a}{\beta_a + f(1 - \beta_a)} \tag{4.7}$$

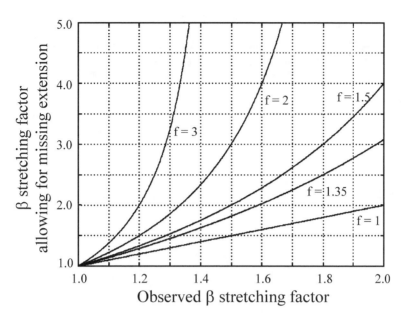

Figure 4.13 "True" (i.e., corrected) stretching factor against apparent (observed) stretching factor as a function of the missing extension factor f as described in the text.

For $f = 1.35$ (35% missing extension; Walsh et al. 1991), faulting beneath the resolution of seismic reflection imaging cannot explain the large extension discrepancies at rifted margins. Explanation of our observations using this principle alone requires unreasonable missing extension fractions and seems unlikely.

Fault Geometry

In this chapter, the calculation of upper-crustal extension assumes that upper-crustal faults have a planar geometry. The use of a listric geometry, however, makes little difference in practice to extension estimates, because the fault angle changes little with depth in the upper part of the seismic reflection sections where fault offsets are measured. Studies of actively extending regions in continents (Turkey, Greece, and western United States) suggest that active basement faults are planar to at least depths of 10 km (Jackson and McKenzie 1988; Jackson 1987; Stein and Barrientos 1985). Although basement faults may appear to be listric on time sections, several authors have demonstrated that typical basement faults are planar in depth (Kusznir and Ziegler 1992) as confirmed by prestack depth migration on the outer Galicia (Reston 1996).

Second-Generation Faults

If extension of the lithosphere at rifted margins is associated with multiple generations of faults that cut earlier generations and mask their detection, our method of estimating upper-crustal extension underestimates upper-crustal extension. Stretching of the upper crust has been determined using seismic reflection data, which allows good imaging of the syn- and prebreakup stratigraphy required to determine fault-controlled extension. In the Goban Spur, Galicia Bank, and the outer parts of the Exmouth Plateau and Pearl River Basin, these margins are sediment starved with little postbreakup deposition obscuring deeper stratigraphy. No second-generation faulting has been observed on the seismic data used in this study. Whilst contributions from second-generation faulting cannot be ruled out, it is notable that the fault-derived stretching factor profiles at rifted margins are relatively constant, with typical maximum stretching factors around $\beta \sim 1.3$, and are considerably less than those observed in many intracontinental rift basins where second-generation faulting is not believed to occur (Roberts et al. 1993).

Nonbrittle (Aseismic) Extension

Upper-crustal extension could be underestimated if upper-crustal extension occurred aseismically rather than by faulting. Upper-crustal extension might conceivably occur through dyking or nonbrittle (plastic) deformation. While dyking is possible, it is difficult to see how it may occur up to ~ 100 km from the continent-ocean boundary. Furthermore, at nonvolcanic rifted margins the transitional region between rotated continental fault blocks in thinned continental crust and unequivocal oceanic crust is surprisingly devoid of significant basaltic volcanism. margins (e.g., Iberian Margin; Whitmarsh et al. 2001). Present-day analogues suggest that the majority of upper-crustal deformation is accommodated by seismic slip on faults (Jackson 1987). Studies of the actively extending Aegean region place good constraints on both geodetic and upper-crustal fault-related strain. That the seismically determined velocity field in the Aegean is comparable with the estimate predicted from simple kinematic arguments (i.e., plate-motion studies) suggests that the majority of the deformation in the upper crust is accommodated by seismic slip on faults (Jackson and McKenzie 1983).

Intrusion of Dense Melt into the
Lithosphere and Phase Transitions

Emplacement and cooling of ultramafic melts may thin the seismically defined crust, giving rise to errors in our determination of crustal thinning and stretching. However, the addition of crustal material generates uplift, not subsidence, unless underplating occurs in the eclogite stability field, which is unlikely (McKenzie 1984). The anomalously high postrift subsidence observed at the outer ~ 100 km of rifted continental margins is therefore not readily explained by melt intrusion. Phase transformation from gabbro to eclogite or garnet granulites would raise the Moho by thinning the crust, and load the lithosphere and generate subsidence, but is unlikely to occur at depths shallower than ~ 50 km.

Mantle Serpentinization

Serpentinization of mantle reduces both its seismic velocity and density such that it could apparently thicken the crust. Such a process would therefore serve to mask rather than enhance the observed increase in thinning and stretching with depth. While many rifted continental margins are associated with a transitional region consisting of serpentinized mantle between unequivocal continental and unequivocal oceanic crust (Pickup et al. 1996), this explanation cannot explain the observed discrepancy between whole-crustal and upper-crustal stretching factors.

Rifting above Sea Level

Rifting above sea level is likely to erode upper-crustal fault blocks, thereby destroying the fault record, leading to an underestimation of upper-crustal stretching from fault-heave measurements. Whereas the uniform lithosphere stretching model (McKenzie 1978) predicts uplift where the initial crustal thickness <18 km, isostatically compensated lithosphere with thin crust should be associated with significant prerift bathymetry, which buffers the subsequent uplift. Alternatively, rifting might occur above sea level if the lithosphere suffers depth-dependent stretching. Analytical models of two-layer stretching demonstrate that uplift is only predicted if the mantle-thinning factor (ε) is greater than 2.2 times the crustal thinning factor (Royden and Keen 1980) and that subsidence is always generated (regardless of mantle-stretching factor) provided that the stretching factor $\beta > 1.5$. While some erosion of fault block crests is observed on all four rifted margin profiles presented in this chapter, the amount of upper-crustal extension "lost" through fault-block crest erosion is believed to be small.

Influence of a Mantle Plume on Subsidence

Anomalous subsidence patterns on rifted margins may sometimes be attributable to mantle-plume influences (Nadin et al. 1997). Whereas plume collapse-driven subsidence may occur simultaneously with post-rift thermal subsidence (e.g., the Norwegian margin), it alone cannot account for the observed subsidence anomaly (see also Roberts et al. 1997). Depth-dependent stretching is also observed at "nonvolcanic" margins that are not associated with mantle-plume activity (e.g., Galicia, Goban Spur).

Flexural Coupling

Unstretched continental crust that is strongly coupled mechanically to adjacent thermally subsiding oceanic lithosphere is expected to experience postbreak subsidence greater than that predicted from its own thinning and stretching. Provision has been made on our analysis to account for the enhanced subsidence of continental crust adjacent to oceanic crust in the reverse postrift modeling analysis by requiring the stretching factor to ramp to $\beta = \infty$ at the conventional continent-ocean boundary. Sensitivity analysis to T_e used to define this coupling has also been performed. Anomalously large lithosphere-stretching estimates for the continental margin lithosphere from subsidence backstripping due to flexural coupling to oceanic lithosphere is therefore ruled out.

Lithosphere Extension on a Low-Angle Extensional Detachment (Simple Shear)

All rifted margins studied in this chapter show greater thinning and stretching of the whole crust and lithosphere than that observed for the upper crust, consistent with similar reports at other margins (Driscoll and Karner 1998). Driscoll and Karner (1998) have noted that all margins appear to be derived from the "upper plate" of a possible simple-shear model (Wernicke 1985; Lister et al. 1991), giving rise to the "upper-plate paradox." The "upper-plate paradox" and simple-shear models of lithosphere extension are discussed further in Discussion and Summary. Although instances of active low-angle faults accommodating lithosphere extension (Wernicke 1985) have been reported (Abers 1991), their existence has been disputed. Jackson and McKenzie (1983) note that there is "not one single example of seismic activity on a subhorizontal fault [dip <20°]"; indeed, the vast majority of seismically active normal faults have dips in from 30–60°. Therefore, if active low-angle normal faults exist, they must occur beneath the earthquake nucleation depth. Although several authors claim to have imaged a detachment either in the field or on seismic reflection profiles (Torres et al. 1993), very little seismic evidence exists for their presence in extensional settings (Collier and Watts 1997). Probably the best example of a possible candidate for such a detachments is a the so-called S-reflector off Galicia (Reston 1996; Charpal et al. 1978).

Lower-Crustal Buoyancy-Driven Flow

Buoyancy-driven flow in the lower crust could give rise to enhanced apparent thinning and stretching of the continental crust at rifted margins. Lower-crustal flow appears not to occur during the formation of intracontinental rift basins since the uniform stretching model is consistent with equal extension observed in the upper crust, the whole crust, and lithosphere (McKenzie 1978; White 1990; Marsden et al. 1990; Roberts et al. 1993). Although thickening of thinned continental crust by buoyancy-driven lower-crustal flow has been proposed (Buck 1991; Davis and Kusznir 2002), if it were to exist it would serve to counteract rather than enhance the observation of crustal thinning and stretching increasing with depth. While lower-crustal flow may occur at rifted continental margins, it is unlikely to contribute to the total extension estimates presented in this paper unless the lower-crust flows into the domain which is conventionally believed to be oceanic.

Model Predictions of Depth-Dependent Stretching at Rifted Margins

We use simple analytical solutions and two-dimensional finite-element models of the divergent lithosphere flow that is expected to occur during early seafloor spreading to explore the mechanism responsible for depth-dependent lithosphere stretching observed at rifted continental margins. The models aim to investigate the effects of early seafloor spreading on the newly formed continental margin lithosphere; they do not investigate the prebreakup rifting of continental lithosphere. The objectives are not to simulate the actual development of a rifted continental margin, but rather to investigate the physics of the conditions under which depth-dependent stretching may occur.

Analytical Solution for Corner Fluid Flow Applied to Young Seafloor Spreading

Insight into the origin of depth-dependent stretching observed at rifted continental margins may be derived from examination of the fluid-flow field at mid-ocean ridge spreading centers. The results for the analytical solution (Batchelor 1967) for corner flow for isoviscous incompressible mantle in which fluid upwells beneath a ridge before spreading laterally as the lithospheric plates diverge is illustrated in figure 4.14. While the two-dimensional corner-flow model for constant viscosity shown in figure 4.14 is highly simplistic, fluid-flow fields predicted by complex models that incorporate temperature-dependent viscosity and melt generation (Spiegelman and McKenzie 1987; Cordery and Phipps-Morgan 1992; Spiegelman and Reynolds 1999) are not dissimilar. The divergent velocity field has been computed for horizontal surface and vertical ridge axis velocities of 5 cm/yr; flow vectors are plotted normalized to this boundary condition.

In the reference frame of the ocean ridge we see a divergent horizontal flow balanced by upwelling beneath the ridge axis (figure 4.14). Only one half of the flow field is plotted because the flow field is symmetric about the ocean ridge axis. The same flow field may be transformed into the reference frame of the adjacent ocean basin or continental interior by subtracting the far-field ocean basin velocity vector. This flow field referenced to the ocean basin is shown in figure 4.14b. The depth dependency of the ocean ridge divergent flow field can be clearly seen in figure 4.14b. The flow field shows oceanward flow of the continental lithosphere mantle and lower crust of the adjacent newly formed rifted continental margin. Figure 14b demonstrates that the divergent motion of the adjacent young seafloor spreading center drives depth-dependent stretching in young rifted continental margin lithosphere.

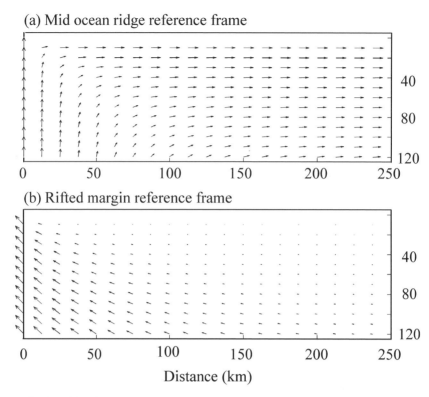

(a) Mid ocean ridge reference frame

(b) Rifted margin reference frame

Distance (km)

Figure 4.14 Fluid-flow velocity vectors for a simple two-dimensional analytical corner flow model of a mid-ocean ridge-spreading center using an isoviscous fluid: (a) in the reference frame of the mid-ocean ridge; and (b) in the reference frame of the far-field ocean basin or continental interior. Depth-dependent flow can be seen clearly in (b) where deeper material flows oceanward relative to upper lithospheric material.

Finite Element Modeling of Young Seafloor Spreading

The analytical corner-flow solution described above assumes constant viscosity, whereas the real viscosity structure associated with a young oceanic spreading center and adjacent continental-margin lithosphere is expected to be highly heterogeneous because of both temperature and compositional variations. While a full dynamic thermal and temperature-dependent viscosity solution is beyond the scope of this study, we have used finite-element modeling to examine depth-dependent stretching in an Earth model with a horizontally and vertically heterogeneous viscosity distribution. We emphasize that the aim of this simple finite element model is to investigate the effects of early seafloor spreading on the young rifted margin lithosphere and is not to investigate the effects of prebreakup intracontinental rifting that usually precedes continental breakup and the initiation of seafloor spreading.

The viscosity structure of this simple schematic model is shown in figure 4.15a. The finite-element model assumes that the cooler continental lithosphere has a greater viscosity than the hotter asthenosphere and that the continental lithosphere

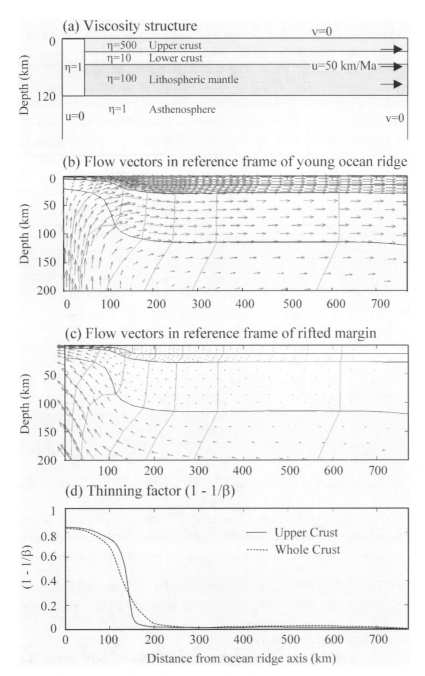

Figure 4.15 Simplified finite-element fluid-flow model of young seafloor spreading and adjacent rifted margin. (a) Viscosity structure and boundary conditions used in the finite-element model. Viscosities: upper crust = 5×10^{22} Pas; lower crust = 1×10^{21} Pas; lithospheric mantle = 1×10^{22} Pas; and asthenosphere = 1×10^{20} Pas; a vertically defined weak zone is present with viscosity = 1×10^{20} Pas. (b) Velocities predicted by the finite-element model in the reference frame of the young ocean ridge. (c) Velocities predicted by the finite-element model in the reference frame of the adjacent rifted margin. (d) Predicted thinning factors for the upper and whole crust.

is rheologically layered with a quartz-feldspathic crust above an olivine mantle (Kusznir and Park 1987). The young seafloor-spreading center is represented by a prescribed low-viscosity weak zone that is given the same viscosity as the asthenosphere. The viscosity of the topmost mantle is assumed to be higher than that of the lower crust, because olivine rheology of the mantle is stronger than that of the quartz-feldspathic crust at the same temperature; and the viscosity of the lower crust is assumed to be lower than that of the upper crust because it is hotter (Kusznir and Park 1987; Kusznir 1991). The finite-element model is run with a constant-velocity far-field boundary condition of 50 km Ma^{-1} (5 cm/yr) for 2.5 Ma, and it is symmetric about an axis of symmetry at $x = 0$ km corresponding to the young seafloor-spreading axis. The top and bottom boundary conditions of the finite-element fluid model have zero vertical fluid velocity, and the model is 660 km deep corresponding to the base of the upper mantle.

The fluid-flow pattern predicted by the finite-element model in the reference frame of the young seafloor-spreading center is shown in figure 4.15b. The divergent motion of the flow field away from the spreading axis can be seen. The same flow field, transformed into the continental interior reference frame, is shown in figure 4.15c. This flow field shows oceanward flow of the mantle and lower crust of the continental margin lithosphere, similar to the analytical solution described in the previous section, and demonstrates that depth-dependent stretching occurs in the continental lithosphere adjacent to the young seafloor-spreading center.

Upper-crustal and whole-crustal thinning factors $(1 - 1/\beta)$ predicted by the finite-element modeling are plotted in figure 4.15d. At distances greater than 150 km from the spreading axis in the postextension reference frame (corresponding to 40 km in the preextension reference frame), the stretching and thinning of the upper crust is predicted to be less than that of the whole crust with the discrepancy decreasing toward the continental interior. These predictions of the finite-element modeling are consistent with observations of thinning at rifted margins. Oceanward of 150 km (and in the region not covered by our observations of rifted margin thinning), the modeled stretching of the upper crust is predicted to exceed that of the whole crust. We identify that the form of predicted depth-dependent stretching is similar to that observed (and inferred) at rifted continental margins, provided that: (a) the lithosphere is horizontally rheologically layered with viscosity varying as a function of depth, and (b) the lithosphere possesses a vertically defined weak zone. In addition, a low-viscosity lower crust is required to generate depth-dependent stretching at the crustal level.

The viscosity structure of real oceanic and continental lithosphere at a young rifted margin is highly temperature- and time dependent due to conductive diffusion and advection of heat. A coupled transient thermal and fluid-flow model incorporating temperature-dependent viscosity and compositionally dependent rheology is required, but it is numerically difficult because of the large dynamic range of viscosity required and is beyond the scope of this chapter. However, preliminary models that incorporate conductive and advective contributions to the fluid-flow modeling of a temperature-dependent and compositionally varying rheology with a viscosity range of 10^{20}–5×10^{22} Pas (a factor of 250) have been performed and show flow vectors and thinning profiles with depth-dependent stretching similar to that of figure 15. This exploratory model also suggests that depth-dependent stretching is small for small amounts of stretching but rapidly increases for $\beta > 2$ and may explain why intracontinental rift basins generally are consistent with the uniform stretching model (McKenzie 1978; Cordery and Phipps-Morgan 1992) and rifted margins are not.

Discussion and Summary

Area Conservation during Depth-Dependent Stretching

Depth-dependent stretching has been observed at a number of rifted continental margins, whereby stretching of the upper continental crust appears to be much less than that of the whole crust or lithosphere. Assuming that this observation is real, how is it possible for stretching to occur without violating area and hence volume conservation? To conserve volume, we require the upper-crustal, lower-crustal, and lithospheric mantle all to suffer the same total extension. In the stretched reference frame the horizontal integral of the thinning factor (ε) along the margin should be a constant:

$$E_{tot} = \int_0^\infty \varepsilon_{uc} dx = \int_0^\infty \varepsilon_{lc} dx = \int_0^\infty \varepsilon_{ml} dx \tag{4.8}$$

The analytical and finite-element models described in the previous section obey the principle of volume conservation and also, under the conditions described, show depth-dependent stretching behavior (figures 4.14 and 4.15). The apparent extension discrepancy, generated where continental upper-crustal stretching factors are less than whole-crustal and lithosphere mantle-stretching factors, may be resolved by the presence of a region farther toward the ocean with an extension discrepancy of opposite polarity (figure 4.16a) so that the equality of the horizontal integral of extension with depth is maintained (equation 4.8). This would require that somewhere oceanward the upper continental crust is strongly stretched and thinned by an amount greater than deeper material. The existence of an oceanward region where upper crustal extension exceeds that of the whole crust and lithosphere is supported by the crossover of upper and lower lithosphere thinning factors predicted by the finite-element model and shown in figure 4.15d.

Evidence for Mantle Exhumation and a Transitional COB

Evidence for the exhumation of continental mantle at rifted continental margins is derived from drilling, dredging, seismic refraction, and wide-angle reflection

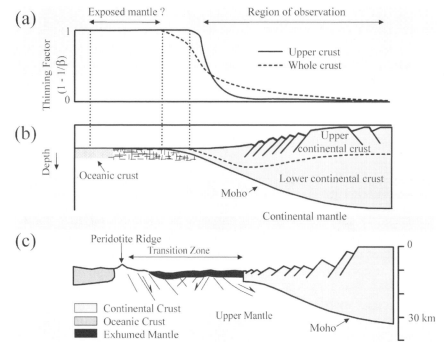

Figure 4.16 Conceptual diagram illustrating depth-dependent stretching and mantle exhumation at rifted continental margins. (a) Thinning factor for upper and whole crust illustrating depth-dependent stretching. In the region we observe, upper-crustal extension is less than that of the whole crust (and mantle); however, mass conservation requires upper-crustal extension oceanward to exceed that of the whole crust (and mantle). (b) Depth-dependent stretching and oceanward flow of continental mantle may lead to exhumation of continental mantle oceanward of thinned continental crust. (c) Schematic diagram of Iberian rifted margin structure showing mantle exhumation between oceanic crust and the rotated fault blocks of thinned continental crust, modified after Pickup et al. (1996).

surveys, normal-incidence seismic reflection, field analogs, and geochemistry (Pickup et al. 1996; Whitmarsh et al. 2001). In addition to field observations, mantle exhumation is also predicted by laboratory analogue models (Brun and Beslier 1996).

Evidence for Transitional Crust from Drilling and Dredging

Prominent ridges drilled in the region between unequivocal oceanic and unequivocal continental crust at the Iberian Abyssal Plain margin consist of highly serpentinized peridotite that appears to have undergone limited melt extraction and may represent exhumed continental mantle (Discovery 215 Working Group 1998; Whitmarsh et al: 2001). Whitmarsh et al. (2001) report that the exhumed mantle on the Iberian margin is "magma-poor" and contains relatively little basaltic material. This paucity of basaltic material within exhumed mantle may be consistent with mantle exhumation occurring during early seafloor spreading such that that basaltic melt is focused (or sucked) into the ridge axis by the matrix pressure field associated with the divergent flow in the lithosphere and asthenosphere at the young ocean ridge (Spiegelman and McKenzie 1987; Spiegelman and Reynolds 1999).

Evidence from Seismic Refraction, Wide-Angle Reflection, and Normal Incidence Reflection

Seismic modeling of refraction and wide-angle reflection data from Galicia Bank reveals that the seismic Moho is frequently poorly defined in the "oceanic" domain adjacent to continental crust and that high-velocity (possibly serpentinitic) bodies are found at shallow depths (Pickup et al. 1996). Throughout the transitional zone, the seismic velocity is not typical of either oceanic or continental crust, because high velocities are found at shallow depths and normal mantle velocities are sometimes not detected. Although there are several possible alternative explanations for the seismic characteristics of the transition zone (ultraslow seafloor spreading, highly thinned and intruded continental crust, and unusually cool mantle), it is likely that the zone comprises continental upper mantle exhumed along a detachment zone and later serpentinized by seawater (Pickup et al. 1996)—a hypothesis consistent with the absence of true crust, and a reflection and refraction Moho. High-velocity zones are frequently noted to bound rifted margins and include the south Australia margin and the Tyrrhenian Sea (Decandia and Elter 1969) and may represent serpentinite belts, although the possibility remains that these zones may represent underplating where drilling has not occurred. Similar transition zones between unequivocal "oceanic" and "continental" crust have also been recognized at rifted continental margins including the Labrador Sea, Galicia, the Lincoln Sea, the Red Sea, and the Gulf of Guinea.

Field Analogs Provide Support for Mantle Exhumation Models

Preserved rift and seafloor-spreading structures in the Austroalpine and Penninic nappes of Graubünden, eastern Switzerland, provide evidence of detachment faulting and mantle exhumation. In particular, absence of normal components of oceanic crust (layers 2A and 2B, i.e., pillow lavas, sheet lava flows, and dykes) in some complexes have led several authors to propose that subcontinental mantle material was tectonically denuded and exposed at the seafloor during rifting (Manatschal and Nievergelt 1997; Manatschal and Bernoulli 1999). The similarities between the mode of extension inferred at both fossil and modern rifted margins provide support for the notion of depth-dependent stretching.

Additional Comments on Mantle Exhumation

Figure 4.16c shows the configuration of exhumed mantle at the Iberian continental margin proposed by Pickup et al. (1996). The exhumed mantle lies between the rotated fault blocks of the extensionally thinned continental crust and unequivocal oceanic crust out to the west. The existence of an oceanward region where upper-crustal extension and thinning exceeds that of the lower crust and lithospheric mantle, as required by volume conservation (equation 4.8) and predicted by finite-element modeling (discussed under Area Conservation during Depth-Dependent Stretching), does not necessarily require the thinning of the upper crust to zero and the resulting exhumation of lower continental crust or continental mantle. However, the observation of exhumed continental mantle at rifted continental margins between the extensionally thinned continental crust and normal oceanic crust suggests that the continental crust may be thinned to zero thickness by the depth-dependent stretching process, allowing the continental mantle to be pulled out oceanward from under the continental crust and exhumed at the sea bed (figure 4.16b).

The "Upper-Plate Paradox"

Two end-member models of lithosphere extension have been proposed (figure 4.17a and b): the pure-shear-stretching model (McKenzie 1978), in which lithosphere stretching is uniform with depth, and the simple-shear model (Wernicke 1985), in which lithosphere extension is controlled by a low-angle lithosphere scale extensional detachment. For intracontinental extension and the formation of intracontinental rift basins, lithosphere extension is observed to be uniform with depth (White 1990; Roberts et al. 1993; Marsden et al. 1990) and the pure-shear (uniform) lithosphere extension model appears to apply (McKenzie 1978). Depth-dependent stretching as observed at rifted continental margins (Driscoll and Karner 1998; Roberts et al. 1997; Baxter et al. 1999; this study) is not consistent however with the "pure-shear" (uniform) lithosphere-extension model. Depth-dependent stretching is, however, predicted by the simple-shear lithosphere extensional detachment model.

The observations of depth-dependent stretching presented in this chapter for the Goban Spur, Galicia Bank, Vøring, and South China Sea margins show that stretching increases with depth, which is consistent with an " upper-plate" location for these margins in the context of the simple-shear lithosphere extensional detachment model (Wernicke 1985). This supports the observation of Driscoll and Karner (1998) that all rifted continental margins correspond to the "upper plate" of a lithosphere " simple-shear" extension model. Driscoll and Karner (1998) note that " upper-plate" scenarios have been proposed for both the Newfoundland and Galicia continental margins (figure 17c and d) yet these margins are conjugate to each other so that they cannot both be "upper plate" within the context of a simple-shear lithosphere extensional detachment model. Driscoll and Karner (1998) name this the "upper-plate paradox."

A possible explanation of the "upper-plate paradox" consistent with depth-dependent stretching and mantle exhumation at rifted continental margins is presented in figure 4.18. If depth-dependent stretching and mantle exhumation are generated by the early seafloor-spreading process, then depth-dependent stretching and mantle exhumation should occur simultaneously and symmetrically on both conjugate margins. Depth-dependent stretching is probably achieved by decou-

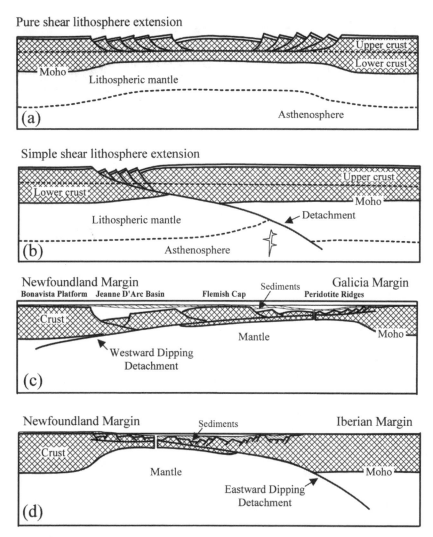

Figure 4.17 (a) Pure-shear (uniform) lithosphere-stretching model (McKenzie 1978) and (b) simple-shear lithosphere-stretching model (Wernicke 1985), adapted from Lister et al. (1986). The observation of increasing stretching with depth at rifted continental margins (Driscoll and Karner 1998; Roberts et al. 1997; Baxter et al. 1999; this study) is consistent with an "upper-plate" location (in the context of simple-shear lithosphere-stretching model; Wernicke 1985) for rifted margins. Driscoll and Karner (1998) named this the "upper-plate paradox." (c and d) Inconsistent interpretations of the form of lithosphere stretching at the Newfoundland and Galicia rifted continental margins identified by Driscoll and Karner (1998); (c) shows a westward-dipping lithosphere extensional detachment, while (d) shows an eastward-dipping lithosphere extensional detachment. Both "upper-plate" and "lower-plate" scenarios have been proposed for the Newfoundland and Galicia margins. If the Newfoundland and Galicia margins are conjugate margins, then the interpretations shown in (c) and (d) are inconsistent.

Conjugate rifted margins

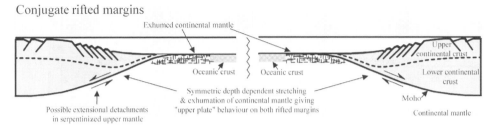

Figure 4.18 Conceptual diagram showing symmetric depth-dependent stretching and exhumation of continental mantle on conjugate rifted margins. The "upper-plate paradox," whereby all rifted continental margins appear to correspond to the hanging-wall of a lithosphere scale extensional detachment, may be explained by depth-dependent stretching and mantle exhumation on both conjugate margins as illustrated.

pling of the upper and lower crust from the continental mantle, either at discrete shear zones or more likely, since detachments are rarely observed, diffuse detachment zones. The conjugate margins would therefore be symmetrical with either discrete or diffuse detachment zones dipping downward toward the continent (figure 4.18), and consequently, both conjugate margins would suffer depth-dependent stretching consistent with the "upper plate" of a lithosphere simple-shear extensional detachment model. The timing of depth-dependent stretching and mantle exhumation with respect to the onset of seafloor spreading is however unknown and unconstrained at present. We suggest that the determination of the timing of depth-dependent stretching with respect to the onset of seafloor spreading is an important priority for future work.

Acknowledgments

We thank Mike Cheadle, Neal Driscoll, Garry Karner, Eric Lundin, Tim Minshull, Alan Roberts, Nicky White, and Bob Whitmarsh for helpful and stimulating discussions. Eric Lundin (Statoil) and Dave Roberts (BP) are thanked for providing seismic data. We also thank Uri ten Brink, Olaf Svenningsen, Neal Driscoll, and David Kohlstedt for constructive reviews. This work was conducted with support to MJD from a NERC Ph.D. studentship GT4/95/165/E.

References

Abers, G. 1991. Possible seismogenic shallow-dipping normal faults in the Woodlark-D'Entrecasteaux extensional province, Papua New Guinea. *Geology* 19:1205–1208.

Batchelor, G. K. 1967. *An Introduction to Fluid Dynamics*. Cambridge, UK: Cambridge University Press.

Baxter, K., G. T. Cooper, K. C. Hill, and G. W. O'Brian. 1999. Late Jurassic subsidence and passive margin evolution in the Vulcan Sub-basin, north-west Australia: Constraints from basin modelling. *Basin Res.* 11:97–111.

Bertram, G. and N. Milton. 1989. Reconstructing basin evolution from sedimentary thickness: The importance of palaeobathymetric control, with reference to the North Sea. *Basin Res.* 1:247–257.

Blystad, P., H. Brekke, R. B. Faerseth, B. T. Larson, J. Skogseid, and B. Torudbakken. 1995. *Structural Elements of the Norwegian Continental Shelf.* NPD Bulletin 8. Oslo: Norwegian Petroleum Directorate.

Boillot, G., G. Feraud, M. Recq, and J. Girardeau. 1989a. Undercrusting by serpentinite beneath rifted margins. *Nature* 341:523–525.

Boillot, G., D. Mougenot, J. Girardeau, and E. Winterer. 1989b. Rifting processes on the West Galicia Margin, Spain. In A. Tankard and H. Balkwill, eds., *Extensional Tectonics and Stratigraphy of the North Atlantic Margins,* Memoir, vol. 46, pp. 363–377. Tulsa, OK: American Association of Petroleum Geologists.

Bott, M. and A. Watts. 1970. Deep structure of the continental margin adjacent to the British Isles. In F. Delaney, ed., *The Geology of the East Atlantic Continental Margin,* pp. 89–109. London: British Geological Survey.

Briais, A., P. Patriat, and P. Tapponnier. 1993. Updated interpretation of magnetic anomalies and seafloor spreading stages in the South China Sea: Implications for the Tertiary tectonics of Southeast Asia. *J. Geophys. Res.* 98:6299–6328.

Brodie, J. and N. J. White. 1995. The link between sedimentary basin inversion and igneous underplating. In J. G. Buchanan and P. G. Buchanan, eds., *Basin Inversion*, Special Publication 88, pp. 21–38. London: Geological Society.

Brun, J. P. and M. O. Beslier, 1996. Mantle exhumation at passive margins. *Earth Planet. Sci. Lett.* 142:161–173.

Buck, W. R. 1991. Modes of continental lithospheric extension. *J. Geophys. Res.* 96:20161–20178.

Charpal, O, P. Guennoc, L. Montadert, and D. Roberts. 1978. Rifting, crustal attenuation and subsidence in the Bay of Biscay. *Nature* 275:706–711.

Cheadle, M., S. McGeary, M. Warner, and D. Blundell. 1987. Extensional structures on the western UK continental shelf: a review of deep seismic reflection profiling. In M. Coward, J. Dewey, and P. Hancock, eds., *Continental Extensional Tectonics,* Special Publication 28, pp. 445–465. London: Geological Society.

Chenet, P., L. Montardet, H. Gairaud, and D. Roberts. 1982. Extension ratio measurements on the Galicia, Portugal and Northern Biscay continental margins: Implications for evolutionary models of passive continental margins. In J. S. Watkins and C. L. Drake, eds., *Studies in Continental Margin Geology*, Memoirs, vol. 34, pp. 703–715. Tulsa, OK: American Association of Petroleum Geologists.

Collier, J. S. and A. B. Watts. 1997. Seismic-reflection constraints on lithospheric extension: Pure shear versus simple shear. *Geophys. J. Int.* 129:737–748.

Cook, D. R. 1987. The Goban Spur: Exploration in a deep water frontier basin. In J. Brooks and K. Glennie, eds., *Petroleum Geology of North West Europe*, pp. 623–632. London: Graham and Trotman.

Cordery, M. and J. Phipps-Morgan. 1992. Melting and mantle flow beneath a mid-ocean spreading centre. *Earth Planet. Sci. Lett.* 111:493–516.

Cordoba, D., E. Banda, and J. Ansorge. 1987. The Hercynian crust in northwestern Spain: A seismic survey. *Tectonophysics* 132.

Davis, M. J. 1999. Unpublished Ph.D. thesis, University of Liverpool, UK.

Davis, M. J. and N. J. Kusznir. 2002. Are buoyancy forces important during the formation of rifted margins? *Geophys. J. Int.* 149:524–533.

Decandia, F. and P. Elter. 1969. Riflessioni sul problema delle ophioliti nell'Appennino Settentrionale (Nota preliminare). *Atti Soc. Toscana Sci. Nat.* 76:1–19.

de Graciansky, P. C. and C. W. Poag. 1985. *Site 550, Initial Reports of the DSDP*, vol. 80, pp. 251–289. Washington, DC: U.S. Government Printing Office.

Discovery 215 Working Group (T. A. Minshull, S. Dean, R. B. Whitmarsh, S. Russell, K. E. Louden, and D. Chian). 1998. Deep structure in the vicinity of the ocean-continent transition zone under the southern Iberia Abyssal Plain. *Geology* 26:743–746.

Dore, A. G. and E. R. Lundin. 1996. Cenozoic compressional structures on the NE Atlantic margin: Nature, origin and potential significance for hydrocarbon exploration. *Petroleum Geosci.* 2:299–311.

Dore, A. G., E. R. Lundin, O. Birkeland, P. E. Eliassen, and L. N. Jensen. 1997. The NE Atlantic Margin: Implications of late Mesozoic and Cenozoic events for hydrocarbon prospectivity. *Petroleum Geosci.* 3:117–131.

Driscoll, N. and G. Karner. 1998. Lower crustal extension across the Northern Carnarvon basin, Australia: Evidence for an eastward dipping detachment. *J. Geophys. Res.* 103:4975–4991.

Edwards, P. 1992. Structural evolution of the western Pearl River Mouth Basin. *Mem. Am. Assoc. Petroleum Geol.* 53:43–52.

Eldholm, O., J. Skogseid, S. Planke, and T. P. Gladczenko. 1995. Volcanic margin concepts. In E. Banda, ed., *Rifted Ocean-Continent Boundaries*, pp. 1–16. Dordrecht, The Netherlands: Kluwer.

Gage, M. and A. Dore. 1986. *A Regional Geological Perspective of the Norwegian Offshore Exploration Provinces. Habitat of Hydrocarbons on the Norwegian Continental Shelf*, pp. 21–38. London: Graham and Trotman.

Hall, B. D. and N. J. White. 1994. Origin of anomalous Tertiary subsidence adjacent to North Atlantic continental margins. *Mar. Petroleum Geol.* 11:702–714.

Hinz, K. 1981. A hypothesis on terrestrial catastrophes: wedge of very thick ocean-dipping layers beneath passive continental margins. *Gelo. Jahrb. (Ser. E)* 22:3–28.

Horsefield, S. J., R. B. Whitmarsh, R. S. White, and J. C. Sibuet. 1993. Crustal structure of the Goban Spur rifted continental margin, northeast Atlantic. *Geophys. J. Int.* 119:1–19.

Jackson, J. A. 1987. Active normal faulting and crustal extension. In M. P. Coward, J. F. Dewey, and P. L. Hancock, eds., *Continental Extensional Tectonics*, Special Publication 28, pp. 3–18. London: Geological Society.

Jackson, J. A. and D. P. McKenzie. 1983. The geometrical evolution of normal fault systems. *J. Struct. Geol.* 5:236–255.

Jackson, J. A. and D. P. McKenzie. 1988. The relationship between plate motions and seismic moment tensors and the rates of active deformation in the Mediterranean and Middle East. *Geophys. J.* 93:45–73.

Jackson, J., J. Haines, and W. Holt. 1992. The horizontal velocity field in the deforming Aegean Sea region determined from the moment tensors of earthquakes. *J. Geophys. Res.* 97:17657–17684.

Joppen, M. and R. S., White. 1990. The structure and subsidence of Rockall Trough from 2-ship seismic experiments. *J. Geophys. Res.* 95:19821–19837.

Joy, A. M. 1992. Right place, wrong time: Anomalous post-rift subsidence in sedimentary basins around the North Atlantic Ocean. In B. C. Storey, T. Alabaster, and R. J. Pankhurst, eds., *Magmatism and the Causes of Continental Break-up,* Special Publication 68, pp. 387–393. London: Geological Society.

Klemperer, S. and R. Hobbs. 1991. *The BIRPS Atlas: Deep Seismic Reflection Profiles from around the British Isles.* Cambridge, UK: Cambridge University Press.

Kusznir, N. J. 1991. The distribution of stress with depth in the lithosphere: thermo-rheological and geodynamic constraints. *Philos. Trans. R. Soc. Lond. A Math. Phys. Sci. A* 337: 95–110.

Kusznir, N. J. and R. G. Park. 1987. The extensional strength of the continental lithosphere: its dependence on continental gradient, and crustal composition and thickness. In M. P. Coward, J. F. Dewey, and P. L. Hancock, eds., *Continental Extension Tectonics*, Special Publication 28, pp. 35–52. London: Geological Society.

Kusznir, N. J. and P. A. Ziegler. 1992. The mechanics of continental extension and sedimentary basin formation: A simple-shear pure-shear flexural cantilever model. *Tectonophysics* 215:117–131.

Lee, T. Y. and L. A., Lawver. 1994. Cenozoic plate reconstruction of the south china sea region. *Tectonophysics* 235:149–180.

Le Pichon, X. and J. C. Sibuet. 1981. Passive margins: A model of formation. *J. Geophys. Res.* 86:3708–3720.

Lister, G. S., M. A. Etheridge, and P. A. Symonds. 1991. Detachment models for the formation of passive continental margins. *Tectonics* 10:1038–1064.

Louden, K. E., J.-C. Sibuet, and J.-P. Foucher. 1991. Variations in heat flow across the Goban Spur and Galicia continental margins. *J. Geophys. Res.* 96:16131–16150.

Louvel, V., J. Dyment, and J.-C. Sibuet. 1997. Thinning of the Goban Spur continental margin and formation of early oceanic crust: Constraints from forward modelling and inversion of marine magnetic anomalies. *Geophys. J. Int.* 128:188–196.

Manatschal, G. and D. Bernoulli. 1999. Architecture and tectonic evolution of nonvolcanic margins: Present day Galicia and ancient Adria. *Tectonics* 18:1099–1119.

Manatschal, G. and P. Nievergelt. 1997. A continent-ocean transition recorded in the Err and Platta nappes (Eastern Switzerland). *Eclogae Geol. Helv.* 90:3–27.

Marrett, R. and R. W. Allmendinger. 1992. Amount of extension on "small" faults: An example from the Viking Graben. *Geology* 20:47–50.

Marsden, G., A. Roberts, G. Yielding, and N. J. Kusznir. 1990. Application of a flexural cantilever simple-shear/pure-shear model of continental lithosphere extension to the formation of the northern North Sea Basin. *Tectonics of the North Sea Rift*, pp. 240–261. Oxford, UK: Oxford Science Publications.

Masson, D. G., L. Montadert, and R. A. Scrutton. 1985. Evolution of the Goban Spur: History of a starved passive margin. In P. C. de Graciansky and P. C. Poag, eds., *Initial Reports of the DSDP*, vol. 80, pp. 1115–1139. Washington, DC: U.S. Government Printing Office.

Minshull, T. A., S. M. Dean, R. S. White, and R. B. Whitmarsh. 1998. Restricted melting at the onset of seafloor spreading: Ocean-continent transition zones at non-volcanic rifted margins. *Trans. Am. Geophys. Union* 79:906.

McKenzie, D. P. 1978. Some remarks on the development of sedimentary basins. *Earth Planet. Sci. Lett.* 40:25–32.

McKenzie, D. P. 1984. A possible mechanism for epeirogenic uplift. *Nature* 307:616–618.

McKenzie, D. P. and M. J. Bickle. 1988. The volume and composition of melt generated by extension of the lithosphere. *J. Petrol.* 29:625–679.

Mjelde, R., S. Kodaira, P. Digranes, H. Shimamura, T. Kanazawa, H. Shiobara, E. W. Berg, and O. Riise. 1997a. Comparison between a regional and semi-regional crustal OBS model in the Vøring Basin, mid-Norway margin. *Pure Appl. Geophys.* 149:641–665.

Mjelde, R., S. Kodaira, H. Shimamura, T. Kanazawa, H. Shiobara, E. W. Berg, and O. Riise. 1997b. Crustal structure of the central part of the Vøring Basin, mid-Norway margin, from ocean bottom seismographs. *Tectonophysics* 277:235–257.

Mooney, W. and R. Meissner. 1992. Multi-genetic origin of crustal reflectivity: A review. *Dev. Geotectonics* 23:45–79.

Moorcraft, H. and S. Roberts. 1991. *Report of the 1990 CNOOC/BP Joint Study of CA 27/31, Pearl River Mouth Basin, South China Sea.* London: BP Internal Publication.

Murillas, J., D. Mougenot, G. Boillot, M. C. Comas, E. Banda, and A. Mauffret. 1990. Structure and evolution of the Galicia Interior Basin (Atlantic western Iberian continental margin). *Tectonophysics* 184:297–319.

Mutter, J. C., M. Talwani, and P. L. Stoffa. 1982. Origin of seaward dipping reflectors in oceanic crust off the Norwegian margin by "subaerial sea floor spreading." *Geology* 10:353–357.

Nadin, P. and N. J. Kusznir. 1995. Palaeocene uplift and Eocene subsidence in the northern North Sea from 2-D forward and reverse stratigraphic modelling. *J. Geol. Soc. Lond.* 152:833–848.

Nadin, P. A., N. J. Kusznir, and M. J. Cheadle. 1997. Early Tertiary plume uplift of the North Sea and Faeroe-Shetland Basins. *Earth Planet. Sci. Lett.* 148:109–127.

Nissen, S. S., D. E. Hayes, B. C. Yao, W. J. Zeng, Y. Q. Chen, and X. P. Nu. 1995. Gravity, heat-flow, and seismic constraints on the processes of crustal extension: Northern margin of the south-china-sea. *J. Geophys. Res.* 100:22447–22483.

Parsons, B. and J. G. Sclater. 1977. An analysis of the variation of ocean floor bathymetry and heat flow with age. *J. Geophys. Res.* 82:803.

Pautot, G. and C. Rangin. 1989. Subduction of the South China Sea axial ridge below Luzon (Philippines). *Earth Planet. Sci. Lett.* 92:57–69.

Peddy, C., B. Pinet, D. Masson, R. Scrutton, J. C. Sibuet, M. R. Warner, J. P. Lefort, and I. J. Shroeder. 1989. Crustal structure of the Goban Spur continental margin, Northeast Atlantic, from deep seismic reflection profiling. *J. Geol. Soc. Lond.* 146:427–437.

Perez-Gussinye, M., T. Reston, D. Sawyer, C. Ranero, and E. Flueh. 1998. The structure of the Galicia Interior Basin, West of Iberia. *Trans. Am. Geophys. Union* 79:906.

Pickup, S.L.B., R. B. Whitmarsh, C.M.R. Fowler, and T. J. Reston. 1996. Insight into the nature of the ocean-continent transition off West Iberia from a deep multichannel seismic reflection profile. *Geology* 24:1079–1082.

Pitman, W. I. 1978. Relationship between eustasy and stratigraphic sequences of passive margins. *Geol. Soc. Am. Bull.* 89:1389–1403.

Planke, S. and O. Eldholm. 1994. Seismic response and construction of seaward dipping wedges of flood basalts: Vøring volcanic margin. *J.Geophys. Res.* 99:9263–9278.

Reston, T. J. 1996. The S reflector west of Galicia: the seismic signature of a detachment fault. *Geophys. J. Int.* 127:230–244.

Roberts, A. M., N. J. Kusznir, G. Yielding, and P. Styles. 1998. 2D flexural backstripping of extensional basins: The need for a sideways glance. *Petroleum Geosci.* 4:327–338.

Roberts, A., E. R. Lundin, and N. J. Kusznir. 1997. Subsidence of the Vøring Basin and the influence of the Atlantic continental margin: *J. Geol. Soc. Lond.* 154:551–557.

Roberts, A. M., G. Yielding, N. J. Kusznir, I. M. Walker, and D. Dorn-Lopez. 1993. Mesozoic extension in the North Sea: constraints from flexural backstripping, forward modelling and fault populations. In J. R. Parker, ed., *Petroleum Geology of Northwest Europe*, Proceedings of the Fourth Conference, pp. 1123–1136. London: Geological Society of London.

Royden, L. and C. E. Keen. 1980. Rifting process and thermal evolution of the continental margin of Eastern Canada determined from subsidence curves. *Earth Planet. Sci. Lett.* 51:343–361.

Ru, K. and J. D. Pigott. 1986. Episodic rifting and subsidence in the South China Sea. *Bull. Am Assoc. Petroleum Geol.* 70:1136–1155.

Sandwell, D. and W. Smith. 1997. Marine gravity from Geosat and ERS-1 altimetry. *J. Geophys. Res.* 102:10039–10054.

Sclater, J. G. and P.A.F. Christie. 1980. Continental stretching: An explanation of the post mid-Cretaceous subsidence of the central North Sea Basin. *J.Geophys. Res.* 85:3711–3739.

Scrutton, R. A. 1984. Modelling of magnetic and gravity anomalies at Goban Spur, northeast Atlantic. In P. C. de Graciansky and C. W. Poag, eds., *Initial Reports of the Deep Sea Drilling Project 80*. Washington, DC: U.S. Government Printing Office.

Sibuet, J. 1992. Formation of non-volcanic passive margins: A composite model applies to the conjugate Galicia and southeastern Flemish Cap margins. *Geophys. Res. Lett.* 19:769–772.

Spiegelman, N. and D. McKenzie. 1987. Simple 2-D models for melt extraction at mid-ocean ridges and island arcs. *Earth Planet. Sci. Lett.* 83:137–152.

Spiegelman, M. and J. R. Reynolds. 1999. Combined dynamic and geochemical evidence for convergent melt flow beneath the East Pacific Rise. *Nature* 402:282–285.

Stein, R. and S. Barrientos. 1985. Planar high-angle faulting in the Basin and Range: Geodetic analysis of the 1983 Borah Peak, Idaho, earthquake. *J. Geophys. Res.* 90:11355–11366.

Su, D., N. J. White, and D. P. McKenzie. 1989. Extension and subsidence of the Pearl River Mouth Basin, northern South China Sea. *Basin Res.* 2:205–222.

Taylor, B. and D. Hayes. 1983. Origin and history of the South China Basin. In Hayes, D., ed., *Tectonic and geologic evolution of Southeast Asian seas and islands. Geophys. Monogr.* 27:23–56.

Taylor, B. and D. E. Hayes. 1980. The tectonic evolution of the South China basin: The tectonic and geologic evolution of Southeast Asian Seas and Islands. *Geophys. Monogr.* 2:89–104.

Torres, J., S. Bois, and J. Burrus. 1993. Initiation and evolution of the Valencia trough (western Mediterranean): Constraints from deep seismic profiling and subsidence analysis. *Tectonophysics* 228:57–80.

Walsh, J., J. Watterson, and G. Yielding. 1991. The importance of small-scale faulting in regional extension. *Nature* 351:391–393.

Watts, A. B. and J. D. Fairhead, 1997. Gravity anomalies and magmatism along the western continental margin of the British Isles. *J. Geol. Soc.* 154:523–529.

Wernicke, B. 1985. Uniform-sense simple shear of the continental lithosphere. *Can. J. Earth Sci.* 22:108–125.

Westaway, R. 1994. Re-evaluation of extension across the Pearl River Mouth Basin, South China sea: Implications for continental lithosphere deformation mechanisms. *J. Struct. Geol.* 16:823–838.

Westaway, R. and N. Kusznir. 1993. Fault and bed 'rotation' during continental extension: Block rotation or vertical shear? *J. Struct. Geol.* 15:753–770.

White, N. 1990. Does the uniform stretching model work in the North Sea? In D. Blundell and A. Gibbs, eds., *Tectonic Evolution of North Sea Rifts*, pp. 217–239. Oxford, UK: Oxford University Press.

White, N. J. and B. Lovell. 1997. Measuring the pulse of a plume with the sedimentary record. *Nature* 387:888–891.

White, N. J. and D. P. McKenzie. 1988. Formation of the "Steer's Head" geometry of sedimentary basins by differential stretching of the crust and mantle. *Geology* 16:250–253.

White, R. S. and D. McKenzie. 1989. Magmatism at rift zones: The generation of volcanic continental margins and flood basalts. *J. Geophys. Res.* 94:7685–7729.

Whitmarsh, R. B., G. Manatschal, and T. A. Minshull. 2001. Evolution of magma-poor continental margins from rifting to sea-floor spreading. *Nature* 413:150–153.

Whitmarsh, R. B., R. S. White, S. J. Horsefield, J. C. Sibuet, M. Recq, and V. Louvel. 1996. The ocean-continent boundary off the western continental-margin of Iberia: Crustal structure west of Galicia Bank. *J. Geophys. Res.* 101:28291–28314.

CHAPTER FIVE

Limits of the Seismogenic Zone

Larry J. Ruff

The seismogenic zone is where earthquakes occur. Because earthquakes are rapid shear slip on nearly planar faults, a region with just one active fault has a planar seismogenic zone. However, in many instances we think of the seismogenic zone as a volume that includes the active fault and surrounding rock. The horizontal bounds on this volume are quite flexible—it may extend across entire continents or plates. In this context, the seismogenic zone is the rock volume *capable* of seismicity. We usually place more emphasis on the depth limits of the seismogenic zone—especially the deeper depth limit. This emphasis follows from empirical studies where we see that small earthquakes occur across wide regions of a continent, but that most of these events are confined to the uppermost 15 km of the crust. The primary exception to this rule is subduction-zone seismicity. Subduction-zone plate-interface earthquakes occur down to about 40 km depth, and intraplate events within the subducted lithosphere occur down to about 700 km depth. We also find seismicity deeper than 15 km in regions where continent–continent collisions and mountain building are underway. Other examples of deeper seismicity are some intraplate events in oceanic lithosphere and some peculiar events in certain continental settings. The upper limit to the seismogenic zone is also of great interest from the theoretical perspective, but it is more difficult to determine this limit for global seismicity. Detailed microseismicity studies in continents typically show that earthquake depths are not evenly distributed all the way up to the surface. On the other hand, many larger earthquakes in these same places show coseismic surface faulting; hence, faults in the uppermost few kilometers of the crust are capable of coseismic slip. This upper limit is particularly important for subduction zones because of tsunami generation by slip near the trench. These basic facts about seismogenic zone limits are depicted in figure 5.1. Since some of the earliest investigations, temperature is thought to explain most aspects of the depth limits of the seismogenic zone. In detail, it appears that rock type and other variables, such as pore pressure, also play a role in some situations.

138 /

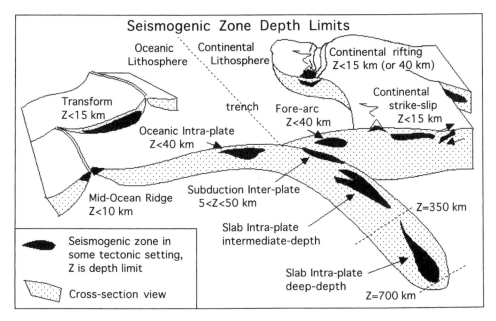

Figure 5.1 Schematic summary of basic seismological results for seismogenic zone limits in various tectonic settings. The upper limit is difficult to determine, and even where there is no small-magnitude seismicity (e.g., the shallowest part of the plate interface in subduction zones), coseismic slip may occur during great earthquakes. The lower limit is about 15 km in most tectonic environments; subduction zones are the significant exception to this rule. Minor but puzzling exceptions are the occasional deeper earthquakes in continental rifts.

The seismogenic zone is difficult to define because of practical and conceptual problems. The main practical problem is the seismological determination of earthquake depth; this chapter focuses on that topic. One conceptual problem is that the seismogenic zone limits might depend on earthquake magnitude. As a hypothetical example, suppose that the largest earthquake in a particular region ruptures down to 15 km depth, but that smaller-magnitude events occur down to 25 km depth; is the seismogenic zone depth 15 or 25 km? Seismologists generally place much greater emphasis on the largest earthquakes that occur in a particular region, and this emphasis is justified for many applications. Limits of the seismogenic zone is one application where small earthquakes provide valuable information, though many of the summary statements about the seismogenic zone limits in this chapter are biased toward the results from larger earthquakes.

This chapter reviews the process of seismogenic behavior and the seismological aspects of seismicity distribution; it then focuses on some of the notable exceptions to the basic rules that offer further insight to the physical and geological controls on the seismogenic zone. Given the broad scope of the topic, this chapter must be regarded as an overview rather than a comprehensive review. Its main theme is to review and combine the contributions from seismology and thermal modeling to test the notion that a critical temperature determines the deeper depth limit on the seismogenic zone. This chapter first gives a brief treatment of the mechanical aspects of seismogenesis, then turns to the main two topics which are the seismological details of earthquake depth and thermal models in various tectonic settings. Wadati-Benioff zone seismicity is discussed in the narrow context of a hypothesis test of a critical temperature. The upper limit of the seismogenic zone, though quite important, receives sparse treatment in this chapter for some practical reasons that are described in the next section. The spirit of this overview is to expose the uncertainties, problems, and challenges in the study of the seismogenic zone.

Seismogenic Zone: Mechanical Considerations

Earthquakes are just one of the various deformation mechanisms available in the Earth (see Scholz 2000 for review and detailed discussion). Figure 5.2 schematically shows the competition between deformation mechanisms to reduce stress. Suppose that some deviatoric stress is applied to a rock volume. There is always the immediate elastic response, and that might be the only response. But if the stress is increased or we wait for a long time, most rock will exhibit other deformation styles. At the scale depicted in figure 5.2, one key competition is between bulk deformation distributed across the entire volume versus concentrated deformation at one major fault zone. There may be intermediate scales of concentrated deformation due to material heterogeneity or nonlinear rheologies, but here the focus is on the end-members of these deformation styles. Also note that bulk deformation could include folding or slip on crack systems widely distributed throughout the volume. The extreme end-member behaviors can be characterized as uniform viscous flow versus pure elasticity throughout the volume, except for fault slip at one fault. There might be some circumstances where both deformation styles play significant roles, but we usually assume that one mechanism dominates at a particular place and time. If one of the mechanisms exhibits feedback behavior that can be described as "work-softening," then this mechanism has a distinct advantage since it can accommodate the deformation at a decreasing stress level. If one mechanism can decrease the stress level such that all other mechanisms have vanishing small deformation, then this one mechanism defeats the others. This competition between deformation mechanisms can be whimsically described as: "Only the weak survive." If earthquakes on the main fault zone accommodate most of the regional deformation, then this main fault must be weaker than the surrounding rock.

Seismic Versus Aseismic Fault Slip

Let us suppose that fault slip is the dominant deformation mechanism. We must probe deeper into fault behavior to see if the slip occurs as steady creep or as

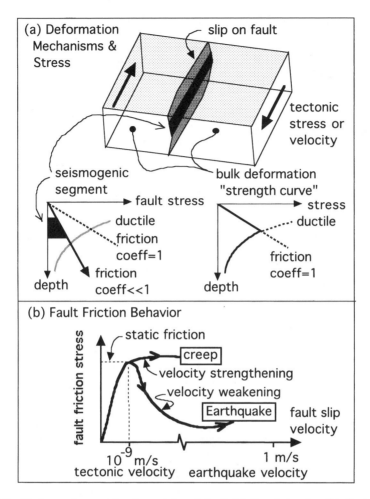

Figure 5.2 Deformation mechanisms (a) and fault-friction behavior (b). In (a), the tectonic stress and velocity will be accommodated either by distributed bulk deformation or by slip at one major fault. Bulk deformation is usually characterized by "strength curves"; fault strength depends on the effective static friction coefficient, which is apparently much less than 1 (see text). If the tectonic deformation is concentrated at one major fault, it can either occur as creep or as earthquakes. The velocity dependence of the dynamic friction coefficient is the key physical property (b).

earthquakes. The choice between creep and earthquakes depends on an obscure yet vitally important aspect of interface behavior: the slip-velocity dependence of the dynamic friction coefficient (figure 5.2b). Both the seismogenic and creeping portions of a fault may share the same static coefficient of friction; the different behaviors are determined by the changes in the coefficient of friction after slip begins. "Velocity strengthening" is when the dynamic coefficient of friction increases as slip velocity increases, whereas "velocity weakening" is when the dynamic coefficient of friction decreases as slip velocity increases. These two behaviors are often referred to as "creep" and "stick-slip," respectively. While the latter terms aptly describe system behavior, the "velocity-strengthening/weakening" terminology focuses on the key aspects of the friction law that produces the creep/ stick-slip behavior.

If the fault-slip velocity is faster than the tectonic loading rate, then shear stress acting on the fault decreases due to elastic unloading of the surrounding rock volume. Hence, rapid fault slip can only occur if the resistive dynamic friction is able to decrease faster than the stress unloading. Thus, seismogenic faults require a velocity-weakening behavior. If the friction coefficient remains constant, or even increases with faster slip velocity, then the fault can only creep slowly as stress is supplied to the fault through tectonic motion. Although there is a tremendous background of experimental work for the static coefficient of friction (see Scholz 2002 for discussion and references), there are much fewer data available for the dynamic friction coefficient. Tracking the temporal evolution of the dynamic friction coefficient is a more difficult experiment than static friction coefficient tests. This is one area of materials science where the Earth Sciences have made key contributions (see Dieterich 1986 and references therein). The efforts of several investigators to parameterize the results of dynamic friction experiments has led to the "rate- and state-dependent friction law," which produces both velocity-strengthening and velocity-weakening behavior. The parameter values for this friction law have been determined for several different fault interface conditions, and some experimental results indicate that temperature can change the velocity-dependent behavior (see Scholz 2002 for references).

There is one other qualitative behavior of the friction law that needs to be discussed: "conditional velocity weakening." This behavior is intermediate between velocity strengthening and velocity weakening. It requires a finite stress jump or an externally forced velocity step before the weakening behavior commences. Among others, Scholz (2002) has advocated that most faults have this behavior in a certain depth range at the edges of the seismogenic fault segments. This "conditional velocity-weakening" fault portion is sometimes referred to as "transitional." An earthquake cannot nucleate in a transitional fault-zone segment, but the rupture front of a large earthquake in the adjacent segment may be able to propagate into this fault segment and induce significant coseismic slip. No aftershocks would occur in this transitional fault segment, even if there were meters of coseismic slip. The possible existence of this behavior in fault segments adjacent to the seismogenic portion greatly complicates many aspects of seismology. In particular, it complicates the definition and interpretation of the seismogenic zone. Observations of coseismic slip in great earthquakes might include this "conditionally velocity-weakening" fault portion as part of the seismogenic zone. On the other hand, observations of hypocentral locations would define the seismogenic zone by where earthquakes nucleate. To further examine this issue, the various waveform inversion methods for the overall average faulting depth should see the co-seismic slip contribution from a deeper "conditional velocity-weakening" fault segment. The best situation to resolve coseismic slip in an otherwise aseismic fault portion would be to combine seismological and geodetic observations for a large earthquake on a fault that has a small dip angle; plate-interface subduction-zone earthquakes are good candidates.

While the rate- and-state-dependent friction law may be a good parameterization of friction, there is great interest in understanding the physical and chemical properties and processes that are responsible for this behavior. This chapter is largely focused on a single physical variable (i.e., temperature) that controls the deeper depth limit of stick-slip behavior. This narrow focus ignores many other variables that may control the seismogenic zone, but it is not possible for this chapter to review all the inferences from geologic studies of faults and rock mechanics (see other chapters in this special volume). One other variable that must be mentioned here is water. The following sections discuss the roles that water pore pressure plays in lowering the effective pressure in fault zones. But the presence of water in fault zones introduces many other complex interactions (Sibson 1973; Rice 1992). One of the important roles is that water may directly control whether a fault zone shows creep or stick-slip behavior (Dieterich and Conrad 1984; Karner et al. 1997). Even if water directly controls seismogenesis, we might still see a correlation between temperature and the seismogenic zone limits due to the indirect influence of temperature on the physical and chemical state of water. This discussion gives a glimpse into how complex the complete description of seismogenesis may be, but this chapter is still focused on testing the "working hypothesis" that temperature controls the seismogenic zone limits. From an observational perspective, temperature is a good variable to test because we can construct thermal models for the various tectonic settings. In this sense, the entire endeavor to correlate earthquake depths with temperature is merely one step in a long process to identify the variables that control seismogenesis.

Is There a Connection Between the Brittle-Ductile Transition and the Interface Friction Law?

The bulk strength of lithospheric rocks is commonly described by the lithosphere strength envelopes (sometimes referred to as the Brace-Goetze curves; see other chapters in this special volume, and Brace and Kohlstedt 1980; Goetze and Evans 1979; Molnar 1992). The basic character of these curves is based on the competing effects of pressure and temperature on the two primary deformation mechanisms of brittle and ductile failure (depicted in figure 5.2a). The shallow brittle portion of the strength curves is based on frictional failure on fault planes of favorable orientation and linearly increases with effective pressure. Because the static coefficient of friction obeys "Byerlee's Law" (Byerlee 1978), brittle strength is nearly independent of rock type and temperature. Variability in the brittle strength is due to the particular choice of the friction coefficient (the typical range is 0.5 to 1.0) and whether or not a hydrostatic pore pressure is assumed. The ductile-brittle transition is the depth where the ductile strength curve falls below the nearly linearly increasing brittle-strength curve. Because the parameters for the ductile-strength curve depend on the geotherm and flow-law variables that depend on rock type and environment, this deeper portion of the strength envelope is somewhat uncertain and subject to interpretation. The upper portion of the strength envelope is based on the stress limit for frictional fault slip. Because earthquakes are a fault-slip phenomenon, it is reasonable to assume that the seismogenic zone coincides with the portion of the strength envelope where brittle failure determines bulk-rock strength. This assumption is probably not exactly correct, as discussed briefly in the following text.

Earthquakes are not just slip on a fault, they are *rapid* slip on a fault. As emphasized in the preceding section, fault slip can occur either as creep or as earthquakes. Hence, bulk deformation of the upper lithosphere may occur as brittle failure but with no seismicity. Thus the deep limit of seismicity can be shallower than the brittle-ductile transition. The companion question is whether the seismogenic zone can extend deeper than the brittle-ductile transition point in the surrounding rock volume? At first glance, the answer to this question should be "no," because ductile deformation is weaker than the fault below the transition depth. Hence, the tectonic deformation should be accommodated by distributed deformation rather than fault slip—be it seismic or aseismic. On the other hand, if we adopt the view that major faults are substantially weaker than expected from Byerlee's Law (for examples of the development and arguments about weak faults, see Brune et al. 1969; Lachenbruch and Sass 1992; Sass et al. 1997; Byerlee 1990; Rice 1992; Scholz 2000; also see articles in special issues, Zoback and Lachenbruch 1992; Hickman et al. 1995), then the answer to the question mentioned previously could be "yes." While the brittle-ductile transition depth is based on a high friction coefficient for the small faults distributed throughout the lithosphere, the one major fault may have an effective friction coefficient that is 0.1 or less such that it is weaker than ductile deformation in the surrounding rock volume (figure 5.2a). This possibility can be tested by observations: the occasional random seismicity distributed through the crust should define a lower limit to the seismogenic zone that is shallower than seismicity on the weaker major fault zone. In most places where accurate hypocentral depths are available, it seems that the seismogenic zone limit is approximately the same for "off-fault" seismicity and along major faults. Thus, the simplest notion that the brittle-ductile transition for bulk strength also determines the deep limit of the seismogenic zone is still regarded as a good default model.

This brief review shows the complexity of considerations in the mechanics of the seismogenic zone and its interactions with other physical processes and rock properties. One way to make progress in such a complex problem is to turn to the observations and look for some general trends that suggest a simple hypothesis that can be tested. One hypothesis that emerges from this approach is that the depth limit of earthquakes is controlled by a critical temperature. The rest of this chapter is focused on testing this hypothesis. This test requires accurate earthquake depths and accurate thermal models. The next three sections are critical examinations of the accuracy of earthquake depths and temperatures inside the Earth.

Methods to Determine Earthquake Depths

Epicentral locations in the National Earthquake Information Center (NEIC) and International Seismological Centre (ISC) global seismicity catalogs have an accuracy of ten kilometers or so. In comparison, hypocentral depth is much less certain. This is evidenced by the fact that NEIC routinely assigns a default depth of 33 km to events that have a depth somewhere between 0 and 70 km. The NEIC will occasionally choose a different default depth, e.g., 10 km, in strike-slip or rifting tectonic environments. Given the central importance of earthquake location to seismology, there is a vast literature on all methods for both experts and novices. For readers that seek a more complete treatment of the basic arrival time methods, many seismology books cover this topic (Gubbins 1990). At a more advanced level in terms of geophysical implications, Stein and Wiens (1986) review the seismological methods and results for depth determination. This section contains an abbreviated review of the three main methods: near-source first arrivals, surface waves, and body wave depth phases. Problems and difficulties with these techniques will be emphasized.

Near-Source *P*-Wave First Arrival Times

Variations in epicentral location (i.e., latitude and longitude) directly affect the pattern of *P*-wave first arrivals at both near-source and teleseismic stations. In contrast, variations in hypocentral depth of "shallow" earthquakes can be largely compensated by variations in origin time for teleseismic observations (i.e., arrival times at seismographic stations located at distances greater than 3,000 km from the epicenter). To obtain reliable depths for crustal events on the basis of first arrivals, we need a regional station network. Figure 5.3 schematically shows the travel-time curves for *P* waves out to an epicentral distance of a few hundred kilometers and it shows that accurate depth estimates require seismographic stations within 10 km of the epicenter so that the depth-dependent near-source curvature in the P_g arrival branch is resolved. Because most seismicity is shallower than about 15 km, the ideal station spacing is less than 10 km in a seismically active region—such a high-density seismographic network is available in only a few places around the world. Figure 5.3 displays another technique to estimate depth from sparse regional networks. If the *P*-wave arrivals resolve both the P_g and P_n travel-time branches, then hypocentral depth appears as a shift in the P_n branch intercept time with respect to the linear portion of the P_g branch. This shift is approximately equal to the hypocentral depth divided by crustal velocity; hence, it is about two seconds for midcrustal seismicity compared with a surface source. The largest obstacle in using this method is that unmodeled lateral variations in crust-mantle velocities and Moho topography also introduce shifts in the apparent P_n-intercept time. Thus, crustal structure across the region must be well characterized before we can have confidence in the hypocentral depths. On the other hand, the depth differences between events in a seismicity cluster can be accurate

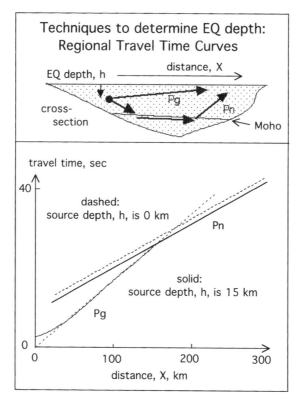

Figure 5.3 Hypocentral depth can be resolved by P arrival times at stations within a few hundred kilometers of the epicenter. Upper part shows the geometry and parameters; lower part shows two different travel-time curves for two different hypocentral depths. "*Pg*" is the direct crustal *P* wave arrival, while "*Pn*" is the Moho refracted *P* wave.

even if the overall absolute depth of the entire cluster is uncertain. Thus, relative earthquake locations may be accurate to 1 km or better, but the entire seismicity pattern can shift up and down by a few kilometers.

The previously described basic effect of relative shifts in apparent intercept times as a function of ray path takeoff angle is also the basis for teleseismic *P*-wave depth determination. To resolve depth, the intercept time relative shifts across the observed distance range must be greater than the scatter in arrival times. On the basis of the experience and procedures of the NEIC, it seems that most events with depths less than 70 km are poorly resolved and thus assigned a depth of "shallow."

Surface Waves

Surface waves can be used to determine hypocenter depth, but unfortunately there are many difficulties. Surface waves contain depth information in the relative excitation of different wave periods. A clear theoretical demonstration of this effect is given in Aki and Richards (1980). In detail, spectral amplitudes also depend on other source characteristics in addition to earth structure and attenuation effects. The most robust observation is the period of a spectral zero due to hypocentral depth (see, e.g., Aki and Richards 1980). Although surface-wave excitation is a complex topic, spectral zeros are basically caused by crustal wave interference, and so their period depends on both earthquake depth and crustal structure. For typical hypocentral depths, the spectral zeros are at periods of 20 s or less. It is challenging to perform quantitative analysis of these "high-frequency" surface waves (see, e.g., Herrmann 1987; Kennett 1989) and the source may introduce additional spectral zeros. Empirical studies of regional seismograms show clear differences in high-frequency surface waves from events at different depths. For the future, the ongoing efforts to fully model both the generation and scattering of high-frequency surface waves promises to provide better crustal structure and hypocentral depths.

Depth Phases

Accurate and robust hypocentral depths can be obtained at teleseismic stations from use of "depth phases" that arrive after the direct P wave. Depth phases are P and S waves that travel up from the hypocenter, reflect off the Earth's surface as P waves, and then follow the direct P wave to far-away seismographic stations (figure 5.4). Because the P and depth phase ray paths are essentially identical except for the segments above the source, the time difference is directly related to hypocenter depth. Of course, to translate the time difference into a depth, we must assume an average velocity above the source. A good choice is to assume an average P-wave velocity of 6.7 km/s because the global variability in average crustal velocity mostly falls between 6.0 and 7.4 km/s; then the hypocentral depth estimates will be within 10% of the "true" value for a particular location. The depth-phase technique is the most accurate and reliable for global studies of hypocentral depth.

The main problem with the depth-phase technique is proper identification of the depth phases. There are many wiggles in seismograms after the first arrival; which one is the depth phase? It is easier to "pick" the depth phases when their time delay behind the P wave is larger. For this reason, the "hand-picked" depth-phase technique works better for deeper earthquakes. For example, the pP-P delay time (see figure 5.4) is about 4 s for hypocentral depth of 15 km, but it is about 30 s for hypocentral depth of 100 km. Because the earthquake rupture duration is less than 10 s for earthquakes smaller than magnitude 7 (Tanioka and Ruff 1997), the pP arrival is seen as a distinct phase on seismograms for most events deeper than 100 km. For crustal seismicity of moderate size, depth phases arrive too soon after the P wave to be seen as separate distinct arrivals. To determine the arrival time of depth phases from shallow seismicity, we must model the entire P waveform—which includes the P, pP, and sP arrivals—to determine the earthquake focal mechanism, source-time function, and depth. This methodology is widely

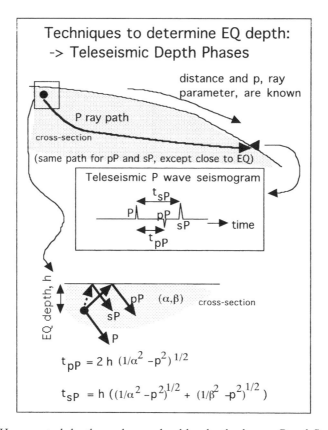

Figure 5.4 Hypocentral depth can be resolved by depth phases. *P* and *S* waves radiated up from the source reflect from the free surface and follow closely behind the direct *P* wave for teleseismic stations. The formulas show that the conversion from observed (*pP-P*) and (*sP-P*) times to hypocentral depth (*h*) depends on average velocity above the source (*α* and *β* are the *P*- and *S*-wave velocities), but the greatest uncertainties come from phase misidentification and modeling trade-offs. The depth-phase delay times also depend on the ray parameter (p), but this parameter is well known for teleseismic waves.

utilized, and several articles have discussed the robustness and resolution of hypocentral depth (Stein and Wiens 1986; Tichelaar and Ruff 1993). A different approach to accurate depth determination is to reprocess all the "hand-picked" depth-phase data in arrival-time catalogs to eliminate misidentifications and other problems; Engdahl et al. (1998) did this reprocessing as part of their effort to recalculate global seismicity hypocenters, and the remarkable clustering in some of their results demonstrates the effectiveness of this approach. For the future, perhaps the combination of better phase reporting and systematic application of statistical techniques will produce more accurate depths for a greater fraction of the shallow seismicity. Here, we turn back to waveform inversion for large shallow earthquakes.

To illustrate the waveform-inversion methodology, we examine an earthquake that is of particular interest to the theme of this chapter. The 2 October 2000 earthquake occurred beneath the southern part of Lake Tanganyika, which lies in the western part of the East African rift system. Previous investigations have found unusually deep earthquakes in the East African rift system (Shudofsky et al. 1987; Wagner and Langston 1988; Jackson and Blenkinsop 1993; Doser and Yarwood 1994) and we shall return to the significance of these observations later. The 2 October 2000 earthquake presents a clear example of how waveform inversion and depth phases can resolve depth. Figure 5.5 shows the overall best focal mechanism, and the synthetic seismograms for the best depth of 41 km ("best" in the sense of the best match between observed and synthetic seismograms). The as-

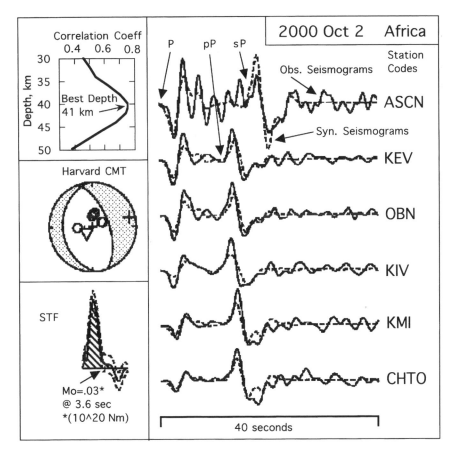

Figure 5.5 Teleseismic *P*-wave analysis of the 2 October 2000 East Africa Rift system earthquake. Observed (solid) and synthetic (dashed) *P*-wave seismograms from stations scattered around the world are shown in the right panel. Synthetic seismograms are produced for the best model: focal mechanism from the Harvard CMT solution, hypocentral depth is 41 km for a *P*-wave velocity above the source of 6.7 km/s, and the source time function (STF) has a duration of 3.6 with a seismic moment of 0.03×10^{20} Nm (M_w of 6.3). Fit between synthetic and observed seismograms is measured by a correlation coefficient. Graph at top left shows the correlation coefficient as a function of assumed hypocentral depth. The best depth of 41 km is well resolved because the depth phases are seen as separate distinct arrivals in the seismograms (see *pP* and *sP* picks).

sumed average *P*-wave velocity between the hypocenter and the surface is 6.7 km/s. We measure the match between observed and synthetic seismograms by a correlation coefficient, which is shown in figure 5.5 as a function of assumed depth. With respect to the typical results found for moderate-sized shallow earthquakes, the depth for the 2 October 2000 Lake Tanganyika event is well resolved. In an earlier review article, Stein and Wiens (1986) show excellent agreement between waveform-inversion depth determinations of shallow events by different research groups; they quote a standard deviation in hypocentral depth of 2.4 km. The Lake Tanganyika earthquake provides an excellent example where the hypocentral depth estimate from *P*-waveform inversion is accurate to within a few kilometers.

There are some problems with waveform inversion for depth phases. Several papers have explored the resolution and trade-off difficulties (see Tichelaar and Ruff 1993 and references therein). While waveform inversion for just the moment tensor (the moment tensor includes the focal mechanism and seismic moment) or just the source-time function (the time history of seismic moment change) is a linear inverse problem, inversion for the best point-source depth or the centroid depth is a nonlinear inverse problem. The simultaneous inversion for moment tensor, source-time function, and depth can be tricky. Waveform inversion becomes more difficult as earthquake size increases and depth decreases. Although waveform inversion is designed to unravel the interference of the source-time function and depth phases, it becomes more difficult to resolve depth as the source-time function duration exceeds the (*pP-P*) delay time. In addition, larger earthquakes rupture over a significant finite depth range. The best-fit point source depth is biased toward the down-dip edge of a finite fault for a dip-slip focal mechanism, and the best-case formal uncertainty in depth is about 2 km (Tichelaar and Ruff 1993). One basic conclusion is that point-source hypocentral depths are more reliable for shallow earthquakes in the 6-to-7 magnitude range than for great earthquakes (magnitude > 8) in the same region.

Comparison of Hypocentral Depths from Waveform Inversion Catalogs

This section contributes new information to the subject of global hypocentral depth determination. The Harvard Centroid Moment Tensor (CMT) and U.S. Geological Survey (USGS) Moment Tensor (MT) catalogs are valuable resources that provide systematic determination of earthquake size and focal mechanism for global seismicity (see basic descriptions in Dziewonski et al. 1981; Sipkin 1986). The moment tensor is the main focus and use of these catalogs, but hypocentral depth must also be determined as part of the waveform-inversion procedure. Many investigators have used these catalog depths, thus it is important to have some independent assessment of depth reliability. There are many specific studies where waveform inversion is performed to obtain accurate hypocentral depths. We must presume that these "special-study" depths are better than those in the "near-real-time" Harvard and USGS catalogs. We focus on comparing depths between the "near-real-time" global catalogs, where the depths from the Michigan Source Time Function (STF) catalog will also be included. The Michigan STF catalog began in June 1994 with systematic inversion of broad-band *P* and *SH* waves for the source-time function of all earthquakes with M_w (M_w is the moment magnitude scale; Kanamori 1977) of 7 or larger, although smaller events are included. The catalog has now been updated so that it is complete for all events with $M_w > 6.7$ since 1996. As of December 2000, there are just over 300 earthquakes in the Michigan STF catalog. The catalog and some basic results are described by Tanioka and Ruff (1997) and Ruff (1999). Note that the Harvard and USGS moment tensor catalogs are complete for earthquakes with magnitude smaller than 6.7, thus these moment tensor catalogs have many more earthquakes than the Michigan STF catalog during the same period. While the primary focus of the Michigan STF catalog is to add source-time functions to the set of earthquake parameters available in "near-real-time," earthquake depth must be determined as part of the waveform-inversion procedure. Theoretically, the depths in the Michigan STF catalog should be more accurate than the depth estimates in the near-real-time moment tensor catalogs. Also note that the Harvard and USGS depths may be revised before inclusion in their final catalogs. The Harvard CMT, USGS MT, and Michigan STF catalogs are compared to reveal the agreement or discord between depths in these catalogs. Recall that there should be no serious disagreements on hypocentral depth for deep events, so our attention is focused on the crucial depth range from 0 to 70 km.

Statistical Comparison

Figure 5.6a shows the Harvard CMT and USGS MT depths plotted as a function of the Michigan STF depth. There are about 180 common events in the 0- to 70-km depth range. A prominent feature of figure 5.6a is the Harvard CMT default shallow minimum depth of 15 km. Both the Michigan STF and USGS MT depths are allowed to approach 0 km depth—though resolution is poor for depths less than 5 km. It is easier to view the details in the plot of depth differences, identified as (HarvCMT-MichSTF) and (USGSmt-MichSTF), versus depth in figure 5.6b. The Harvard CMT 15-km default depth is now displayed by the cutoff line slanting across the plot. There is an apparent cutoff line for USGS MT depths that is a simple consequence of a zero kilometer bound on depth. We will first consider the summary statistics for the depth differences, then discuss some specific aspects of variability in catalog depths. The overall mean value of the Harvard CMT-Michigan STF depth difference distribution is +5.9 km, but this is largely due to the Harvard CMT depth assignment of 15 km to events that have Michigan STF depths between 0 and 15 km. The standard deviation of the (Harvard CMT-Michigan STF) distribution is 8.7 km with maximum and minimum values of +29 and −25 km. The overall mean value of the (USGS MT-Michigan STF) distribution is +2.4 km and the standard deviation is 11.9 km with maximum and

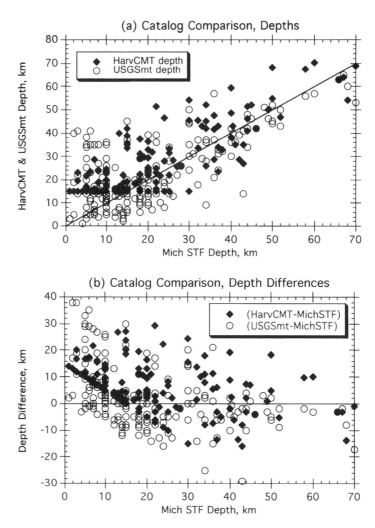

Figure 5.6 Comparison of depths from three catalogs that perform near-real-time wave-form inversion for various earthquake parameters. This comparison focuses on the crucial depth range from 10 to 70 km. In (a), depths for the Harvard CMT and USGS MT catalogs are plotted versus the Michigan STF depths (see text for detailed description). To better examine the details, the depth differences are plotted as a function of the Michigan STF depths in (b). The overall statistical variability in these "rapid" depth estimates is about ± 10 km.

minimum values of +38 and −29 km. While the overall bias between the USGS MT and Michigan STF depths is less than 3 km, the scatter is slightly larger than for the (Harvard CMT–Michigan STF) comparison. One question to ponder is whether the scatter is less if we directly compare the USGS MT and Harvard CMT catalogs. The answer is no: the (Harvard CMT–USGS MT) distribution has a standard deviation of 12.6 km. Furthermore, there is no significant correlation between the (Harvard CMT–Michigan STF) and (USGS MT–Michigan STF) depth differences. In other words, a deeper Harvard CMT depth with respect to the Michigan STF depth does not imply a deeper USGS MT depth as well.

Depth Difference Extremes

Now lets look at specific examples of depth difference extremes. For the Harvard CMT catalog, the largest "shallow" anomaly is for the 30 September 1999 Mexico earthquake ($M_w = 7.4$) where the "quick" CMT solution was at 17 km while the Michigan STF depth was 42 km. It is worth noting that the final CMT catalog depth for this event is 46.8 km, in much better agreement with the Michigan STF value. In general, the depth adjustments between the quick and final Harvard CMT catalog are less than 5 km. The largest "deep" anomaly in the (Harvard CMT-Michigan STF) comparison is for the 21 April 1997 Santa Cruz Island earthquake ($M_w = 7.7$) where the Harvard CMT (final) solution is 51 km while the Michigan STF depth is 22 km; the USGS MT depth is 26 km for this large earthquake. If we now look at the largest discrepancies in the (USGS MT-Michigan STF) distribution, the largest "shallow" anomaly is for the small ($M_w = 6.3$) Kyushu event of 24 January 1999 where the USGS MT depth is 14 km and the Michigan STF depth is 43 km; the Harvard CMT depth is 27 km. The (USGS MT-Michigan STF) depth difference is $+38$ km in two events: the 14 November 1994 Philippines event (M_w 7.1) with USGS MT and Michigan STF depths of 40 and 2 km; and the 17 June 2000 Iceland event (M_w 6.4) with USGS MT and Michigan STF depths of 41 and 3 km; the Harvard CMT depths are 15 km for both these earthquakes.

Further analysis of the events with the largest depth differences reveals that there are *a posteri* justifications for most of these discrepancies. Special inversion studies of earthquake depths might eliminate these problematic cases and thus achieve the "target" depth accuracy of 2 km. If the goal of this comparison were to show how well all three catalogs can agree, then we could apply additional filters to the catalogs that would yield smaller standard deviations. However, a more demanding and critical view of these catalog depths is to ask: how good is the depth in one of the "near-real-time" catalogs for any shallow earthquake of any size? A simple answer to this question is that the standard deviation from these intercatalog comparisons is about 10 km. We can use this number as the "rule-of-thumb" standard uncertainty. However, this uncertainty can be reduced—for smaller events—to 2 km or so with more time and effort. As we enter a discussion of temperature and the seismogenic zone in the next section, we shall presume that special studies have defined the seismogenic depth limit to an accuracy of less than 5 km in all tectonic settings.

Does Temperature Explain the Limits of the Seismogenic Zone?

The simplest explanation for the deep limit of the seismogenic zone is that ductile deformation is weaker than frictional fault slip below a certain depth. Alternatively, the fault may be weaker than the surrounding rock volume below the seismogenic zone, but the friction constitutive law only allows fault creep below a certain depth. While it is difficult to decide which one of these explanations is correct, there are reasons to believe that both of them may depend on a critical temperature (see discussion and references in Scholz 2002). Through the years, there have been many empirical studies that correlate temperature with seismogenic zone limits—especially the down-dip edge of the seismogenic zone. Many studies have found that an isotherm of about 300 to 350°C corresponds to the deep limit of the seismogenic zone in crustal rocks (Chen and Molnar 1983). On the other hand, it seems that intra-plate events in oceanic lithosphere mantle require a critical temperature of about 700 to 750°C (Stein and Wiens 1986; Seno and Yamanaka 1996). Use of two critical temperatures is not as elegant as just one, but it is still a remarkably simple explanation. We shall briefly review the seismicity and thermal modeling results for the various tectonic settings. We begin with the oceanic intraplate tectonic setting where thermal models are thought to be accurate.

Oceanic Intraplate Seismicity and Thermal Models

Although most earthquakes occur at plate boundaries, there are enough intraplate events in oceanic lithosphere to test the critical isotherm hypothesis. Earthquake depths range from top of the oceanic crust down to about 40 km. Most of the depth determinations are based on teleseismic depth phases, and the consistent oceanic structure gives us high confidence in the depth estimates. It is more controversial whether coseismic slip extends deeper than 40 km during the rare great earthquake (see references in Spence 1987). For the smaller events with more precise depths, there appears to be a consistent trend where maximum earthquake depth increases with lithosphere age. Because hypocentral depths greater than about 6 km place the earthquake in the mantle, it is thought that the critical isotherm for mantle rocks should control this deep limit.

Thermal models of oceanic lithosphere are based on a cooling plate and have been quite successful in explaining ocean-floor bathymetry and heat flow away from the ridge (see Stein and Stein 1992). Since radiogenic heating is minor in the oceanic lithosphere, the temperature structure is mostly controlled by age and the basal temperature (a constant temperature at a fixed depth). Heat transfer by hydrothermal circulation confuses the heat-flux interpretation, but bathymetry gives us a reasonably reliable picture of integrated heat loss. The cooling half-space model predicts continued cooling and contraction throughout the life of the lithosphere, whereas the plate model produces a nearly flat bathymetry and heat flow for ages older than about 80 Ma. The temperature structure at depths relevant to the seismogenic zone are not too different for the plate and cooling half-space models. The GDH1 model of Stein and Stein (1992) has a fixed basal temperature of 1,450°C at a depth of 95 km; for old lithosphere, this model yields an asymptotic depth of about 45 km for the 700°C isotherm. There is still some argument as to which model is most appropriate for "typical" oceanic lithosphere. Even with this uncertainty, the temperature structure as a function of lithosphere age down to a depth of 50 km or so is widely regarded as the most reliable thermal regime. Thus, the occurrence of seismicity at temperatures up to 700°C in oceanic mantle rocks must be accepted.

Oceanic Transform Faults

Seismicity is quite shallow along ridge-ridge transform faults. While most events are probably shallower than 6 km, and hence likely occur in crustal rocks, there are a few events in the middle of long transforms that may occur below Moho, perhaps down to a depth of 15 km or so (Abercrombie and Ekstrom 2001). Thermal models for transform faults are more complex than for the cooling-plate model. Because there is a contrast in lithosphere age across a ridge-to-ridge transform fault, we should solve a two-dimensional thermal conduction problem with time-variable boundary conditions. Furthermore, the possible role of plate boundary shear heating and fault hydrothermal circulation must be considered. Engeln et al. (1986) used a simple thermal model of an oceanic transform fault—take the average temperature of two superimposed cooling-plate temperature models—to show that the seismogenic limit of transform seismicity corresponds to an isotherm of about 300–400°C. This result has been corroborated by others (Abercrombie and Ekstrom 2001). At the middle of a long transform with slow spreading rate, such as the Romanche Fracture Zone in the Atlantic Ocean, the lithosphere age is about 25 Ma. Hence, the temperature at Moho should be about 200°C, while 700°C is reached at 25 km depth. Yet, it seems that the 10- to 15-km seismogenic depth limit is determined by the 350°C isotherm, not the 700–750°C isotherm as seen for intraplate events (though depth resolution is a concern here). This result complicates the working hypothesis that there is a strict critical isotherm of 350°C for crustal rock or 750°C for mantle rock. Alternatively, we can simply claim that ridge-ridge transform faults are hotter than expected from the simplest thermal models.

We shall now consider oceanic-transform faults that do not connect ridge segments. The Macquarie Ridge Complex is one oceanic-plate boundary where seismicity is deeper than 15 km. There is still some controversy over the depth range of the great 1989 Macquarie Ridge event (see Das 1993, and references therein) but it is possible that coseismic displacement occurred deeper than 20 km. However, because this boundary is now a combination of thrust and strike-slip motion with some crustal thickening, perhaps the thermal structure is "colder" than expected for a strike-slip boundary. The long oceanic-transform boundary between the Azores triple junction and Gibraltar is another tectonic setting where large strike-slip earthquakes have occurred close to regions with shortening (see Lynnes and Ruff 1985), though the hypocentral depth appears to be shallow.

Continental Strike-Slip Faults

The San Andreas Fault in California is one of the best-characterized systems in the world with detailed studies of both seismicity and thermal structure. Indeed, the central portion of the San Andreas fault is a key site for the argument that faults are weak since shear heating must be small, i.e., the shear stress is no greater than 10 MPa (see previous citations for weak faults, in particular, the special issues edited by Zoback and Lachenbruch 1992; Hickman et al. 1995). Pore fluids can explain a weak main fault surrounded by stronger faults (see Byerlee 1990; Rice 1992), but vigorous discussion of these issues still occurs (Scholz 2000).

Many papers that show hypocentral depths to be mostly above 15 km depth (see summaries in Scholz 2002; Doser and Kanamori 1986; Furlong and Atkinson 1993). In addition, the precise hypocenters seem to define an upper limit of the seismogenic zone in some localities. In places where the uppermost few kilometers of the crust are aseismic for small events, this uppermost fault segment must be in the "transitional" stability field, because some large strike-slip earthquakes do have coseismic slip at the free surface. To generalize, continental strike-slip faults in other parts of the world that have been well studied show 15 km as the limit to the seismogenic zone.

Thermal models for continental strike-slip faults must incorporate all the considerations of continental geotherms, plus the possibility that there is a significant shear-heating component. However, results from the central San Andreas show that there is no significant shear heating, which constrains the frictional shear stresses to be less than 10 MPa (see previous citations, e.g., Sass et al. 1997). If fault-zone shear heating is small at all strike-slip faults, then the thermal modeling procedure is greatly simplified so that we can use standard continental geotherms where the lower limit of the seismogenic zone at 15 km depth commonly corresponds to the 350°C isotherm. Complications in the thermal model arise from fault zone fluid flow, contrast in lithologies, and subsidence and erosion in the fault region. Even without these complications, there is still the ambiguity of geotherm construction due to uncertainties in crustal heat production. Many heat-flow province studies conclude that (i) approximately half the surface heat flux is generated by crustal radiogenic heat production; and (ii) estimates for the scale depth of radiogenic production is about half the crustal thickness (see Wong and Chapman 1990 for references). However, the allowed variations in the heat-production distribution impart significant uncertainties to the temperature estimate at 15 km depth. Due in part to the difficulties of translating the observed heat flux into a geotherm, many studies focus on the correlation between the observed parameters of seismicity depth and heat flow, rather than the temperature at depth (see Wong and Chapman 1990 for discussion).

Other Continental Settings

This category includes mixtures of continental rifting, thrusting, and strike-slip settings. Active thrusting should eventually cool the upper crust, whereas rifting and volcanism will heat the upper crust. Despite the variety in focal mechanisms and deformation, most seismicity still occurs in the upper crust (Shimazaki 1986). In some thrusting and mountain-building environments, hypocenters do occur deeper than 15 km, and it seems that seismicity in many rifting, volcanic, and geothermal regions is shallower than 15 km (Ito 1993).

Any region with significant vertical motions—mountain building, erosion, and basin formation—will have much greater uncertainties in the geotherm (Ehlers and Chapman 1999). Given these difficulties in constructing reliable geotherms, most investigators have just tested the variations in seismogenic depth and regional heat flow, as discussed previously. There are many results that seem quite reasonable and offer support for the notion that seismogenic depth is thermally controlled. For example, Doser and Kanamori (1986) show that the Salton Sea region of California has rifting, higher heat flow, and shallower seismicity than other strike-slip regions of California. On a broader scale, Sibson (1982), Smith and Bruhn (1984), and Wong and Chapman (1990) show the correlation of heat flow with seismogenic depth across western North America. Wong and Chapman (1990) discuss the curious anomalies of unusually deep small events in Utah at depths greater than 50 km. Although one might dismiss the Utah events because of their very small magnitude, we cannot ignore the 2 October 2000 Lake Tanganyika earthquake with a M_w of 6.3 and a depth of 41 km. Doser and Yarwood (1994) speculated that these rare deeper events associated with active rifts actually occur off the rift axis and therefore might be in colder crust—but the Lake Tanganyika event clearly violates this notion. If we use typical crustal geotherms (see references mentioned previously), then the minimum temperature at 40 km depth should be 480°C for a mantle contribution of 30 mW/m^2. Hence, these East African Rift events seem to occur at higher temperatures than the crustal critical isotherm of 350°C. One solution to this dilemma is that the temperature at 40 km is in fact only 350°C—this ad hoc solution might be acceptable given all the uncertainties in geotherm construction. Another solution to this dilemma would be to say that the deep African events are occurring just below Moho; thus, the relevant critical isotherm is the hotter one for mantle rocks. However, Shudofsky et al. (1987) studied several deep African events and they found that these earthquakes mostly occur above Moho. Could there be some process that cools the crust beneath active rifts?

At this time, the deep crustal African events present us with an apparent violation of the working hypothesis that 350°C limits the seismogenic zone in crustal rocks. There seem to be three possible solutions to the problem: (i) there is no violation of the working hypothesis because temperature at 40 km depth is 350°C; (ii) there is no violation of the working hypothesis because these deep events occur in mantle rocks where we expect a critical temperature of 700°C or so; (iii) the working hypothesis is violated in continental rifts. Further work should be focused on this issue in the continental rift tectonic setting.

Subduction Zones: Plate Interface

Since subduction zones were explained in terms of plate tectonics (Isacks et al. 1968), we have known that the seismogenic portion of the subduction-plate interface extends no deeper than about 70 km—the depth range associated with the "shallow" designation in global seismicity catalogs. To make progress in delimiting the seismogenic zone, it is necessary to determine both hypocentral depth and focal mechanisms to discriminate between plate interface and intraplate events. Progress has occurred through many regional investigations and by a few global summaries. In their systematic global survey of variations in the seismogenic limiting depth, Tichelaar and Ruff (1993) found that the lower limit of the seismogenic plate interface for most subduction zones is about 40 ± 5 km. There are some subduction-zone segments that show significant variations from this 40-km value. Parts of the Chile and Japan zones are distinctly deeper, about 50–55 km, and the Mexico subduction zone is distinctly shallower, about 20 km. Further studies have modified some of the numbers in Tichelaar and Ruff, while other studies have added more subduction zones to this global survey (see Comte and Suarez 1994; Hyndman et al. 1995). The global variation in seismogenic depth offers valuable insight to the controlling mechanism.

Although the emphasis throughout this review is on the deeper limit of the seismogenic zone, the subduction-plate interface is one of the few plate-boundary environments where we should obtain reliable information on the up-dip edge of the seismogenic zone due to the small dip of the fault plane. In many subduction zones, the plate interface is aseismic from the trench axis down to a depth of perhaps 10 km (Byrne et al. 1988). Unfortunately, this aseismic behavior is difficult to interpret because of the multiple processes occurring at shallow depths in subduction zones. While some investigators attribute this up-dip limit to a critical temperature of about 100–150°C (see Hyndman et al. 1995), there are other viable mechanisms such as distributed deformation in accretionary prisms or pore-pressure effects. An additional complication is the possibility that great earthquake coseismic slip might occur in this otherwise aseismic portion of the plate interface. Tsunami wave inversion is a key tool in testing this notion (see references in Tanioka et al. 1997). Next, we return to the primary focus on the deep limit to the seismogenic zone.

Thermal models of the subduction plate interface must account for advection of the subducting lithosphere thermal structure with the consequent transient effects, add the shear heating along the plate interface, and match this to a compatible thermal structure in the overlying lithosphere. In detail, many parameters contribute to the plate-interface temperature through the seismogenic zone: subducting lithosphere age, convergence rate, dip angle, shear stress, and even radiogenic heat production in the overlying plate. Most of these thermal models have been produced with two-dimensional numerical methods, but Molnar and England (1990) show a quasi-analytical procedure that works well down to depths of 50 km or so. To match heat-flow observations above the seismogenic zone, Tichelaar and Ruff (1993), Wang et al. (1995), and Peacock (1996) all conclude that shear stresses are relatively low (<50 MPa) through the seismogenic zone. This conclusion contrasts with some of the earlier subduction-zone thermal models that merely assumed high levels of shear heating. Given all the complexities, we usually consider subduction-plate interface temperatures to be "poorly" determined in comparison with other tectonic settings.

Tichelaar and Ruff (1993) tested the notion that a critical isotherm explains the global observations of the seismogenic zone-depth limit. They found that the down-dip edge of the seismogenic zone corresponds to a plate-interface temperature close to 350°C for about half of the subduction zones. To explain the other subduction zones, one must invoke either a second critical temperature or some additional mechanism. Ruff and Tichelaar (1996) argued that a change along the plate interface from crustal to mantle rock in the overlying plate may be this additional mechanism. They used the coastline as a proxy for the intersection of Moho in the overlying plate with the plate-interface zone, and showed that this proxy coincides with the down-dip edge of the seismogenic zone. Peacock and Hyndman (1999) have proposed that mantle hydration could explain these observations and the requirement for a second mechanism to help limit the seismogenic zone. This idea can be tested by future work that precisely determines the intersection of Moho with the plate interface (Oleskevich et al. 1999; Ito et al. 2000). At this time, one critical temperature is not an adequate explanation for the deep limit to the seismogenic subduction-plate interface.

Subduction Zones: Intraplate Seismicity

Seismicity within the subducting slab begins in the outer rise and may extend down to a depth of 700 km. The maximum depth of Wadati-Benioff zone seismicity varies from less than 100 km (e.g., Cascadia and Nankai) down to the 700-km limit (e.g., Tonga, Indonesia, Marianas, Japan-Kuriles, and parts of South America). There have been many investigations of the tectonic controls on this maximum seismicity depth. The simplest explanation is that "thermal assimilation" limits the seismicity, though this idea is clouded by the bi-modal distribution of maximum seismicity depth and problems with thermal models (see Wortel and Vlaar 1988, and references therein). One basic idea in all models is that old lithosphere that subducts fast should have a colder interior than young lithosphere that slowly subducts. To quantify this connection, Molnar et al. (1979) correlate the slab-thermal parameter (product of age and rate) to slab length and depth. It seems that there is a critical value for slab-thermal parameter that must be achieved for a subduction zone to have deep earthquakes. Most zones with deep seismicity subduct old (about 120 Ma) lithosphere. This empirical result is consistent with recent ideas that a metastable phase change in the slab may be the physical mechanism for deep seismicity (Green and Houston 1995; Kirby et al. 1996a). Note that this mechanism is quite different from the frictional-slip mechanism relevant to all other seismicity. Here, we test whether the critical isotherm subducts to sufficient depth to explain the existence of intermediate-depth seismicity. Because the minimum temperature within the subducting slab occurs in the mantle below the subducting oceanic crust, 700°C is the relevant critical temperature.

Thermal models are reliable at the outer rise, but our confidence degrades as the slab subducts deeper. The lower portion of the oceanic lithosphere continues to follow the plate-cooling formula as it subducts, with a slight correction for adiabatic compression. In contrast, there are big changes in the upper portion of the slab. Plate-interface temperature history is a transient boundary condition that diffuses into the slab interior. This temperature history eventually controls the temperatures within the slab. We have already discussed the difficulties in determining the plate-interface temperatures down through the seismogenic zone. To probe deeper, the main uncertainties are shear heating in the creep portion of the plate interface and the depth at which the corner flow is encountered. At greater depths, phase changes modify the thermal structure.

Most detailed models of a particular subduction zone are purely numerical models. However, it is possible to extend the quasi-analytical method down to 700 km depth by superimposing the solutions for the plate-interface transient and the corner-flow transient boundary conditions. The GDH1 model (Stein and Stein 1992) provides the initial thermal structure as a function of lithosphere age, and we choose a very low shear heating along the plate-interface zone. At some depth, plate-interface creep terminates as corner flow is viscously coupled to the downgoing slab. An open question still exists as to the depth of the apex of the corner flow—modelers have assumed depths that range from 50 to 100 km. The corner flow is significant for the thermal structure of the slab as hot mantle rocks contact the top of the slab at the corner flow apex. For the "very cold" model calculated here, the corner flow apex is placed at 100 km depth.

Since the seismogenic zone limit is about half the thickness of the lithosphere, some value for shear stress must be assumed for the deeper creeping portion of the interface. Most models assume a fairly low value for creep shear stress, and a recent speculative "measurement" (Ruff 2001) of this creep shear stress gives values that range from 0.1 to 1.7 MPa at the Nankai, Alaska, and Chile subduction zones. This low level of shear stress produces very little shear heating.

The corner flow then couples the hot basal temperature of the GDH1 model (1,450°C) to the plate interface. There is an immediate adjustment to half the temperature difference between the plate-interface temperature and the basal temperature, followed by heat flow from the corner mantle into the slab at a decreasing rate. Hence, the plate-interface temperature is about 800°C at the corner-flow apex, but the slab interior can keep a temperature less than 750°C for a considerable depth of subduction. To demonstrate this effect, figure 5.7 plots the penetration depth of the 700°C isotherm within the slab as a function of sinking rate for three different ages of subducted lithosphere. Slab interior temperatures in the 100–350 depth range show a strong dependence on age of the subducting lithosphere for ages less than 60 Ma. For lithosphere older than 60 Ma, some part of the slab interior is colder than 700°C to depths beyond 350 km. Note that intermediate-depth seismicity tapers off and disappears at about 350 km—even in subduction zones that subduct very old lithosphere (see Kirby et al. 1996b). Thus, while the existence of the intermediate-depth seismogenic zone can be explained by temperatures less than a critical isotherm of 700–750°C , some mechanism other than temperature is required to explain the seismicity minimum at a depth of 350 km. At the other extreme of slab temperatures, figure 5.7 shows that when very young (less than 20 Ma old) lithosphere subducted, the minimum interior temperature will be hotter than 700°C for depths below about 100 km. Thus, the subduction

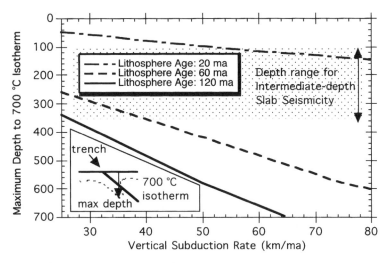

Figure 5.7   Maximum depth of the 700°C isotherm within subducted lithosphere for three different lithosphere ages and for a range in vertical subduction rate. As described in the text, these calculations are for a very cold thermal model. With that caveat, we see that part of the slab interior is colder than 700°C down to at least 350 km for old lithosphere (120 Ma) at any rate greater than 25 km/Ma. Also 60-Ma-old lithosphere has a slab interior colder than 700°C down to 350 km for vertical rates greater than 40 km/Ma. In contrast, the minimum interior slab temperature for young lithosphere with age less than 20 Ma will be hotter than 700°C for depths greater than about 100 km.

of young lithosphere provides another test of the critical isotherm hypothesis. Global surveys of subduction zones show that Wadati-Benioff zone seismicity does not extend deeper than about 125 km for zones that subduct oceanic lithosphere younger than 20 Ma (for review and discussion, see Jarrard 1986; Kirby et al. 1996b). This observed characteristic of intermediate-depth Wadati-Benioff zone seismicity is consistent with the working hypothesis that seismicity in mantle rocks requires a temperature less than 700–750°C.

Conclusions

Limits of the seismogenic zone are fundamental to both our understanding of earthquakes and the broader issues of deformation and tectonics. Seismologists found that most earthquakes occur at shallow depths—with the most curious and informative exception of Wadati-Benioff zone earthquakes. While there are many characteristics that change with depth in the Earth, an early supposition was that temperature is the main variable that limits the maximum depth of seismicity. This basic supposition survives to the present, but we must use at least two critical temperatures to encapsulate most of the world's seismicity. In detail, it seems that a critical temperature of 350°C limits the seismogenic zone in crustal rocks and most plate boundaries, while seismicity within oceanic lithosphere mantle can occur for temperatures up to 700–750°C. This latter temperature may explain both outer-rise earthquake depths and the depth limit for intermediate-depth Wadati-Benioff zone seismicity. The population of very deep earthquakes down to 700 km requires some other mechanism. The down-dip limit to the seismogenic subduction-plate interface requires at least two mechanisms: the 350°C critical temperature and either another critical temperature or some rock composition mechanism. Some of the most baffling observations are the occasional occurrence of a deep event in a continental setting where we expect the seismogenic limit to be at about 15 km depth. Perhaps the most outstanding example of this puzzle is the East Africa Rift, as evidenced by the 2 October 2000 Lake Tanganyika event with $M_w = 6.3$ at a depth of 41 km. This odd event, and other aberrations around the world, give us valuable clues as to what controls the limits to the seismogenic zone—if only we can make sense of this evidence!

Acknowledgments

Thanks to the organizers of the Snowbird 2000 workshop; this meeting provided interesting discussions on various issues of lithosphere deformation. This research was partially supported by the National Science Foundation (EAR-9725175 and OCE-9905503 to LJR).

References

Abercrombie, R. and G. Ekstrom. 2001. Earthquake slip on oceanic transform faults. *Nature* 410:74–77.

Aki, K. and P. G. Richards. 1980. *Quantitative Seismology*. New York: W. H. Freeman.

Brace, W. F. and D. Kohlstedt. 1980. Limits on lithospheric stress imposed by laboratory experiments. *J. Geophys. Res.* 85:6248–6252.

Brune, J., T. Henyey, and R. Roy. 1969. Heat flow, stress, and rate of slip along the San Andreas fault, California. *J. Geophys. Res.* 74:3821–3827.

Byerlee, J. D. 1978. Friction of rocks. *Pageoph* 116:615–626.

Byerlee, J. D. 1990. Friction, overpressure, and fault normal compression. *Geophys. Res. Lett.* 17:2109–2112.

Byrne, D., D. Davies, and L. Sykes. 1988. Loci and maximum size of thrust earthquakes and the mechanics of the shallow region of subduction zones. *Tectonics* 7:833–857.

Chen, W. and P. Molnar. 1983. Focal depths of intracontinental and intraplate earthquakes and their implications for the thermal and mechanical properties of the lithosphere. *J. Geophys. Res.* 88:4183–4214.

Comte, D. and G. Suarez. 1994. An inverted double seismic zone in Chile: Evidence of phase transformation in the subducted slab. *Science* 263:212–215.

Das, S. 1993. The Macquarie Ridge earthquake of 1989. *Geophys. J. Int.* 115:778–798.

Dieterich, J. 1986. A model for the nucleation of earthquake slip. In S. Das, J. Boatwright, and C. Scholz, eds., *Earthquake Source Mechanics*, AGU Monograph 37, pp. 37–49. Washington, DC: American Geophysical Union.

Dieterich, J. and G. Conrad. 1984. Effect of humidity on time- and velocity-dependent friction in rocks. *J. Geophys. Res.* 89:4196–4202.

Doser, D. and H. Kanamori. 1986. Depth of seismicity in the Imperial Valley region (1977–1983) and its relationship to heat flow, crustal structure, and the October 15, 1979 earthquake. a *J. Geophys. Res.* 91:675–688.

Doser, D. and D. Yarwood. 1994. Deep crustal earthquakes associated with continental rifts. *Tectonophysics* 229:123–131.

Dziewonski, A., T. Chou, and J. Woodhouse. 1981. Determination of earthquake source parameters from waveform data for studies of global and regional seismicity. *J. Geophys. Res.* 86:2825–2853.

Ehlers, T. and D. S. Chapman. 1999. Normal fault thermal regimes; conductive and hydrothermal heat transfer surrounding the Wasatch Fault, Utah. *Tectonophysics* 312:217–234.

Engdahl, E. R., R. D. van der Hilst, and R. P. Buland. 1998. Global teleseismic earthquake relocation with improved travel times and procedures for depth determination. *Bull. Seism. Soc. Am.* 88:3295–3314.

Engeln, J., D. Wiens, and S. Stein. 1986. Mechanisms and depths of Atlantic transform earthquakes. *J. Geophys. Res.* 91:548–577.

Furlong, K. and S. Atkinson. 1993. Seismicity and thermal structure along the northern San Andreas Fault system, California USA. *Tectonophysics* 217:23–30.

Goetze, C. and B. Evans. 1979. Stress and temperature in the bending lithosphere as constrained by experimental rock mechanics. *Geophys. J. R. Astron. Soc.* 59:463–478.

Green, H. and H. Houston. 1995. The mechanics of deep earthquakes. *Annu. Rev. Earth Planet. Sci.* 25:169–215.

Gubbins, D. 1990. *Seismology and Plate Tectonics.* Cambridge, UK: Cambridge University Press.

Herrmann, R. 1987. Broadband Lg magnitude. *Seismol. Res. Lett.* 58:125–133.

Hickman, S., R. Sibson, and R. Bruhn. 1995. Introduction to special section: Mechanical involvement of fluids in faulting. *J. Geophys. Res.* 100:12831–12840.

Hyndman, R., K. Wang, and M. Yamano. 1995. Thermal constraints on the seismogenic portion of the southwestern Japan subduction thrust. *J. Geophys. Res.* 100:15373–15392.

Isacks, B., J. Oliver, and L. Sykes. 1968. Seismicity and the new global tectonics. *J. Geophys. Res.* 73:5855–5899.

Ito, K. 1993. Cutoff depth of seismicity and large earthquakes near active volcanoes in Japan. *Tectonophysics* 217:11–21.

Ito, S., R. Hino, S. Matsumoto, H. Shiobara, H. Shimamura, T. Kanazawa, T. Sato, J. Kasahara, and A. Hasegawa. 2000. Deep seismic structure of the seismogenic plate boundary in the off-Sanriku region, northeastern Japan. *Tectonophysics* 319:261–274.

Jackson, J. and T. Blenkinsop. 1993. The Malawi earthquake of March 10, 1989: Deep faulting within the East African Rift system. *Tectonics* 12:1131–1139.

Jarrard, R. 1986. Relations among subduction parameters. *Rev. Geophys.* 24:217–284.

Kanamori, H. 1977. The energy release in great earthquakes. *J. Geophys. Res.* 82:2981–2987.

Karner, S. L., C. J. Marone, and B. Evans. 1997. Laboratory study of fault healing and lithification in simulated fault gouge under hydrothermal conditions. *Tectonophysics* 277:41–55.

Kennett, B. 1989. Lg wave propagation in heterogeneous media. *Bull. Seism. Soc. Am.* 79:860–872.

Kirby, S., S. Stein, E. Okal, and D. Rubie. 1996a. Metastable phase transformations and deep earthquakes in subducting oceanic lithosphere. *Rev. Geophys.* 34:261–306.

Kirby, S., E. R. Engdahl, and R. Denlinger. 1996b. Intermediate-depth intraslab earthquakes and arc volcanism as physical expressions of crustal and uppermost mantle metamorphism in subducting slabs. In G. Bebout, D. Scholl, S. Kirby, and J. Platt, eds., *Subduction: Top to Bottom,* AGU Geophysics Monograph 96, pp. 195–214. Washington, DC: American Geophys Union.

Lachenbruch, A. H. and J. H. Sass. 1992. Heat flow from Cajon Pass, fault strength, and tectonic implications. *J. Geophys. Res.* 97:4995–5015.

Lynnes, C. and L. Ruff. 1985. Source process and tectonic implications of the great 1975 North Atlantic earthquake. *Geophys. J. R. Astron. Soc.* 82:497–510.

Molnar, P. 1992. Brace-Goetze strength profiles, the partitioning of strike-slip and thrust faulting at zones of oblique convergence, and the stress-heat flow paradox of the San Andreas fault. In B. Evans and T. Wong, eds., *Fault Mechanics and Transport Properties of Rocks,* pp. 435–459. London: Academic Press.

Molnar, P. and P. England. 1990. Temperatures, heat flux, and frictional stress near major thrust faults. *J. Geophys. Res.* 95:4833–4856.

Molnar, P., D. Freedman, and J. Shih. 1979. Lengths of intermediate and deep seismic zones and temperatures in downgoing slabs of lithosphere. *Geophys. J. R. Astron. Soc.* 56:41–54.

Oleskevich, D. A., R. D. Hyndman, and K. Wang. 1999. The updip and downdip limits to great subduction earthquakes; thermal and structural models of Cascadia, South Alaska, SW Japan, and Chile. *J. Geophys. Res.* 104:14965–14991.

Peacock, S. M. 1996. Thermal and petrologic structure of subduction zones. In G. Bebout, D. Scholl, S. Kirby, and J. Platt, eds., *Subduction: Top to Bottom,* AGU Geophysics Monograph 96, pp. 119–133. Washington, DC: American Geophys Union.

Peacock, S. and R. Hyndman. 1999. Hydrous minerals in the mantle wedge and the maximum depth of subduction thrust earthquakes. *Geophys. Res. Lett.* 26:2517–2520.

Rice, J. R. 1992. Fault stress states, pore pressure distributions, and the weakness of the San Andreas fault. In B. Evans and T. Wong, eds., *Fault Mechanics and Transport Properties of Rocks,* pp. 475–504. London: Academic Press.

Ruff, L. 1999. Dynamic stress drop of recent earthquakes: variations within subduction zones. *Pageoph* 154:409–431.

Ruff, L. 2001. Stress on the seismogenic and deep creep plate interface during the earthquake cycle in subduction zones. *Earth, Planets, Space* 53:307–320.

Ruff, L. and B. Tichelaar. 1996. What controls the seismogenic plate interface in subduction zones? In G. Bebout, D. Scholl, S. Kirby, and J. Platt, eds., *Subduction: Top to Bottom,* AGU Geophysics Monograph 96, pp. 105–111. Washington, DC: American Geophysical Union.

Sass, J., C. Williams, A. Lachenbruch, S. Galanis, and F. Grubb. 1997. Thermal regime of the San Andreas fault near Parkfield, California. *J. Geophys. Res.* 102:27575–27586.

Scholz, C. H. 2000. Evidence for a strong San Andreas fault. *Geology* 28:163–166.

Scholz, C. H. 2002. *The Mechanics of Earthquakes and Faulting*, 2nd ed. Cambridge, UK: Cambridge University Press.

Seno, T. and Y. Yamanaka. 1996. Double seismic zones, compressional deep trench outer rise events, and superplumes. In G. Bebout, D. Scholl, S. Kirby, and J. Platt, eds., *Subduction: Top to Bottom*, AGU Geophysics Monograph 96, pp. 347–355. Washington, DC: American Geophysical Union.,

Shimazaki, K. 1986. Small and large earthquakes: The effects of the thickness of the seismogenic layer and the free surface. In S. Das, J. Boatwright, and C. Scholz, eds., *Earthquake Source Mechanics*, AGU Monograph 37, pp. 209–216. Washington, DC: American Geophysical Union.

Shudofsky, G., S. Cloetingh, S. Stein, and R. Wortel. 1987. Unusually deep earthquakes in east Africa: Constraints on the thermo-mechanical structure of a continental rift system. *Geophys. Res. Lett.* 14:741–744.

Sibson, R. H. 1973. Interactions between temperature and fluid pressure during earthquake faulting: A mechanism for partial or total stress relief. *Nature* 243:66–68.

Sibson. R. H. 1982. Fault zone models, heat flow, and the depth distribution of earthquakes in the continental crust of the United States. *Bull. Seism. Soc. Am.* 72:151–163.

Sipkin, S. A. 1986. Estimation of earthquake source parameters by the inversion of waveform data: Global seismicity. *Bull. Seism. Soc. Am.* 76:1515–1541.

Smith, R. and R. Bruhn. 1984. Intraplate extensional tectonics in the eastern Basin-Range: Inferences of structural style from seismic reflection data, regional tectonics, and thermo-mechanical models of brittle-ductile deformation. *J. Geophys. Res.* 89:5733–5762.

Spence, W. 1987. Slab pull and the seismotectonics of subducting lithosphere. *Rev. Geophys.* 25: 55–69.

Stein, C. A. and S. Stein. 1992. A model for the global variation in oceanic depth and heat flow with lithospheric age. *Nature* 359:123–129.

Stein, S. and D. Wiens. 1986. Depth determinations for shallow teleseismic earthquakes: Methods and results. *Rev. Geophys.* 24:806–832.

Tanioka, Y. and L. Ruff. 1997. Source time functions. *Seismol. Res. Lett.* 68:386–400.

Tanioka, Y., L. Ruff, and K. Satake. 1997. What controls the lateral variation of large earthquake occurrence along the Japan trench? *The Island Arc* 6:261–266.

Tichelaar, B. W. and L. J. Ruff. 1993. Depth of seismic coupling along subduction zones. *J. Geophys. Res.* 98:2017–2037.

Wagner, G. S. and C. A. Langston. 1988. East African body wave inversion with implications for continental structure and deformation. *Geophys. J.* 94:503–518.

Wang, K., T. Mulder, G. Rogers, and R. Hyndman. 1995. Case for very low coupling stress on the Cascadia subduction fault. *J. Geophys. Res.* 100:12907–12918.

Wong, I. G. and D. S. Chapman. 1990. Deep intraplate earthquakes in western United States and the relationship to lithospheric temperatures. *Bull. Seism. Soc. Am.* 80:589–599.

Wortel, M. J. and N. J. Vlaar. 1988. Subduction zone seismicity and the thermo-mechanical evolution of downgoing lithosphere. *Pageoph* 128:625–659.

Zoback, M. D. and A. H. Lachenbruch. 1992. Introduction to special section on the Cajon Pass scientific drilling project. *J. Geophys. Res.* 97:4991–4994.

CHAPTER SIX

Controls on Subduction Thrust Earthquakes: Downdip Changes in Composition and State

R. D. Hyndman

Introduction

In this chapter we discuss changes downdip on the subduction thrust interface or in adjacent rocks that control whether or not portions of the thrust are seismogenic. The main seismogenic zone appears to be continuous between upper and lower slip stability transitions (Scholz 1990). Therefore, of special importance are the controls of the updip limit and downdip limits to the seismogenic zone (Shimamoto et al. 1993; Tichelaar and Ruff 1993; Hyndman and Wang 1993; Hyndman et al. 1997). The updip or seaward limit is important for tsunami generation. Larger tsunamis are generated if the great earthquake rupture extends updip close to the seafloor. The downdip limit usually represents the nearest approach of the seismic source zone to nearby coastal cities, so is an important factor for earthquake hazard. The total downdip extent of the seismogenic zone (i.e., distance between the updip and downdip limits) is the primary control to the maximum-magnitude earthquake. The thrust seismogenic zone commonly extends between about 10 and 40 km depth for many subduction zones beneath continental areas (Tichelaar and Ruff 1993; Ruff and Tichelaar 1996; Pacheco et al. 1993), but there are some continental subduction zones such as southwest Japan where both limits are significantly shallower (Hyndman et al. 1995). The seismogenic portion of the subduction thrusts beneath most island arcs is limited to much shallower depths (Pacheco et al. 1993). An especially wide seismogenic zone, $\sim$200 km, allows great earthquakes such as the south Alaska $M \sim 9$ event of 1964. A narrow seismogenic zone allows only intermediate-magnitude thrust earthquakes. An extreme example may be the Mariana arc where there are few events larger than about $M = 7$.

A related topic is the strength and long-term regional stress transmission of the subduction thrust between the incoming plate and forearc. The downdip width of the seismogenic zone is an important control on the seismic coupling factor, that is, the fraction of the plate convergence that appears to be accommodated in thrust earthquakes. Many controls on the coupling factor have been discussed, including stress regime (Scholz and Campos 1995) and seamount subduction (Cloos and Shreve 1996; Scholz and Small 1997; Kodaira et al. 2000).

In analysis of seismogenic limits, it is important to recognize the complexities in rupture, which include (1) slip rate. Rapid slip generates seismic energy, moderate slip generates tsunamis, and slow slip is seen only in geodetic data. (2) postearthquake afterslip, updip and downdip of the main rupture zone; and (3) bulk shortening and out-of-sequence thrust faulting in the updip region where large accretionary sedimentary prisms occur.

The Limits to the Seismogenic Part of the Subduction Thrust

The convergence between plates at subduction zones appears to be concentrated in a narrow zone at the thrust interface based on seismic reflection data. In this summary we assume that the composition and physical state of the material along this interface are the primary controls to the seismic behavior, i.e., for the occurrence of great earthquakes and for the stress transmitted across the margin. However, a model has been presented by Scholz and Campos (1995) explaining the common weak seismic coupling in west Pacific arcs compared with strong coupling in North and South American subduction zones in terms of the difference in regional stress regime.

An important change in conditions appears to be the increase in temperature downdip (Hyndman et al. 1997). Although there is not enough data on downdip effective pressure changes to make accurate correlations, the seismogenic limits do not appear to be correlated simply with depth (Oleskevich et al. 1999). There are several obvious changes recognized in composition and state of the rocks in the fault zone, especially in the hanging wall, that may affect seismic behavior. The composition of the overlying rocks changes downdip from varying amounts of accreted sediments in the updip region near the trench, to forearc crustal rocks, and then to forearc mantle rocks. In most areas the footwall of mafic oceanic crust is probably stronger than the overlying fault-zone material. For continental subduction zones, the overlying crustal rocks and accreted sediments are probably more felsic and thus weaker at the same temperature. If the fault-zone material is strong, the fault may cut up into the hanging wall. For oceanic subduction zones, the overlying forearc crust may have similar composition and high strength as the downgoing crust, and it is likely that deformation is constrained to the fault shear zone material. Downdip on the thrust, below the forearc Moho, we conclude that the overlying weak serpentinized forearc mantle material deforms (see discussion later in this chapter). At great depths where the overlying mantle is dehydrated ($\sim 600°C$), the footwall subducting oceanic crust may be weaker than the overlying mantle, but this is much deeper than the limit of subduction-thrust earthquakes.

Within these zones, the downdip increase in temperature and pressure results in a number of changes that may affect seismic behavior. These include changes in mineralogy due to metamorphic reactions, upward expulsion of large amounts of water, and high formation pressures due to porosity collapse and dehydration reactions (Peacock 1993), and temperature-controlled changes in rock rheology from "brittle" to "viscous/plastic" (Scholz 1990), or velocity weakening to velocity strengthening (Blanpied et al. 1991). Other important observations on the behavior of the subduction thrust are, first, the thrust is very weak producing little or no detectable heat-flow anomaly due to frictional heating, and there is only small stress transmission to the forearc region as inferred from earthquake data (Wang et al. 1995a, and references therein). Second, thrust earthquakes, including great events, occur only at depths less than about 50 km (Pacheco et al. 1993) (figure 6.1). There is an aseismic updip zone of variable width ranging from 0 to 100 km, and depth up to 10 km.

The Updip Aseismic Zone

Many explanations have been given for the updip aseismic to seismic transition. These possibilities include: (1) changes in physical properties with depth, especially the downdip change from overlying, unconsolidated accretionary prism sediments to forearc crustal rocks (Byrne et al. 1988); (2) changes in pore pressure with depth (discussion of Moore and Saffer 2001); (3) temperature-controlled changes in mineralogy, especially the dehydration of stable-sliding clays (Vrolijk 1990), or the stable-sliding clays (Vrolijk 1990). As discussed in the following text, examination of a number of subduction zones has resulted in the conclusion that the updip limits usually correspond to a critical temperature of 100–150°C, not to depth or to changes in the composition of the overlying material. For some

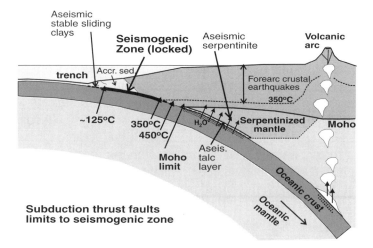

Figure 6.1 Composition and temperature limits to the subduction-thrust seismogenic zone that generates great earthquakes.

subduction zones, the updip seismogenic limit lies under accretionary prism sediments (e.g., southwest Japan, south Alaska, Cascadia); for others it lies beneath forearc crustal rocks (many island arcs such as Marianas, Izu-Bonin). This critical temperature occurs on the thrust near the trench for hot subduction zones such as Cascadia and as much as 100 km landward for margins subducting old, cold oceanic crust at shallow angle, such as south Alaska and northeast Japan.

A number of possible temperature-controlled chemical/mineralogical changes occur at 100–150°C, notably the breakdown of sheet-structure smectite to illite-chlorite. The presence of stable-sliding clays has been proposed to explain the aseismic behavior of the upper part of the San Andreas transform fault (Marone and Scholz 1988; Saffer et al. 2001). Smectite clays dehydrate to chlorite/illite at 100–150°C (Jennings and Thompson 1986; Pytte and Reynolds 1989; discussion by Moore and Vrolijk 1992) (Figure 6.2). Chlorite/illite are probably seismic because they lack the smectite sheet structure, but more laboratory data are needed. In subduction zones the clays may occur either in underthrust sediments or may form by shearing in the fault-zone gouge. No detailed study of differences in seismic behavior with the amount of clay in incoming sediments has yet been performed, but Logan and Rauenzahn (1987) suggest that 30 to 50 percent of smectite is required to appreciably weaken the frictional characteristics of a shear zone. Also, some correlations between deformation character in accretionary prisms and clay content has been suggested (Underwood 2002), so sediment composition may be important where there are large accretionary prisms.

Clay dehydration may be too simple a description for many diagenetic to low-

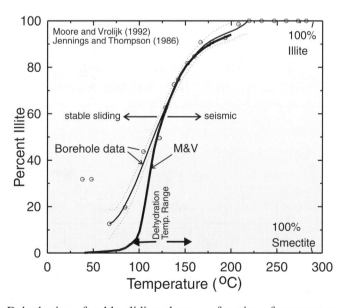

Figure 6.2   Dehydration of stable-sliding clays as a function of temperature, showing the transition at 100–150°C (after Moore and Vrolijk 1992; Jennings and Thompson 1986).

grade metamorphic and consolidation changes occurring at 100–150°C that may result in stick-slip seismic behavior (Moore and Saffer 2001). Moore and Saffer (2001) describe the changes associated with declining fluid production and decreasing fluid-pressure ratio, with active clay, carbonate, and zeolite cementation, and with the transition to pressure solution and quartz cementation. Thus, the temperature for the updip limit is expected to vary somewhat with local conditions.

The Downdip Aseismic Zone

Subduction thrusts are aseismic below a depth that varies from about 10 to 50 km in different subduction zones. We have examined two downdip changes that should affect seismic behavior. First, a temperature-controlled change from seismic (velocity weakening) to aseismic (velocity strengthening) slip occurs at about 350°C for crustal composition rocks. Second, serpentinite and related stable-sliding minerals form through hydration of the forearc mantle by fluids driven off the downgoing plate.

The Maximum Temperature Limit

At some depth, temperatures are reached on the subduction-thrust fault where rocks behave plastically. Ductile behavior at temperatures above a critical limit is often taken as the reason most continental crustal earthquakes are confined to depths less than about 25 km. More precisely, above the critical temperature, the fault zone no longer exhibits frictional instability, rather than there being a change in bulk rheology of the surrounding rocks (e.g., Scholz 1990). The critical transition is between velocity weakening (seismic) to velocity strengthening (aseismic stable sliding). The maximum depth of crustal earthquakes commonly corresponds to where the temperature is estimated to be about 350°C (Scholz 1990). Laboratory measurements on quartzofeldspathic crustal rocks indicate that the critical temperature is about 350°C for wet or dry conditions (Tse and Rice 1986; Blanpied et al. 1991, 1995) (figure 6.3). Laboratory data are not yet available giving the variation in transition temperature for different crustal composition rocks. There may be a transition zone from 350 to 450°C within which rupture, initiated at shallower depth, continues downdip with decreasing slip offset. There is a rapid rise in instantaneous shear strength above about 450°C for felsic rocks (e.g., discussion by Tse and Rice 1986). The transition from "brittle" to "ductile" deformation of crustal rocks occurs at approximately 450°C, not at the velocity-weakening seismic limit of about 350°C. The 350°C temperature and the seismic limit occur on the thrust as shallow as 15 km for very hot subduction zones such as Cascadia (Hyndman and Wang 1995), and as deep as 100 km for cold subduction zones such as northeast Japan (Peacock and Wang 1999).

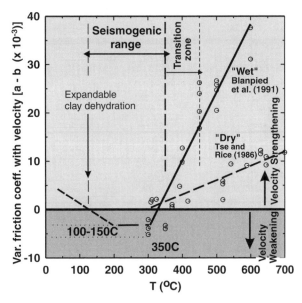

Figure 6.3 The transition from seismogenic velocity weakening to stable-sliding velocity strengthening above ~350°C based on laboratory data for quartzofeldspathic rocks (after Blanpied et al. 1991; Tse and Rice 1986). Above the horizontal line the friction increases with faster sliding (velocity strengthening); below the line, friction decreases with faster sliding (velocity weakening). There may be stable sliding at temperatures below 100–150°C.

Stable-Sliding Hydrated Mantle Rocks

The subduction thrust comes in contact with the forearc mantle below the forearc Moho. This part of the mantle is inferred to contain concentrations of serpentinite and related minerals (Peacock and Hyndman 1999). The sliding behavior of such hydrated minerals is complex, but they are expected to be weak and generally behave aseismically (Reinen et al. 1991) (figure 6.4). Large amounts of fluid are driven off the downgoing sediments and porous upper oceanic crust with increasing temperature and pressure (Peacock 1993). There also are geochemical arguments for a thin layer of very weak talc at the thrust boundary with the overlying forearc mantle (Peacock and Hyndman 1999) (figure 6.5). Hydrated mantle serpentinite will be transformed to talc by the infiltration of SiO_2 in fluids driven from underthrust siliceous sediments and from the underlying altered oceanic crust with increasing temperature and pressure (figure 6.6).

For subduction beneath continents, the forearc crust is commonly 35–40 km thick, so aseismic behavior is expected on the thrust below this depth. In the island arcs, the forearc crust usually is much thinner and the aseismic forearc mantle is reached by the thrust at much shallower depths. As an example, the Mariana forearc crust may be only ~10 km thick and the landward limit of the seismogenic zone may be only a short distance from the trench (Hyndman et al. 1997). If there

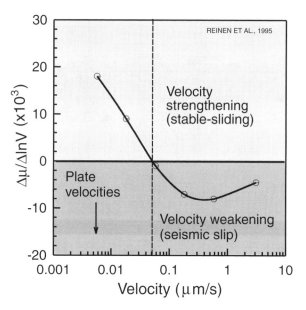

Figure 6.4 Laboratory data suggesting that serpentinite may exhibit stable sliding at low deformation rates, but move seismically if driven by rapid earthquake slip initiated in an adjacent region (data from Reinen et al. 1991). Above the horizontal line the friction increases with faster sliding (velocity strengthening); below the line, friction decreases with faster sliding (velocity weakening).

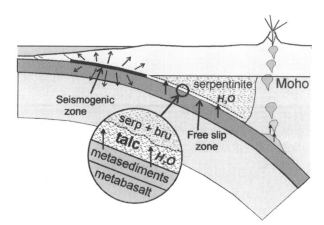

Figure 6.5 The postulated layer of talc along the subduction-thrust contact with the forearc mantle that may exhibit aseismic stable sliding (after Peacock and Hyndman 1999).

is a broad updip aseismic zone, as suggested by the clay dehydration hypothesis for such cold subduction zones, there will be, at most, only a very narrow seismic zone between the two limits (figure 6.7). The shallow depth of the downdip seismogenic limit provides an explanation for why such subduction zones do not have great thrust earthquakes.

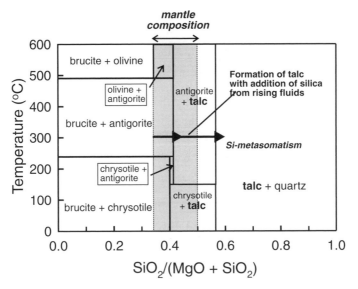

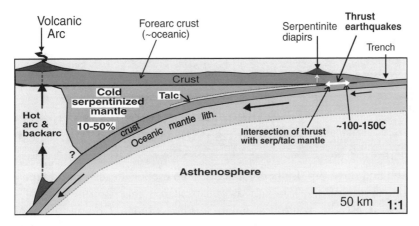

Figure 6.6   Phase diagram illustrating the formation of talc at subduction-thrust interfaces from rising silica-rich fluids (after Peacock and Hyndman 1999). The arrows indicate the change in forearc mineral composition with the addition of silica to the hydrated mantle.

Figure 6.7   Cross section of an island arc with a thin forearc crust such as Mariana, showing that there may be only a short section of the subduction thrust between updip stable-sliding clays and downdip stable-sliding serpentinite/talc.

Subduction Thrust Temperatures

Because two of the three suggested seismogenic zone limits are temperature controlled, temperatures have been calculated on profiles across a number of subduction zones including Cascadia, southwest Japan, Mexico, northeast Japan, Chile, south Alaska, and the Aleutian arc (Hyndman and Wang 1995; Hyndman et al.

1995; Peacock and Wang 1999; Oleskevich et al. 1999; Peacock and Hyndman 1999; Currie et al. 2002). There are large variations among subduction zones and along strike for individual subduction zones. Analytic solutions to the thermal regime have been used by Tichelaar and Ruff (1993), who found that there might be several seismogenic zone thermal limits for different areas. However, the accuracy of analytic solutions is limited by the required approximations. We therefore have used finite-element modeling with careful specification of local model parameters for each subduction-zone cross section (Hyndman and Wang 1993; Wang et al. 1995b). The important model parameters include: (1) the curved thrust dip profile (not planar approximations), (2) the age of the subducting plate (and its age history), (3) the subduction rates (and rate history), (4) the effect of insulating sediment on incoming crust (for example at Cascadia, just landward of the deformation front, the thrust is ~250°C), (5) overlying forearc crust radioactive heat generation, and (6) the thermal properties of each of the main geological elements (accreted sediments, crust, mantle). Corner flow in the backarc asthenosphere is concluded to have an insignificant effect for the depths of interest but has been included in some models (Peacock and Wang 1999; Currie et al. 2002). An example of the cross section and the model thrust temperatures for the hot north Cascadia subduction zone is given in figure 6.8b. The predicted heat flow across the margin is compared with observations in figure 6.8a. The model temperatures for the cold subduction zones such as northeast Japan are very much lower, with 350°C being reached at depths as great as 100 km (Peacock and Wang 1999). The locations of 100–150°C on the thrusts usually are not well constrained because the isotherms intersect the subduction thrust at a shallow angle. The locations of 350°C are usually better defined.

Subduction Zone Examples: Comparison of Predictions with Observed Rupture Limits

The updip and downdip limits predicted by the models described previously have been compared with the limits of the seismogenic zone in a number of subduction zones—Cascadia, southwest Japan, northeast Japan, Chile, south Alaska, the Aleutian arc, and Mexico (Hyndman and Wang 1995; Hyndman et al. 1995; Oleskevich et al. 1999; Peacock and Hyndman 1999; Currie et al. 2002). The seismogenic limits are estimated from (1) waveform modeling of great earthquake seismic data, (2) the updip and downdip limits of great earthquake aftershocks, (3) the seaward updip limit from modeling of tsunami wave data, (4) the downdip limit of intermediate magnitude thrust earthquakes, (5) the downdip limit from elastic dislocation modeling of geodetic measurements of coseismic deformation, and (6) dislocation modeling of interseismic data to define the interseismic locked zone that may rupture in future thrust earthquakes. These constraints all measure different limits, but the estimates from the different methods are usually very similar. The updip temperature limit of 100–150°C corresponds approximately to the subduction thrust updip seismogenic limits for all of the subduction zones

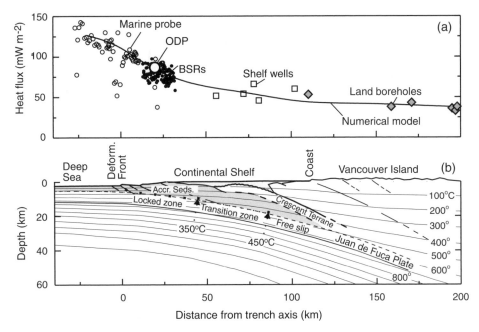

Figure 6.8 (a) Heat-flow data across the margin compared with the prediction of a numerical thermal model (after Hyndman and Wang 1995). The heat-flow data are from marine probes, gas hydrate bottom-simulating reflector depths, Ocean Drilling Program (ODP) downhole measurements, continental shelf petroleum exploration wells, and land boreholes. (b) Cross section of the north Cascadia subduction zone showing the positions of 350°C and 450°C on the subduction thrust from a numerical thermal model.

studied, based on earthquake, tsunami, and geodetic data. Obana et al. (2001) also found that the updip limit of current shallow seismicity corresponded to the updip limit of the great 1946 earthquake of southwest Japan. However, the uncertainties are quite large in both the updip part of the thermal models and the estimated updip seismogenic limits. Also, for the updip limit, where large accretionary sedimentary prisms occur, there is an uncertain role of prism shortening and subsidiary (out-of-sequence) thrust faulting rising to the seafloor landward of the trench.

The study of these margins suggests that the downdip limit of thrust earthquake rupture corresponds to whichever is reached first, the 350°C crustal thermal limit or the intersection of the thrust with aseismic hydrated forearc mantle. Figure 6.9 compares the thermal limits with the limits of the locked and transition zone on the basis of geodetic data for Cascadia. For the hydrated forearc mantle limit, the forearc Moho is at a depth of 30–50 km for subduction beneath most continents but is as shallow as 10 km for some island arcs such as the Marianas.

The general weakness of subduction thrusts may result from clays in the updip zone, high pore pressure at intermediate depths, and serpentinite and talc at greater depths. On a larger scale, although the forearc crust is strong because of the very low temperatures, the forearc mantle may be weak due to the presence of hydrated mineral assemblages.

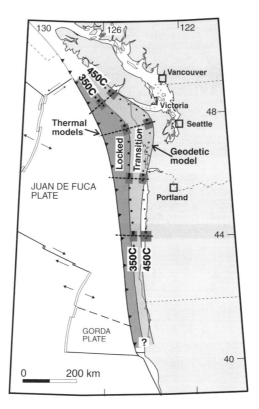

Figure 6.9 Comparison of positions of 350°C and 450°C, with uncertainty estimates, on the Cascadia subduction thrust from numerical thermal models with the locked and transition zones from dislocation models of geodetic data (after Hyndman and Wang 1995).

Summary

An important constraint to the earthquake and tsunami hazard from great subduction zone earthquakes is the updip and downdip limits to rupture. The updip limit is especially important for the size of tsunamis generated. The downdip limit usually provides the closest approach of the seismic source zone to coastal cities so is important for earthquake hazard. This chapter summarizes studies on factors that control these limits. Seismic slip requires stick-slip or velocity weakening material characteristics. For the updip limit, there is evidence for temperature control at 100–150°C. Our preferred mechanism for the onset of seismic behavior is the dehydration of stable-sliding smectite clay to stick-slip chlorite/illite, either in overlying sediments, or in the fault zone gouge. Other possible controlling downdip changes are the landward limit of weak accretionary prism sediments, the slate transformation, mechanical consolidation, and changes in pore pressure and changes in the stress regime. However, most of these changes do not have the inferred temperature dependence. For the downdip limit, our preferred controlling changes are, first, temperature dependence of slip characteristics (i.e., stick-slip to stable-sliding) in the fault zone material and second, downdip change in the overlying material composition. The most important composition change downdip occurs at the fault intersection with the forearc Moho, from overlying forearc crust to either overlying serpentinite in the forearc mantle or talc along the thrust plane. Testing the predictions of these hypotheses requires that the seismogenic zone be delineated by: (1) great earthquake rupture area from seismic modeling, (2) the extent of aftershocks, (3) modeling of great earthquake tsunamis (constrains seaward extent), and (4) modeling of coseismic geodetic data (constrains landward extent). Other constraints include the extent of smaller thrust events interpreted to be on the subduction thrust and models of interseismic geodetic data that constrain the thrust "locked" zone. Detailed analyses of a number of subduction zone (Cascadia, S.W. Japan, N.E. Japan, Chile, Aleutians, and Mexico) so far indicate seismogenic zones consistent with the limits suggested, although the uncertainties are large both in the location of the controlling factors (i.e., temperature and composition boundaries, etc.) and in the extent of past great earthquake rupture.

References

Blanpied, M. L., D. A. Lockner, and J. D. Byerlee. 1991. Fault stability inferred from granite sliding experiments at hydrothermal conditions. *Geophys. Res. Lett.* 18:609–612.

Blanpied, M. L., D. A. Lockner, and J. D. Byerlee. 1995. Frictional slip of granite at hydrothermal conditions. *J. Geophys. Res.* 100:13045–13065.

Byrne, D., D. Davies, and L. Sykes. 1988. Loci and maximum size of thrust earthquakes and the mechanics of the shallow region of subduction zones. *Tectonics* 7:833–857.

Cloos, M. and R. L. Shreve. 1996. Shear zone thickness and the seismicity of Chilean and Marianas-type subduction zones. *Geology* 24:107–110.

Currie, C. A., R. D. Hyndman, K. Wang, and V. Kostoglodov. 2002. The seismogenic zone of Mexican megathrust earthquakes. *J. Geophys. Res.* 107(B12):2370, 15-1 to 15-13.

Hyndman, R. D. and K. Wang. 1993. Thermal constraints on the zone of major thrust earthquake failure: The Cascadia subduction zone. *J. Geophys. Res.* 98:2039–2060.

Hyndman, R. D. and K. Wang. 1995. The rupture zone of Cascadia great earthquakes from current deformation and the thermal regime. *J. Geophys. Res.* 100:22133–22154.

Hyndman, R. D., K. Wang, and M. Yamano. 1995. Thermal constraints on the seismogenic portion of the south-west Japan subduction thrust. *J. Geophys. Res.* 100:15373–15392.

Hyndman, R. D., M. Yamano, and D. A. Oleskevich. 1997. The seismogenic zone of subduction thrust faults. *Island Arcs* 6:244–260.

Jennings, S. and G. R. Thompson. 1986. Diagenesis of Plio-Pleistocene sediments of the Colorado River delta, southern California. *J. Sediment. Petrol.* 56:89–98.

Kodaira, S., N. Takahasi, J.-O. Park, K. Mochizuki, M. Shinohara, and S. Kimura. 2000. Western Nankai Trough seismogenic zone: Results from a wide-angle ocean bottom seismic survey. *J. Geophys. Res.* 105:5887–5905.

Logan, J. and J. Rauenszahn. 1987. Frictional dependence of gouge mixtures of quartz and montmorillonite on velocity, composition, and fabric. *Tectonophysics* 144:87–108.

Marone, C. and C. Scholz. 1988. The depth of seismic faulting and the upper transition from stable sliding to unstable slip regimes. *Geophys. Res. Lett.* 15:621–624.

Moore, J. C. and P. Vrolijk. 1992. Fluids in accretionary prisms. *Rev. Geophys.* 30:113–135.

Moore, J. C. and D. Saffer. 2001. Updip limit of the seismogenic zone beneath the accretionary prism of southwest Japan: An effect of diagenetic to low-grade metamorphic processes and increasing effective stress. *Geology* 29:183–186.

Obana, K, S. Kodaira, K. Mochizuki, and M. Shinohara. 2001. Micro-seismicity around the seaward updip limit of the 1946 Nankai earthquake dislocation area. *Geophys. Res. Lett.* 28:2333–2336.

Oleskevich, D., R. D. Hyndman, and K. Wang. 1999. The updip and downdip limits of subduction earthquakes: Thermal and structural models of Cascadia, south Alaska, S.W. Japan, and Chile. *J. Geophys. Res.* 104:14965–14991.

Pacheco, J. F., L. R. Sykes, and C. H. Scholz. 1993. Nature of seismic coupling along simple plate boundaries of the subduction type. *J. Geophys. Res.* 98:14133–14159.

Peacock, S.M. 1993. Large-scale hydration of the lithosphere above subducting slabs. *Chem. Geol.* 108:49–59.

Peacock, S. M. and R. D. Hyndman. 1999. Hydrous minerals in the mantle wedge and the maximum depth of subduction thrust earthquakes. *Geophys. Res. Lett.* 26:2517–2520.

Peacock, S. M. and K. Wang. 1999. Seismic consequences of warm versus cool subduction metamorphism: Examples from southwest and northeast Japan. *Science* 286:937–939.

Pytte, A. M. and R. C. Reynolds. 1989. The thermal transformation of smectite to illite. In N. D. Naeser and T. H. McCulloh, eds., *Thermal History of Sedimentary Basins: Methods and Case Histories*, pp. 33–40. New York: Springer-Verlag.

Reinen, L. A., J. D. Weeks, and T. E. Tullis. 1991. The frictional behavior of serpentinite: Implications for aseismic creep on shallow crustal faults. *Geophys. Res. Lett.* 18:1921–1924.

Reinen, L. A. 1995. Microstructural evidence of strain localization and distributed strain in serpentine friction experiments. *EOS Trans. Am. Geophys. Union* 76:560.

Ruff, L. J. and B. W. Tichelaar. 1996. What controls the seismogenic plate interface in subduction zones? In G. Bebout, D. W. Scholl, S. H. Kirby, and J. P. Platt, eds., *Subduction: Top to Bottom*, pp. 105–111. Washington, DC: American Geophysical Union.

Saffer, D., K. M. Frye, C. Marone, and K. Mair. 2001. Laboratory results indicating complex and potentially unstable frictional behavior of smectite clay. *Geophys. Res. Lett.* 28:2297–2300.

Scholz, C. H. 1990. *The Mechanics of Earthquakes and Faulting*. New York: Cambridge University Press.

Scholz, C. H. and C. Small. 1997. The effect of seamount subduction on seismic coupling. *Geology* 25:487–490.

Scholz, C.H. and J. Campos. 1995. On the mechanism of seismic decoupling and back arc spreading at subduction zones. *J. Geophys. Res.* 100:22,103–22,115.

Shimamoto, T., T. Seno, and S. Uyeda. 1993. A simple rheological framework for comparative subductology. In K. Akai and R. Dmowska, eds., *Relating Geophysical Structure and Processes: The Jeffreys Volume*. American Geophysical Union Monograph Series 76:39–52.

Tichelaar, B. W. and L. J. Ruff. 1993. Depth of seismic coupling along subduction zones. *J. Geophys. Res.* 98:2017–2037.

Tse, S. T. and J. R. Rice. 1986. Crustal earthquake instability in relation to the depth variation of frictional slip properties. *J. Geophys. Res.* 91:9452–9472.

Underwood, M. B. 2002. Strike-parallel variation in clay mineralogy, fault vergence, and up-dip limits to the seismogenic zone. *Geology* 30:155–158.

Vrolijk, P. 1990. On the mechanical role of smectite in subduction zones. *Geology* 18:703–707.

Wang, K., T. Mulder, G. C. Rogers, and R. D. Hyndman. 1995a. Case for very low coupling stress in the Cascadia subduction thrust fault. *J. Geophys. Res.* 100:12907–12918.

Wang, K., R. D. Hyndman, and M. Yamano. 1995b. Thermal regime of the southwest Japan subduction zone: Effects of age history of the subducting plate. *Tectonophysics* 248:53–69.

CHAPTER SEVEN

Thermo-Mechanical Models of Convergent Orogenesis: Thermal and Rheologic Dependence of Crustal Deformation

Sean D. Willett and Daniel C. Pope

Introduction

Preface

Convergent orogens present a particularly difficult problem for Earth scientists interested in the mechanics of crustal deformation. The deformation associated with plate convergence is intense and widespread, in many cases deforming the lithosphere for hundreds of kilometers from a plate boundary. In addition, lithospheric deformation associated with convergent orogenesis is diverse and no single model for the mechanical processes of orogeny is applicable to all convergent systems. The challenge to mechanical modelers is to find or adapt numerical tools capable of simulating these mechanical processes. Finite-element methods have proven to be a versatile and powerful method of analysis and have seen broad application to the study of orogenesis (Bird and Piper 1980; Villotte et al. 1982, 1986; Bird and Baumgartner 1984; England and Houseman 1986; Houseman and England 1986; Bird 1989, 1996; Willett 1992, 1999a, 1999b; Beaumont et al. 1992, 1995, 1996a, 1996b, 2000, 2001; Braun 1993; Bird and Kong 1994; Wdowinski and Bock 1994; Willett and Beaumont 1994; Fullsack 1995; Ellis et al. 1995; Batt and Braun 1997; Ellis et al. 1998; Pysklywec et al. 2000). Potential applications of numerical models to orogenic processes are virtually unlimited; in this chapter we present one finite-element formulation of thermo-mechanical deformation of convergent orogens and demonstrate its use in addressing the rheologic dependence of crustal deformation.

Setting of Convergent Orogenesis

Convergent orogens exhibit a spectrum of scales from forearc deformation belts at subduction zones to full continent-continent collisions with the India-Asia collision as a modern example. In all cases, relative convergence between two plates leads to crustal shortening, thickening, and deformation. At oceanic subduction boundaries, this crustal deformation is limited to the forearc where offscraping of material from the subducting plate leads to the formation of an accretionary wedge (Karig and Sharman 1975; Seely 1979). At a mature subduction boundary, this accretion-driven deformation can reach hundreds of kilometers inland from the trench as, for example, occurs on the Cascadia and Alaska margins. Compressional stresses associated with subduction can also lead to shortening of the upper plate; the Andes of South America are the best modern example of this process where several hundred kilometers of east-west shortening of the South American plate is evident from the structurally thickened crust and widespread convergent deformation despite very little accretion of new material from the subducting Nazca plate (Isacks 1988; Sheffels 1995).

Oceanic plate subduction ceases when continental crust on the downgoing plate collides with the accretionary margin of the upper plate. The continental crust on the subducting plate might be an island arc, a continental fragment, or a major continent, and this determines the character of resulting collision (Dewey and Bird 1970). In an arc-continent collision or with the collision of a small continental fragment, the plate boundary typically experiences a short phase of intense orogeny as the continental crust grows by accretion, but subsequently the plate boundary continues to consume oceanic lithosphere by subduction (Dewey and Bird 1970). At the largest scale of collisional orogeny, two continental masses collide. If convergence continues, it is only with massive deformation of one or both continental masses as is presently occurring between India and Asia in the greater Himalaya-Tibetan plateau system (Molnar and Tapponnier 1975; Dewey et al. 1988).

Mechanical Models of Convergent Orogens

The deformational and thermal processes of arc-continent or continent-continent collisions are complex, difficult to study, and thus poorly understood. They are a prime objective for mechanical modeling that can make an important contribution to interpretation of observations consistent with the physics of the processes.

An important class of mechanical models is represented by the thin-sheet models of Bird and Piper (1980), England and McKenzie (1982, 1983), Bird and Baumgartner (1984), England et al. (1985), Houseman and England (1986), Sonder et al. (1986), Bird (1989; 1996), Bird and Kong (1994), and subsequent work by these authors and others. In most of these models, lithospheric deformation is modeled in planform by integrating stress and strain rate over depth, thus implicitly assuming that the entire crust and lithospheric mantle deform together by pure-shear shortening and thickening. These models have been useful for analyzing the planform deformation of large orogens such as the Himalaya-Tibet system, and have been used to demonstrate the importance of competition between gravitational forces arising from crustal thickening and horizontal compressive forces presumably arising from far-field plate forces. Variations on these thin-sheet models include those of Villote et al. (1982, 1986), who used both a plane-stress approach and a plane-strain formulation in planform to approximate the effect of gravity on the thickened lithosphere, and Wdowinski et al. (1989), who used a one-dimensional thin sheet to investigate the role of basal traction exerted on the lithosphere by a subducting slab in an Andean-type orogen.

These models assume that the lithosphere deforms by homogeneous pure shear. That is, the horizontal component of velocity does not vary with depth, so that shear strain rates on horizontal planes are assumed to be negligible. Thus, the crust and mantle deform as a single unit, experiencing only uniform shortening and thickening. This limitation is partially avoided by formulations described by Bird and Baumgartner (1984), who model only deformation of the crust coupled by viscous drag to a fixed mantle, and Bird (1989), in which two thin sheets, representing the brittle upper crust and the deforming lithospheric mantle, are coupled across a weak lower crust. This formulation allows shear of the lower crust to accommodate differential strain in the mantle and upper crust. This was also the principle behind a thin-sheet model proposed by Ellis et al. (1995) in which a thin sheet was used to model crustal deformation with prescribed mantle deformation coupled to the crust through a basal traction. However, this does not avoid the assumption that depth-averaged quantities over each layer accurately capture the physics of the deformation.

The assumption depth-independent horizontal velocity has been called into question by other geodynamic models of continental deformation. Bird (1991), Royden (1996), and Beaumont et al. (2002) demonstrated that locally thick crust drives viscous flow of the lower crust in an extrusive manner. The lower crust thus has a horizontal velocity greater than the underlying mantle or overlying upper crust, clearly not consistent with the assumptions of the thin-sheet models.

Thin-sheet assumptions are also not likely to be valid for small orogens or accretionary complexes at oceanic subduction boundaries where deformation is driven by shear tractions of a subducting plate. In these settings crustal deformation clearly shows the importance of simple-shear strain on near-horizontal detachments within the crust, likely driven by underthrusting of one lithospheric plate beneath the other (Suppe 1981; Davis et al. 1983; Mattauer 1986). Observations of crustal structure in fold-and-thrust belts led to another important class of models, the critical-wedge models (Chapple 1978; Stockmal 1983; Davis et al. 1983; Dahlen 1984, 1990; Dahlen et al. 1984; Zhao et al. 1986). Critical-wedge models are based on the theory that convergent orogens comprise crustal material that forms a deforming wedge analogous to a pile of sand in front of an advancing bulldozer blade. The surface slope of such a (frictional) plastic wedge will steepen until reaching the minimum angle at which gravitational stresses balance the basal traction while keeping the material within the wedge everywhere at its yield stress, a condition necessary to accommodate the accretion of new material. Critical-wedge solutions provide static solutions for the geometry and stress states of wedges, but do not make predictions of the deformation or kinematics within deforming wedges; these require additional assumptions (Dahlen and Barr 1989; Barr and Dahlen 1989; Willett 1992).

Critical-wedge theories were initially applied to accretionary wedges and fold-and-thrust belts, but the concept of a mountain belt attaining a critical topographic profile in balance with gravitational forces can be incorporated into more general models (Willett et al. 1993). In principle, provided crustal deformation is driven by convergence and basal tractions are significant, the deforming crust attains a "critical" balance between gravitational stresses, basal tractions, and the strength of the crust at the scale of the entire crust, thus resulting in the formation of an "orogenic wedge" of crustal scale (Platt 1986).

The Role of Numerical Modeling

Mechanical models are constructed to help provide a physical explanation for observations of crustal structure and deformation. Where these physical models become too complex for analytical treatment, numerical modeling techniques become an important tool. In continuum problems involving heat transfer or deformation, complexity is typically the result of either complicated geometry or nonlinear material behavior. Geologic problems rarely involve simple geometry, so although insight can be gained from simple problems involving, for example, layered media, thin-sheet approximations, or simple wedges, most applications require methods capable of representing complex geometric domains. During deformation, geologic materials exhibit complex, often nonlinear rheologic behavior (Kohlstedt et al. 1995), and this also generally requires numerical methods for solution. Finite-element methods are well suited for arbitrary domain shapes, have great versatility in incorporating complex mechanical behavior, and thus are frequently used in geologic applications.

As a tool, numerical models can be applied to mechanical problems in a variety of ways. One method is through regional simulation. Regional simulation consists of posing a mechanical boundary-value problem applied to some specific regional domain. If the problem is well constrained, with boundary conditions and domain properties reasonably well determined, a model can be constructed to estimate remaining unknown parameters. For a quasi-static mechanical problem, this typically requires specification of a deformational mechanism and constitutive law at all material points, the physical constants in that constitutive law and boundary displacements or velocities. The boundary-value problem is then solved; the predictions of deformation, structure, topography, etc., are compared to observations and in the event of misfit, the assumptions of material properties or boundary conditions are changed and the process repeated. This type of simulation is useful for determining physical properties or boundary conditions, but, in practice, requires that the system be very well constrained before the exercise is attempted (Bird 1996; Bird and Kong 1994; Beaumont et al. 1996b; 2000).

A second method of application of numerical models is in the demonstration of fundamental mechanical behavior through simple "numerical experiments." Numerical models are used to test the response of particular physical systems with specific materials and boundary conditions, not to simulate any specific geologic setting, but to illustrate mechanical behavior that might serve as an analog to geologic processes. Such experiments are useful in developing intuition for mechanical processes that are difficult to investigate in isolation or in nature. For example, before simulating contraction of continental crust with multiple layers of differing composition and continuously varying pressure and temperature, it might be preferable to apply the same boundary conditions to a linear, viscous material of uniform viscosity. Having developed some understanding of the response of a linear-viscous material, progressive complexity (nonlinear viscosity, plastic behavior, etc.) can be added to the model with a clearer isolation of specific effects of such complexity. Examples of this "numerical experiment" approach include investigations of thin-sheet models (England and McKenzie 1982, 1983; England et al. 1985; Houseman and England 1986; Ellis et al. 1995), critical wedge models (Davis et al. 1983; Dahlen 1984, 1990; Willett 1992) and larger-scale orogenic wedge models (Braun 1993; Beaumont et al. 1994; Willett 1999a; Pysklywec et al. 2000). It is this approach that we follow in this chapter. Consequently, there is little direct comparison of models and observations. It is assumed that the reader is aware of the general geologic settings to which the models are applicable and detailed comparison is not warranted.

Mechanical Processes Investigated in This Chapter

Two processes are investigated in this chapter. The first is the formation of doubly vergent wedges in subduction and collisional settings. Analog modeling of these wedges suggests a strong dependence of modes of strain on the crustal rheology. In particular, there is an important issue regarding the rheologic dependence of extension of orogenic wedges; surface extension has been associated with accretionary wedges (Platt 1986), and one potential mechanism for extension is gravity spreading as observed in viscous analog models (Buck and Sokoutis 1994). However, this extension has not been observed in comparable sandbox analog models (Davis et al. 1983; Wang and Davis 1996; Buck and Sokoutis 1994), suggesting a rheologic dependence. Willett (1999b) investigated this dependence, and those results are reviewed here.

The second process investigated in this chapter is the formation of orogenic plateaus. The genetic mechanism for orogenic plateaus is controversial, but observations and models suggest that the low-relief of high-elevation plateaus such as the Tibetan Plateau is the result of flow of a thermally weakened lower crust (Bird 1991; Willett and Beaumont 1994; Royden 1996; Nelson et al. 1996; Pope and Willett 1998). This process depends on the thermal and rheologic structure of the crust. These dependencies are investigated through a series of numerical models.

Finite-Element Method

Finite-element methods in engineering applications are typically based on elastic, or elastic-plastic constitutive laws and thus are most easily cast in a Lagrangian (material) reference frame in which the equations of motion are formulated relative to the initial configuration of the medium (Zienkiewicz 1977). In many geologic applications the deformation is so extensive that the initial configuration is difficult to track. This is particularly true for high-temperature deformation dominated by viscous flow of rock. In this case the mechanical problem is more easily formulated in an Eulerian (spatial) reference frame, so that material moves relative to the reference frame and finite-element grid. In the method employed here, a second, Lagrangian grid is defined to track material properties. Points in this grid are advected with the velocity field from the Eulerian grid. This permits interpolation of material properties back to the Eulerian grid for subsequent velocity calculations, resulting in what is referred to as an "Arbitrary Lagrangian, Eulerian" or ALE method (Zienkiewicz and Godbole 1975; Villotte et al. 1982; Dunbar and Sawyer 1989; Willett 1992; Fullsack 1995).

Nonlinear Stokes Flow

In a Eulerian reference frame, two-dimensional force equilibrium is expressed as a quasi-static equation of motion. Assuming low Reynolds number (Stokes flow) and a Newtonian viscous material, this is given by

$$-\frac{\partial \bar{\sigma}}{\partial x_j} + \mu \frac{\partial}{\partial x_i}\left(\frac{\partial v_i}{\partial x_j} + \frac{\partial v_j}{\partial x_i}\right) + \rho g_j = 0 \quad j = 1,2 \tag{7.1}$$

where v_i, $i = 1, 2$ are the components of velocity, μ is the viscosity, ρ is the density, g_j is the acceleration of gravity in the x_j direction, and $\bar{\sigma}$ is the mean stress or pressure defined in plane-strain as

$$\bar{\sigma} = -\frac{\sigma_{kk}}{2} \tag{7.2}$$

Note that summation is implied over repeated indices. Strain rate, or more specifically, the rate-of-deformation is defined in terms of the partial derivatives of velocity:

$$\dot{\varepsilon}_{ij} = \frac{1}{2}\left(\frac{\partial v_i}{\partial x_j} + \frac{\partial v_j}{\partial x_i}\right) \tag{7.3}$$

Conservation of mass in an incompressible material can be written in terms of the divergence of the velocity field:

$$\frac{\partial v_k}{\partial x_k} = 0 \tag{7.4}$$

This condition is required to solve for the two components of velocity and pressure.

For geologic applications, it is important to include more complex constitutive behavior such as nonlinear viscous and plastic deformation. In general, this requires specification of a new constitutive equation, resulting in a different form of the equation of motion (equation 1). However, many complex rheologies can be modeled with a viscous constitutive relationship with the appropriate selection of a nonlinear viscosity.

High-temperature deformation of rock is typically described by a first-order rate equation with temperature dependence through an activation energy and nonlinear stress dependence through a power-law relationship. These functional dependencies can be combined into a temperature and strain-rate-dependent (nonlinear) viscosity:

$$\mu_{nlv} = I_2'^{((1-n)/2n)} A^{-1/n} \exp\left(\frac{Q}{nRT}\right) \tag{7.5}$$

where Q is an activation energy, A and n are constants, T is temperature, and R is the gas constant. Material constants, A, n, and Q can be determined experimentally (Kohlstedt et al. 1995). The strain-rate dependence comes through I_2', which is the second invariant of the deviatoric rate-of-deformation tensor defined as

$$I_2' = \frac{1}{2} \dot{\varepsilon}_{ij}' \dot{\varepsilon}_{ij}' \tag{7.6}$$

Equation (7.5) defines a nonlinear viscosity, so that solution of equation (7.1) requires a nonlinear solution method, but these are easily included in finite-element methods.

Plastic deformation can also be treated as a nonlinear viscous flow problem. Plastic deformation is fundamentally different from viscous deformation in that there is no deformation at stress levels less than the yield stress. For stress equal to the yield stress, strain rate is undefined, although many theories have been proposed for postyield deformation (Zienkiewicz 1977; Bathe 1996). Because the problem here has been formulated as a Eulerian, velocity-based equilibrium problem, a natural formulation of plastic deformation is provided by the Levy-Mises theory (Malvern 1968). The Mises yield stress is defined as a limit on the second invariant of deviatoric stress,

$$\sqrt{J_2'} = \sigma^Y \tag{7.7}$$

where J_2' is the second invariant of the deviatoric stress and σ^Y is the yield stress, a material property. The second invariant of deviatoric stress is defined as

$$J_2' = \frac{1}{2} \sigma_{ij}' \sigma_{ij}'$$

The yield criterion of equation (7.7) is consistent with a viscous constitutive law if a strain-rate-dependent viscosity is defined as

$$\mu_p = \frac{\sigma^Y}{2\sqrt{I_2'}} \tag{7.8}$$

It is apparent that this definition of a nonlinear viscosity and the viscous constitutive law:

$$\sigma_{ij}' = \mu \dot{\varepsilon}_{ij},$$

are consistent with the Von Mises yield criterion (equation 7.7).

For geologic materials that exhibit pressure-dependent frictional behavior at low temperature, a Coulomb yield criterion is more appropriate than the Mises criterion described previously. A Coulomb yield criterion can also be cast in terms of a viscous constitutive equation by writing it in terms of J_2'. In general, the Coulomb yield criterion is written in terms of the three stress invariants, but for the two-dimensional, plane-strain, incompressible deformation, the third invariant is zero, so that the criterion can be written in terms of the first and second invariants of the deviatoric stress, J_1 and J_2', respectively.

$$\sqrt{J_2'} = c\cos(\phi) + \frac{1}{3}J_1\sin(\phi) = \sigma^Y \tag{7.9}$$

The first invariant, J_1, is the trace of the stress tensor, so that pressure is a third of this quantity. The cohesion, c, and ϕ, the angle of friction, are the two physical constants defining the Coulomb yield criterion. Equation (7.9) is of the same form as equation (7.7) so the right side of equation (7.9) is the Coulomb equivalent of the Mises yield stress, σ^Y, a scalar quantity, and equation (7.8) can be used to define an effective viscosity for a Coulomb plastic material.

To summarize, equations (7.1) and (7.4) provide a statement of quasi-static equilibrium with incompressible flow. The viscosity, μ, in equation (7.1) can be taken as a constant, or as defined by equation (7.5) or (7.8), to reflect the appropriate rheology.

Finite-Element Implementation

The finite-element discretization of equations (7.1) and (7.4) is standard and is not described in detail here; the interested reader is referred to Bathe (1996) or other finite-element texts. The result of element integration over a Eulerian grid is a system of linear equations in velocity, v_i, and pressure, $\bar{\sigma}$:

$$\mathbf{Kv} + \mathbf{K}_T \bar{\sigma} = \mathbf{Mg} \qquad (7.10)$$
$$\mathbf{K}_T \mathbf{v} = 0$$

where $\mathbf{K}$ and $\mathbf{M}$ are the stiffness and mass matrices, respectively, and $\mathbf{v}$ and $\bar{\sigma}$ are the vectors of nodal velocities and pressures, respectively. In practice, the second equation is not solved explicitly but is recast using a penalization method:

$$\frac{\partial v_k}{\partial x_k} = -\frac{1}{\kappa}\bar{\sigma} \qquad (7.11)$$

where κ is a pseudo-bulk modulus, taken to be large with respect to the viscosity, thereby forcing incompressibility. With this expression, the pressure can be condensed out at the element-integration level by solving equation (7.11) for pressure in terms of velocity and substituting this expression in equation (7.10) (Zienkiewicz and Godbole 1975; Fullsack 1995). This yields a system of linear equations in $\mathbf{v}$. Here, the element integration is done using four node elements and bilinear interpolants for velocity and a single degree of freedom for pressure (constant over the element). Specified-velocity boundary conditions are applied by factoring the corresponding equation out of the system. Specified flux conditions are imposed by adding the integrated flux to the right side (data vector) of the matrix equation system. The natural boundary condition is a no flux condition.

Arbitrary Lagrangian-Eulerian Method

The equilibrium expressions described previously are based on a Eulerian formulation that operates in a reference frame fixed in space such that material moves through it. The disadvantage of Eulerian formulations is that one does not naturally track particles (material points) through the course of a deformation problem, and thus cannot track history-dependent quantities such as strain. Material properties associated with material points are also not tracked, making it difficult to model material heterogeneity other than through state variables such as pressure and temperature, which are associated with spatial points. The second problem with Eulerian methods is the inability to track moving boundaries. In most geologic applications the upper surface is a stress-free surface that moves with the deformation; a Eulerian finite-element grid must be redefined to update the position of this, or other, moving boundaries. To address the former problem, a second, Lagrangian mesh can be defined to track material points. An initial mesh is defined to spatially cover the Eulerian domain. Mesh points are subsequently moved by calculating the velocity from the Eulerian grid and advecting the Lagrangian nodes with this velocity at each timestep. The Eulerian grid is also redefined to track moving boundaries, but in a way to maintain as closely as possible its rectilinear structure. Material properties are tracked with the Lagrangian grid and interpolated to the Eulerian grid as needed. The Lagrangian grid can also be used to calculate history-dependent quantities such as strain or pressure, temperature paths.

Applications to Convergent Orogens

In this chapter, we present two applications of finite-element modeling to convergent orogen mechanics. In the first application, we investigate the role of crustal rheology in the formation and structure of orogenic wedges. The general wedge geometry of orogenic belts has long been noted and is clear from sections of modern and ancient orogens. For example, figure 7.1a shows a section of the European Alps with the wedge-shaped crustal stack of nappes and deformed basement that comprise the orogenic belt (Pfiffner 1992; Schmid et al. 1996). In the second example, we demonstrate the role of temperature-dependent flow of the crust in the formation of orogenic plateaus such as the modern Tibetan plateau (figure 7.1b) (Molnar and Tapponier 1975; Dewey et al. 1988; Owens and Zandt 1997; Royden et al. 1997; Beaumont et al. 2001). These geomorphic features represent the culmination of convergent orogenesis with thousands of kilometers of convergence resulting in massive crustal thickening and mountain building. This extreme crustal thickening is clearly an important deformational process, and the rheology of the crust plays a fundamental role in the resultant crustal structure, as demonstrated by the series of numerical models presented later in this chapter.

Orogenic Wedge Mechanics

Geologic Problem

Mechanical wedge models have been applied to orogens at many scales from small accretionary wedges of offscraped sediment (Chapple 1982; Davis et al. 1983) to crustal orogenic belts (Willett et al. 1993). At most scales, orogenic wedges take on a doubly vergent geometry with thrusting over both plates. The orogen is defined as the doubly vergent wedge of deforming crustal material overlying the boundary of the two, effectively rigid, underlying plates (figure 7.2).

The rates and distribution of deformation within the orogenic wedge (figure 7.2) depend on the plate-convergence velocity and the rheology of the deforming crustal material. In its simplest form, the mechanical problem consists of deformation of a crustal layer of some specified rheology driven by the relative velocities of the underlying plates. According to critical-wedge theory, the geometry of the wedge is a function of its internal strength and the strength of the detachment separating the deforming orogen from the undeforming lower plates (Dahlen 1990). Coulomb wedge theory describes that relationship analytically, but, at a crustal scale, the rheology of the deforming orogen is not likely to be purely Coulomb plastic because higher-temperature deformational mechanisms become dominant. It is more likely that the orogen will deform by some combination of

a) European Alps

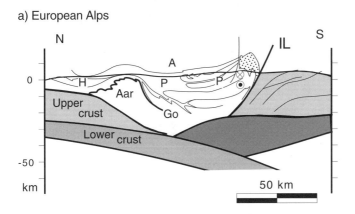

b) Tibetan Plateau

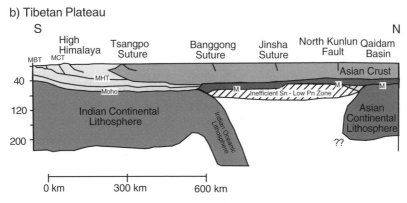

Figure 7.1 Examples of convergent orogens at two scales. (a) The European Alps, modified from Schmid et al. (1996). European lower crust and mantle are observed to underthrust alps to the south. Crust shortens above, forming an orogenic wedge including major features: Helvetic Nappes (H), Penninic Nappes (P), Aar basement massif (Aar), and Insubric Line (IL) denoting southern extent of metamorphic core. (b) Tibetan plateau crustal structure as interpreted by Owens and Zandt (1997).Thrust faults in the Himalaya include the Main Boundary Thrust (MBT) and the Main Central Thrust (MCT), both of which root into the Main Himalayan Thrust (MHT). Mesozoic and early Cenozoic sutures young to the south with the Tsangpo Suture representing final closure of the Tethys and India-Asia collision.

plastic and viscous deformational mechanisms depending on pressure and temperature conditions (Platt 1986). Further complexities in the mechanical problem include spatial heterogeneity, time-dependent thermal evolution of the lithosphere, and the isostatic compensation of the thickened crust. Numerical models are able to address this problem with its complexities as demonstrated in the next sections.

Analog Models

The mechanical problem of development of a doubly vergent wedge as described in figure 7.2 lends itself to analog modeling, provided the rheology of the deform-

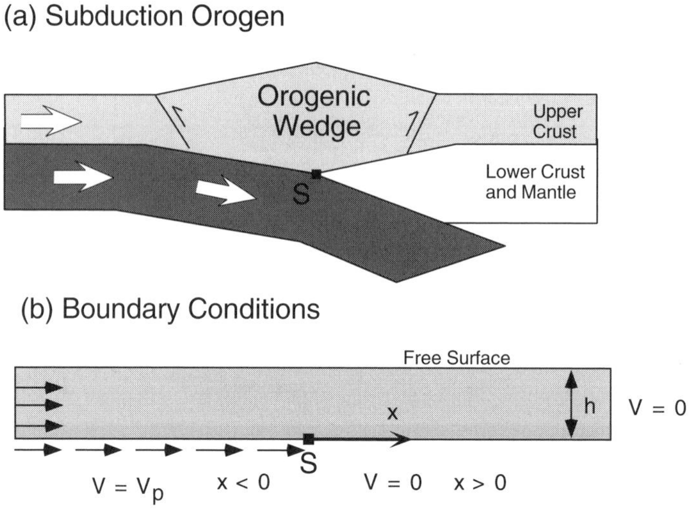

(a) Subduction Orogen

(b) Boundary Conditions

Figure 7.2 (a) Idealized setting for formation of crustal orogenic wedge. Subducting substrate (dark gray) comprises lithospheric mantle and, in some cases, crust. (b) Domain and boundary conditions representing setting of (a) in subsequent analog and finite-element models. Layer of thickness h has material properties as described in the text. Upper surface is traction free. Base and sides have specified velocity. Horizontal component of basal velocity is constant at V_p left of point S, and zero to the right of S. Side-boundary conditions are also constant velocity but are far enough away so that the layer is effectively infinite in horizontal dimension.

ing crustal layer is relatively simple. Analog models provide useful tests of numerical models and thus are reviewed here. Much of Coulomb wedge theory was verified and, in part, motivated by analog models using sand as a plastic material (Davis et al. 1983). Many of these experiments used a vertical wall to deform the sand layer and thus only investigated the mechanics of a wedge with uniform vergence, and therefore modeled conditions of near-uniform state of stress. However, several modeling studies considered the problem of doubly vergent wedges with boundary conditions as in figure 7.2b.

Malavieille (1984) presented an experiment in which a sandpack was deformed by pulling a mylar sheet along the base of the sandbox, then down a vertical slot, thus matching the boundary conditions of figure 7.2b. In a similar experiment, Wang and Davis (1996) slid a mylar sheet beneath a thin, flat, rigid backstop to produce similar velocity conditions (figure 7.3a). The discontinuity in velocity at the base of the sandpack induced deformation initially localized at this point (figure 7.3b). Subsequent convergence in the experiment led to the formation of a doubly vergent wedge with a narrow tapered wedge overlying the moving mylar sheet (pro-wedge) and a steeper wedge tapering in the opposite direction (retro-wedge) (figure 7.3c–e). The analogous structures in subduction zones would include the seaward-tapering accretionary wedge and the landward-verging retro-wedge separated by an outer-arc high. The taper angles of each side of the orogenic

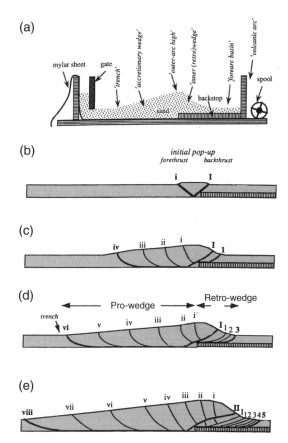

Figure 7.3 Analog, sandbox model of Wang and Davis (1996) for doubly vergent wedge growth consistent with boundary conditions described in figure 7.2. (a) Model setup consisting of a mylar sheet pulled under an initially constant thickness layer of sand. Rheology of sand is often approximated by rigid-plastic behavior with a Coulomb yield criterion. (b)–(e) Sequence of growth of sand wedge with progressive shortening. Note initial symmetric pop-up and subsequent asymmetric development of minimum-taper prowedge and steeper, retrowedge.

wedge in this experiment are explained by critical Coulomb wedge theory. The surface slope of the narrow-taper pro-wedge reaches a constant value at the minimum angle at which plastic failure occurs within the wedge interior and on the basal detachment. Once developed, this surface slope is maintained over the pro-wedge as the wedge grows by convergence and accretion of new material. The retro-wedge that forms by material backthrusting the rigid basal plate is considerably steeper than the minimum critical taper angle. It is difficult to measure the surface slope because of its limited extent, but the surface slope seems to be at or just below the maximum critical-taper angle that represents the steepest slope that can be maintained with the wedge interior and base both at failure (Dahlen 1984). However, as this steep wedge grows and propagates over the rigid backstop, it begins to deform the flat layer of sand that initially overlies the backstop. This deforming material attains the minimum taper angle so that the retro-wedge has two segments, a steeper one just "landward" of the outer-arc high and a lower-angle segment between this and the landward-verging retro-deformation front.

Buck and Soukoutis (1994) presented results from a similar analog experiment (figure 7.4). The boundary conditions in their experiment were identical to those described previously; the important distinction of their model is that the deforming material has a linear viscous rheology. As such, it provides a valuable contrast with the sandbox experiments and their Coulomb plastic rheology. As with the sandbox experiments, a linear viscous material forms a doubly tapered wedge of deforming material (figure 7.4). The deforming region is not defined as distinctly as in the sandbox experiments because of the lack of localization in the viscous fluid. An interesting effect occurs in the viscous wedge that is not observed in the sand models. Once the wedge reaches a characteristic height, the upper surface begins to extend horizontally, in the direction of convergence. This is evident in

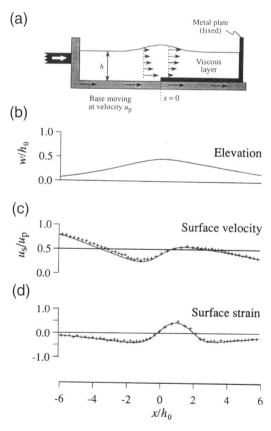

Figure 7.4 Analog, viscous fluid model of Buck and Sokoutis (1994). (a) Model setup of initially constant thickness fluid layer above moving base, consistent with boundary conditions of figure 7.2. Material is linearly viscous fluid characterized by a constant viscosity. Surface elevation (b), velocity (c), and strain (d) are as shown. Note horizontal extension in both strain rate (slope of velocity) and strain (positive value).

the surface velocity profile in figure 7.4 in which a positive slope indicates horizontal extension. There are important geologic implications to this result in that such extension in the high-elevation regions of convergent orogens would result in extensional structures developing as the orogen is undergoing net contraction and convergence. The fact that no comparable extension was observed in the sandbox experiments suggests a rheologic control on this phenomenon.

Parameterization, Scaling of Orogenic Wedges

The problem of formation of a doubly vergent orogenic wedge by imposition of the boundary conditions of figure 7.2 is easily addressed by using the finite-element method for a variety of rheologies (Willett 1999b). It is convenient to express the problem in nondimensional terms, using the intitial layer thickness as a scaling length (figure 7.2). The natural scaling time for the problem comes from the convergence velocity, v_p, so that nondimensional length and time are expressed as: $x^* = x/h$, and $t^* = v_p t/h$, respectively. This nondimensional time reflects total convergence of the system so that a nondimensional time of 1 represents convergence of an amount equal to the layer thickness, h. Using this scaling, the physical parameters combine to produce the Argand number (England and McKenzie 1982; Buck and Sokoutis 1994):

$$Ar = \frac{\rho g h^2}{\mu v_p} \qquad (7.12)$$

with parameters as defined in equation (1) and figure 7.2. England and McKenzie (1982) argued that Ar represented a ratio between gravitational stresses and viscous stresses. Medvedev (2002) demonstrated that in wedge problems the appropriate viscous stress in a scale analysis is the basal shear stress and thus proposes that this nondimensional group be referred to as the Ramberg number to differentiate it from England and McKenzie's (1982) usage in terms of a horizontal compressive stress. In this chapter, we do not interpret the Argand number in terms of stress ratios; it is simply the nondimensional group relevant to this form of the equation of motion and, therefore, we retain the original terminology. For the special cases of nonlinear, power-law viscous flow or plastic deformation, equation (7.5) or (7.8) is used for the viscosity in equation (7.12), respectively. The plastic case is interesting in that if one considers the characteristic strain rate to be V_p/h, the Argand number reduces to a ratio of the gravitational stress, $\rho g h$ and the yield stress:

$$Ar = \frac{2\rho g h}{\sigma^Y} \qquad (7.13)$$

The plastic Argand number (equation 7.13) is independent of velocity, consistent with the rate-independent plastic theory used here. The Argand number is the primary parameter in these mechanical problems. In fact, once boundary conditions are specified as in figure 7.2, for a homogeneous deforming layer, the Argand number is the only physical parameter. As such, it will be used to characterize all models in this section. For the more realistic, geologic case of temperature varying in space, Ar is itself a spatially varying parameter. For example, with a linear and temperature-dependent viscosity (equation 7.5 with $n = 1$), and a constant geothermal gradient, viscosity will vary exponentially with depth and so will Ar. In this case, there is not a constant Argand number to characterize a model system.

Finite-Element Simulation of Sandbox Models

Simulation of a sandbox analog model of a doubly vergent wedge provides an important test of the numerical methods outlined previously (Willett 1999b). Sand is a complex material with strain-hardening and -softening behavior depending on its state of consolidation, but in its rheologic behavior, it is frequently approximated by a Coulomb plastic material. Given that analytical solutions exist for wedge geometry through Coulomb wedge theory, it is important that numerical codes be able to satisfactorily address this problem. The Coulomb plastic rheology is highly nonlinear and strongly localizing and thus difficult to simulate with a continuum, finite-element model. To simulate most sandbox models it is necessary to include two physical parameters, the internal angle of friction of the material (sand), ϕ, and the angle of friction between the sand and the underlying base, ϕ_b. The basal friction must be less than the internal friction; otherwise, the material simply fails internally and does not slide along the base. Model results are thus expressed in terms of these two parameters and nondimensional length (in two dimensions) and time. Results for a simulation with $\phi = 25°$ and $\phi_b = 5°$ are shown in figure 7.5. Total deformation is shown by the Lagrangian mesh and the instantaneous deformation is given by the color-contoured values of strain rate. The horizontal, longitudinal component of rate of deformation is indicated to emphasize horizontal extension, as discussed later. Other components, or the second invariant of the rate of deformation, would show a similar spatial pattern.

The instantaneous strain rate provides a clear picture of the progression of deformation with time (figure 7.5a–c). Deformation is initially localized above the velocity discontinuity, as occurs in the analog model (figure 7.3b). With time, the deformation zone expands in both directions to form a doubly vergent wedge. Deformation is localized into zones that are the continuum equivalent of faults. The shear zones delineating the limits of deformation are denoted pro- and retro-shear zones on the corresponding side of the orogen (figure 7.5a). Shear zones have a strong component of simple shear with a sense of shear consistent with reverse or thrust faulting. Vergence of these structures is dominantly toward the "foreland" toe of the wedge, although there are backthrusts that can have strain rates of nearly the same magnitude. Comparison with the sandbox model (figure 7.3) indicates similarities and differences between numerical and analog models. Both models show the same pattern of outwardly vergent structures propagating in both directions. However, the numerical model cannot resolve discrete zones of slip and thus produces more diffuse deformation. The shear zones that do de-

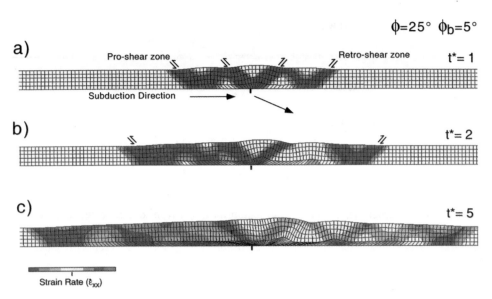

Figure 7.5 Finite-element model of deformation of a uniform layer with a plastic rheology and a Coulomb yield stress. Boundary conditions as in figure 7.2. Material properties are set to simulate sand over a weaker base with internal angle of friction of 25° and basal angle of friction of 5°. Longitudinal component of strain rate in the horizontal direction is shown along with the deformed Lagrangian mesh.

velop in the numerical model have some important differences from the discrete faults in that they do not develop at the angle predicted by Anderson fault theory and Coulomb failure mechanics. They form at 45° to the principle stresses, consistent with the viscous constitutive law in which principal directions of stress and strain rate are the same.

Even though the individual shear zones do not accurately reproduce the faults observed in the sandbox analog model, the overall geometry of the sand wedge is simulated well by the numerical model. The variable taper angles of the pro- and retrowedges are both reproduced and match the analytical Coulomb wedge solutions (Willett 1999b).

The internal strain as predicted by the finite-element model is a continuous representation of what would be slip on a discrete system of faults in a geologic material such as sand. Strain in the numerical model is calculated by the Lagrangian mesh as shown in figure 7.5. The Lagrangian mesh is highly distorted. In contrast, the Eulerian mesh used for the FE calculations retains its roughly rectangular form (figure 7.6). Note that the weak basal detachment in the Coulomb wedge simulation is included as several rows of thin, mechanically weak elements along the base of the model (figure 7.6a). The localization of shear in this layer is also apparent, though not well resolved by the Lagrangian mesh (figure 7.6b).

The important question of whether horizontal extension is present in the Coulomb wedge can be addressed by noting the surface strain. In figure 7.7 the horizontal component of the surface strain and strain rate are shown for the end

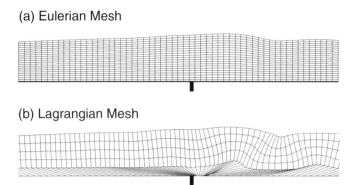

Figure 7.6 Eulerian mesh (a) and Lagrangian mesh (b) from one timestep of model of figure 7.5. Eulerian mesh is distorted only to match free surface. Lagrangian mesh tracks actual material points. Note high basal shear in lower row of elements reflecting shear on base.

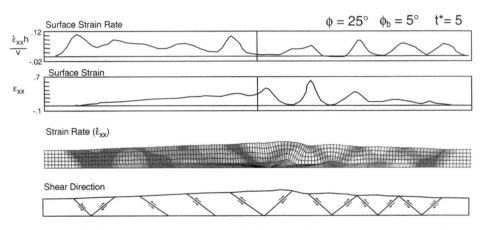

Figure 7.7 Horizontal component of longitudinal strain and strain rate of surface of model in figure 7.5. Results shown after nondimensional time of 5. Bottom frame shows sense of shear localizations. Note that no extension occurs on surface of model.

condition of the model of figure 7.5. Strain rate is normalized by the layer thickness and convergence velocity, but both are calculated as longitudinal shortening in the x direction, so that positive values indicate contraction in the x direction and negative values extension. It is clear that although surface strain and strain rate are highly variable in space, there is no extension in the model. The variations in strain rate have an almost periodic nature, reflecting the spatial scale of the plastic shear zones. The net strain is also influenced by this spatial variation, but the motion of material through the deformation field tends to produce a smoother strain field.

Finite-Element Simulation of a Viscous Wedge

Deformation of a linear viscous material with the boundary conditions of figure 7.2 provides a rheologic contrast to the Coulomb plastic model of the preceding section. Such a model is directly comparable with the analog model of Buck and Soukoutis (1994) and the analytical solution of Royden (1996). The nondimensional form of this problem is characterized by the Argand number (Ar) (equation 7.12) and nondimensional time (t^*). A model with $Ar = 10$ and $t^* = 5$ is shown in figure 7.8. As in the model with a plastic rheology, thickening is localized at the basal velocity discontinuity and a doubly vergent wedge structure develops. However, there are important differences in the strain field. First, there is no strain localization in shear zones; strain is more uniformly distributed through the wedge. Second, there is a zone of horizontal extension in the wedge interior at high elevation, indicated by the red colors in figure 7.8. This region of extension is also clear in the surface deformation indicated by the large negative values of both strain and strain rate. In fact, the stretching strain is locally larger than any contractional strain. The region of extension is limited in its depth extent, being constrained to the upper two thirds of the deforming layer indicating that it is a consequence of the gravitational forces and the free surface.

It is also important to ask whether horizontal extension is a function of the viscous rheology or if it exists only for a limited range of values of the viscosity. Figure 7.9 shows a model result for an Argand number of 1.0, corresponding to an order-of-magnitude increase in the viscosity relative to the model of figure 7.8. With the smaller Argand number, the orogenic wedge is thicker and steeper. The lower Argand number does reduce the surface extension, but it does not eliminate it; a significant extension zone is still present. Although not shown here, models with a higher Argand number (lower bulk viscosity) favor extension and the region of horizontal extension is larger and extends at higher rates (Willett 1999a).

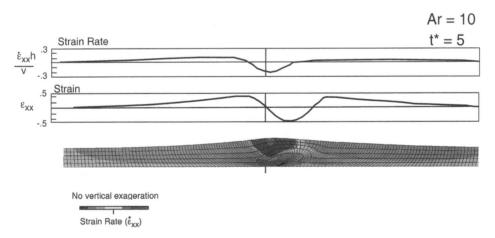

Figure 7.8 Finite-element model of deformation of uniform layer with a linear viscous rheology and an Argand number of 10. Boundary conditions as in figure 7.2. Strain-rate field, Lagrangian mesh and surface strain, and strain rate shown after nondimensional time of 5. Note region of horizontal extension in both surface plots (negative values) and strain-rate field (orange and red colors).

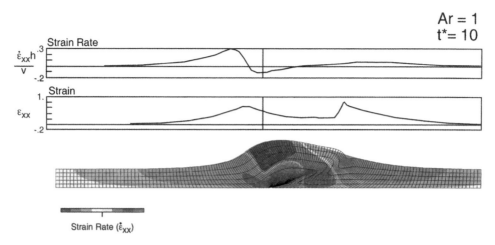

Figure 7.9 Finite-element model of deformation of uniform layer with a linear viscous rheology and an Argand number of 1.0. Boundary conditions as in figure 7.2. Strain-rate field, Lagrangian mesh and surface strain, and strain rate shown after nondimensional time of 10. Note region of horizontal extension in surface strain rate (negative values) and strain-rate field (orange and red colors), but no net surface extension.

Pressure- and Temperature-Dependent Wedge

In a natural orogenic setting, crustal rocks deform by both frictional and viscous deformational mechanisms depending on pressure, temperature, and stress conditions. In general, brittle, plastic deformational mechanisms dominate at the low pressures and temperatures of the upper crust, whereas the temperature dependence of viscous mechanisms supports their dominance at deeper crustal levels. Assuming a Coulomb plastic yield strength for the upper crust, a viscous constitutive law of the form of equation (7.5), and a geotherm permits calculation of a yield-stress envelope to describe the strength of the crust (figure 7.10). Note that "strength" is not defined for a viscous material, so in this case, it is defined to be the differential stress necessary to maintain a constant strain rate.

Multiple deformational mechanisms are easily incorporated into the finite-element model. The temperature-dependent viscous mechanism is included simply by assigning a viscosity to each element depending on its temperature as calculated through an assigned geotherm. This geotherm remains constant in a spatial reference frame, which is equivalent to an assumption of thermal steady state. This viscosity is used as the starting value for the iterative process to determine an effective viscosity (equation 7.8) corresponding to the plastic failure criterion if needed. If stresses remain below the failure criterion, the temperature-dependent viscosity is used. Otherwise, the effective viscosity is found by direct iteration. Models in this section use a linear viscosity ($n = 1$ in equation 7.5), but the nonlinear case, which is not shown here, can be considered by including calculation of viscosity by direct iteration. In either case, the result is an orogenic wedge that tends to deform plastically at low temperature and pressure and viscously at depth, although the deformational mechanism for each element is not prescribed a priori.

A model wedge including this rheology is shown in figure 7.11. In addition to the multiple deformational mechanisms, crustal thickening is also locally isostatically compensated. The deformation field shows the characteristics of both deformational mechanisms. The thermally activated viscous flow in the lower crust leads to distributed shear along the base of the crust and widening of the orogenic wedge. The flattening of the topographic slope in the wedge interior is also a response to deep crustal flow. Plastic deformation leads to strain focusing at each end of the wedge. There is no horizontal extension in this model, suggesting that

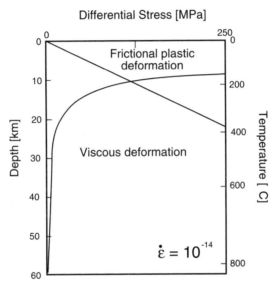

Figure 7.10 Yield-stress envelope for continental crust representing pressure-dependent frictional deformational mechanism with effective angle of friction of 25° and temperature-dependent viscous deformation at depth based on the quartz flow law of Kronenburg and Tullis (1984). These are the physical parameters used in the model of figure 7.13.

Figure 7.11 Finite-element model of deformation of a crustal layer with a rheologic model that includes frictional plastic and linear viscous deformational mechanisms as in figure 7.10. Boundary conditions as in figure 7.2. Strain-rate field, Lagrangian mesh are shown after nondimensional time of 10. Thickening is locally compensated. There is no horizontal extension observed in strain or strain rate.

the plastic deformation in the upper crust inhibits the tendency for the viscous wedge to spread by extension.

Discussion

The rheologic control on the modes of deformation in an orogenic wedge is evident in the contrasts among the models of figures 7.5 and 7.8 or 7.9. The physical parameters for any specific constitutive law affect the deformation patterns in predictable ways. The internal and basal friction angles in the Coulomb models affect taper angle as predicted by critical wedge theory (Davis et al. 1983), and growth at different taper angles affects the distribution of strain. In the viscous models, viscosity expressed through the Argand number affects the overall shape of a wedge and the distribution of deformation that produces that shape. However, there are also important differences in wedge deformation that reflect the constitutive relationship itself. Primary among these are the strain localization inherent to a plastic rheology and the extensional flow field of a viscous wedge.

An extensional flow field has important consequences for the deformation history of rocks in a natural orogenic system. As material is accreted into the deforming wedge from the left (pro-) side of the system it would initially undergo contraction, but as it moves to the right (retro-) side in figure 7.8 with time, a material point moves into the region of extension. The deformational path of this material point would thus exhibit a complex, polyphase deformational history with phases of both contraction and extension.

The existence of normal faults is the evidence often cited for extension in convergent settings (Crespi et al. 1996). This is somewhat paradoxical because the mechanical models suggest that brittle plastic deformation should inhibit extension by this mechanism. One possibility is that the plastic crustal layer is thin and confined to shallow levels of the crust and that the strain field is dominated by the viscous deformation of the deeper crust. The model of figure 7.11 suggests that a Coulomb plastic "lid" to the crust will inhibit extension. However, it is possible that for a weaker or thinner plastic layer, the viscous deformation could dominate and produce extension by this mechanism.

Formation of Orogenic Plateaus

Geologic Problem

Large, mature collisional orogens are frequently characterized by high-elevation, low-relief surfaces of large ($>10^4$ km^2) extent. The modern examples of these orogenic plateaus are the Tibetan plateau associated with the Asia-India collision (Molnar and Tapponnier 1975; Dewey et al. 1988; Nelson et al. 1996; Owens and Zandt 1997) and the Altiplano-Puna plateau of the central Andes (Isacks 1988; Beck et al. 1996). In each of these cases the plateau is characterized by thick (65–80 km) continental crust. This crust may have been thickened in part by the addition of magmatic products, but it appears to be primarily the result of structural thickening (Isacks 1988; Dewey et al. 1988; LePichon et al. 1992). This suggests that orogenic plateaus form as the culmination of convergent processes similar to those presented previously. The early stage of orogenic crustal thickening was modeled in the previous section as an accreting orogenic wedge (figure 7.11). However, these models, as formulated, provide no explanation for the existence of a morphological plateau; continued convergence leads to approximately self-similar growth of either plastic or viscous wedges. With a homogeneous material, there is no limit to the crustal thickness that can develop provided it maintains a wedge geometry, nor is there any reason for regional slopes to decrease and form a plateau. This is not the case, however, for temperature-dependent rheologies as described, for example, by equation (7.5). If viscosity decreases with temperature and temperature increases with depth, as the crust thickens the lower crust will heat and its viscosity will decrease; the lower viscosity will lead to lower surface slope, particularly where the crust is thickest, thereby producing a thick-crust, high-elevation, low-slope plateau (Willett et al. 1993; Willett and Beaumont 1994; Royden 1996; Beaumont et al. 2001).

This process of crustal thickening, heating, and weakening represents a coupled dynamic system where the thermal and mechanical processes are linked. Crustal thickening can lead to cooling of crustal rocks by underthrusting of cold material to depth and by the associated erosion that brings material toward the surface. However, the primary, long-term effect is one of heating, as crustal material is taken to greater depths where long-term, equilibrium temperatures are higher. Radioactive heat production in the crust also leads to heating if the thickened crust contains a greater integrated heat production. Thicker crust thus leads to higher temperatures and lower viscosity in the lower crust. Lower viscosity, however, implies lower surface slope (compare figures 7.8 and 7.9) and thus, as the viscosity decreases, the rate of crustal thickening and heating decrease. Furthermore, if the viscosity becomes low enough, existing topography cannot be supported and the orogen thins, either by collapse (Willett et al. 1993) or by extrusion of lower crust (Bird 1991; Royden 1996; Beaumont et al. 2001). Crustal thinning leads to lower crustal temperatures, higher crustal viscosity, and mechanical stability. This feedback provides damping to the dynamic system so that a long-term stable state of near-constant crustal thickness is expected.

Numerical Model Parameterization

The stable crustal thickness that produces an orogenic plateau and the timescales over which it develops is a function of (1) the rates and processes of lithospheric thickening, (2) the rates and processes of thermal evolution of the lithosphere, and (3) the rheologic constitutive relationships and physical parameters. The thermal processes introduce a timescale to the problem, independent of the convergence velocity, reflecting the rates of thermal equilibration of thickened crust. It is thus necessary to include the transient thermal evolution in numerical models of the orogen development.

The processes of lithospheric thickening are also important to the plateau formation problem and must be dealt with explicitly. In particular, at this, the largest scale of convergent orogenesis, the mechanism by which the lithospheric mantle accommodates convergence is unclear. In the orogenic wedge models the substrate is assumed to remain undeformed and subduct, at least to depths where it does not affect the mechanics of the crustal wedge. This is justified for small orogens associated with oceanic plate subduction or for the periphery of larger orogenic belts where deformation is dominated by underthrusting of one plate beneath the orogen. However, at the larger scale of a full continent-continent collision, the fate of the deeper lithosphere (mantle and lower crust) remains ambiguous. In a full thermo-mechanical problem the motion of the lithospheric mantle deformation is important both to crustal deformation through the induced stresses and to thermal equilibration because the motion of the lithospheric mantle advects heat and thus affects the temperature in the crust.

We consider three models of deformation by which the lithosphere can accommodate convergence (figure 7.12). The first (figure 7.12a) involves subduction of an oceanic plate with shortening of the overriding, continental plate. To accommodate the shortening of the upper plate, some of the lithospheric mantle is dragged down into the deeper mantle with the subducting slab in a process Tao and O'Connell (1992) termed ablative subduction. With lithospheric mantle removed, the continental crust will thicken. This is the setting in which the Andes form. In the central Andes, where convergence and crustal thickening are largest, heating of the lower crust has led to the formation of the Altiplano plateau (Pope and Willett 1998). The second model of lithospheric thickening investigated here is an extension of the wedge models of the previous section (figure 7.12b). Continuous subduction of the deeper lithosphere (mantle and lower crust) with offscraping and accretion of the upper crust initially produces an orogenic wedge; with increased mantle subduction the wedge will grow to the point that lower crustal temperatures are high, viscosity is low, and a plateau forms. The third model for lithospheric deformation assumes that the lithospheric mantle thickens uniformly in response to continental collision (figure 7.12c). In this model the crust thickens in response to the uniform strain of the deforming lithospheric mantle.

Each of these models for lithospheric deformation provides an independent prediction for the stresses and heat applied to the crust. In the models presented here, the focus will be on the crustal response. The lithospheric mantle will be treated kinematically. Deformation will not be predicted for the mantle; its motion will be assumed and these motions will be transmitted into the crust, where the deformation will be calculated dynamically by the finite-element model. No simple assumption for temperature can be made and it is, in part, through heat advection that differences in the mantle kinematics manifest themselves in crustal deformation. It is thus important that the full, time-dependent temperature field be calculated, not just in the crust, but through the entire lithosphere.

The temperature is also calculated using a finite-element method with a combination of dynamic and kinematic velocities used for the advective term. Temperature is determined by solving the advection-diffusion equation:

$$\frac{\partial}{\partial x_i}\left(k\frac{\partial T}{\partial x_i}\right) - \rho c v_i \frac{\partial T}{\partial x_i} + A = \rho c \frac{\partial T}{\partial t}, \tag{7.14}$$

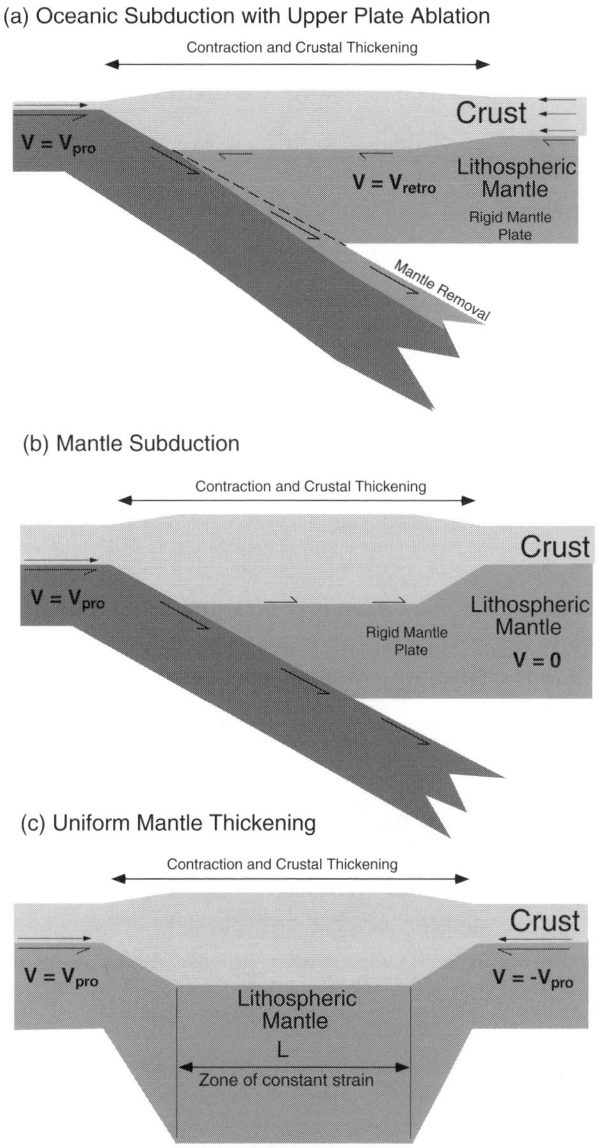

(a) Oceanic Subduction with Upper Plate Ablation

Contraction and Crustal Thickening

Crust

V = V_pro

V = V_retro

Lithospheric Mantle

Rigid Mantle Plate

Mantle Removal

(b) Mantle Subduction

Contraction and Crustal Thickening

Crust

V = V_pro

Lithospheric Mantle

Rigid Mantle Plate

V = 0

(c) Uniform Mantle Thickening

Contraction and Crustal Thickening

Crust

V = V_pro

V = -V_pro

Lithospheric Mantle

L

Zone of constant strain

Figure 7.12 Modes of lithospheric convergence leading to orogenic plateau formation as investigated here. (a) Lithospheric mantle is entrained with subducting oceanic plate, ablating continental plate, and leading to shortening of overlying crust. (b) Lithospheric mantle of one continental plate is subducted beneath the other. Crust is accreted. (c) Lithospheric mantle is shortened uniformly over a zone of width L. Overlying crust shortens in response.

where T is temperature, k is thermal conductivity, ρ, c, and A are the density, specific heat, and heat production of the medium, respectively, and v_i is the velocity of the medium in the x_i direction. In nondimensional form, this becomes

$$\frac{1}{Pe} \cdot \frac{\partial^2 T}{\partial x_i^2} - \frac{v_i}{V_c} \frac{\partial T}{\partial x_i} + N_Q = \frac{\partial T}{\partial t} \qquad (7.15)$$

where the variables are normalized by the characteristic length of the system, h, and the boundary velocity, V_c. The thermal parameters appear as two nondimensional groups, Pe, the Peclet number, defined as

$$Pe = \frac{V_c h \rho c}{k}, \qquad (7.16)$$

and, N_Q, a nondimensional heat production or Damkohler number:

$$N_Q = \frac{Ah}{V_c \rho c}. \qquad (7.17)$$

This differential equation with its boundary conditions is solved by a finite-element method using linear basis functions defined over elements in the Eulerian mesh. In this problem, the Eulerian mesh and the associated velocity field are larger than that used for the mechanical problem because both crust and mantle must be included. The Eulerian mechanical mesh which covers the crust is included as a subset of the thermal mesh, but the thermal mesh also includes the lithospheric mantle. The motion and deformation of the mantle is described kinematically, consistent with the specific model (figure 7.12). Each of the models in figure 7.12 is simple enough to construct an analytic velocity representation and this representation is used to construct the advective component of the heat-transfer problem in the mantle. Thermal boundary conditions consist of an isothermal upper surface and a constant-flux basal surface. The region of the model that we consider as-thenosphere is held to near isothermal conditions by assigning elements a high thermal conductivity.

Equations (7.1), (7.4), and (7.15) define a coupled thermo-mechanical system. The coupling is in both directions because the temperature depends on advection of heat and the deformation is temperature dependent through the rheology. The crustal structure of the resultant orogen thus depends on both the thermal processes and the rheologic parameters. As in the orogenic wedge models of the preceding section, the mechanical processes are characterized by an Argand number. In the thermo-mechanical models there are two additional thermal parameters, the Peclet number and the Damkohler number representing heat production. With temperature dependence to the viscous rheology, Ar is a function of temperature and hence space and time, thereby losing much of its usefulness as a scalar characteristic of the system. However, quantities such as the maximum Ar in the crust or the change of

Ar with time reflect changes in temperature or strength of the crust and will be used as a characteristic metric. Coulomb plastic deformation of the crust is not temperature dependent and therefore retains the same definition of *Ar* (equation. 7.13).

Isostatic compensation of the thickened crust is included in these models. Because elastic deformation has been neglected in the mechanical model, elastic plate flexure must be included explicitly. This is done by calculating the weight of the thickened crust and applying it as a load on an elastic plate supported by an inviscid half-space. The deflection of the plate is calculated analytically and the entire model is deflected accordingly.

To summarize the model, there are three mechanical components: (1) the lithospheric mantle for which a complete velocity field is assumed, (2) a crustal layer for which a rheologic model is assumed and deformation is calculated dynamically, and (3) an isostatic compensation model that consists of a specified flexural rigidity and density structure for the crust and mantle. Temperature is calculated for the entire lithosphere using the velocity field taken from the combination of mantle kinematics, compensation deflection, and dynamic velocity of the crust. Deformation (strain and velocity) is calculated dynamically for the crust using the temperature field for the entire system and velocity boundary conditions taken from the mantle kinematic model.

In the series of models that follow, we use this coupled model to address several questions. First, how do orogenic plateaus develop by crustal thickening and heating processes? Second, what are the rheologic implications and parameter constraints required by the existence of plateaus? Third, how do the rates of heat diffusion affect the timescale over which plateaus develop? Fourth, can we distinguish between different models for mantle kinematics during convergence from crustal structure of orogens that include plateaus? These questions are addressed by a series of models presented next that investigate the thermo-mechanical evolution of crustal orogens under the boundary conditions represented by the three scenarios of figure 7.12.

Ablative Subduction Model

In a tectonic setting with subduction of an oceanic plate, shortening of the crust of the overriding plate due to accretion of material from the subducting plate is minimal. Significant deformation is more likely caused by shortening of the entire upper plate. This can either be accommodated by whole-lithosphere shortening (a case considered later in the chapter) or it can be accommodated by removal of the lithospheric mantle. One means of doing the latter is by entrainment or ablation of the upper plate into the downwelling flow represented by the subduction (Tao and O'Connell 1992) (figure 7.12a). This process is likely to occur where tractions between the subducting slab and the overriding plate are large and is attractive as a shortening mechanism in settings such as the Andes where the relatively shallow angle of subduction presents space problems for whole-lithosphere thickening (Pope and Willett 1998). If the continental crust detaches from the entrained mantle, the crust will be progressively shortened as the plate boundary moves toward the upper-plate interior. A consequence of this kinematic scenario is that the convergent orogenesis will be confined to the crust of the upper plate and the orogen propagates toward the continent interior, progressively accreting crust from the upper plate. This is the situation in the Andes of South America, where the Sub-Andean fold-and-thrust belt is actively shortening the South American crust and is consuming the foreland as the deformation propagates east relative to the stable continent.

A model of this process of crustal growth with ablative subduction is shown in figure 7.13. Results consist of the temperature field as a function of time and the crustal thickness; the internal deformation determined from the Lagrangian mesh is not shown. The initial and boundary conditions include a mantle velocity field representing subduction of an oceanic plate at 67 mm/yr with a slab dip of 30°, values that are characteristic of the modern central Andes (Isacks 1988). This velocity field imposes tractions on the base of the crust that provide the boundary conditions for the mechanical model. The thermal model extends into the upper mantle covering the entire area shown in the figure; the mechanical mesh is confined to the crust, the base of which is indicated by a bold line that initially mirrors the upper surface. The model in figure 7.13 is designed to demonstrate the effects of ablative removal of mantle from the upper plate, but it is important to start the process with the thermal conditions appropriate for a subduction zone. The initial conditions thus include no ablation of the upper plate; ablation is initiated after steady thermal conditions are established. Without ablation, the velocity field of the mantle drives only a small amount of deformation at the plate boundary as a result of accretion of the sedimentary cover of the oceanic plate. The thermal field, however, is highly perturbed by the heat advection of the downgoing slab. After 8 m.y. of subduction, the thermal field is near steady state. At this point in time, ablation is initialized by imposing a trenchward motion to the mantle of the upper plate. Mass is balanced at the plate boundary by subducting the upper-plate material that reaches the slab. This is indicated by the two bold lines defining the top of the slab. The lower line represents the top of the oceanic plate; the area between the two lines represents material removed from the upper plate. With reference to figure 7.2, ablation corresponds to motion of the upper plate mantle toward the point S at a velocity of V_R. In figure 7.13, the ablation velocity, V_R, is imposed at 18 mm/yr. The trenchward motion of the upper plate carries the continental crust toward the plate boundary. As the lithospheric mantle detaches and descends, this crust shortens and thickens (figure 7.13b). The region of thickened crust grows outward with a wedge geometry, dependent on the plastic and viscous properties of the crust. As the crustal wedge thickens the lower crust becomes hotter, both because of the increased insulation of the overlying thick crust, but also because of the internal, radiogenic heat production. The crust has a heat production of 1.5 mW/m^3 in this model and as the crust thickens, this heat production plays an important role in the heating. Note that in figure 7.13c there is actually a temperature inversion at the base of the crust such that the temperature of the lower crust is higher than that of the underlying mantle. This is the result of both heat production in the crust and the underthrusting of the relatively cold mantle.

(a) Oceanic plate subduction

(b) Subduction with upper plate ablation

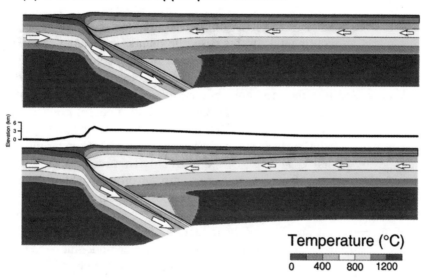

Temperature (°C)

0   400   800   1200

Figure 7.13 Finite-element model of coupled temperature and deformation of lithosphere at an ablative subduction boundary as described in figure 7.12a. Dimensional results shown for initial continental crustal thickness of 30 km, subduction rate of 67 km/Ma, and ablation rate of 18 km/Ma. (a) Initial condition given as near- steady-state thermal model with no ablation. (b) Model following 20 and 30 Ma of ablative convergence. Note crustal heating from heat production and widening of orogen as lower crust weakens. See text for additional details.

The mechanical consequence of this heating is weakening of the crust and decrease in the surface of the orogenic wedge. The weakened wedge cannot maintain a steep surface slope and thus exhibits a low-relief, high-elevation (plateau) surface. In this case (figure 7.13) the surface has a slight slope dipping to the right in response to the underthrusting mantle; depending on the rheologic and isostatic models, a more distinct plateau can develop as in subsequent models, discussed in following sections.

The time of development of the plateau is dictated by the rate of ablation and by thermal diffusion of the internal heating. The ratio of these parameters determines the characteristic Peclet number for the system. The characteristic diffusion time for this problem is determined by the crustal thickness, not the lithospheric thickness, because the mechanical strength is limited by the viscosity of the lower crust. Once the lower crust heats sufficiently, the orogenic wedge loses its strength and the plateau begins to form, thereby limiting any additional heating. Under most geologic conditions, this occurs on the order of 1 to 10 m.y. The model of figure 7.13 was constructed for comparison with the Andes so that the subduction velocity is 67 mm/yr, the ablation velocity is 18 mm/yr, and the upper-plate crustal thickness is 30 km (Pope and Willett 1998). Under these conditions, development of the plateau to the stage represented by figure 13c takes 30 m.y.

Continental Subduction Model

Orogenic plateaus also develop in regions of continental collision where both bounding plates are continental. In this case, the mantle kinematics are likely to be complex and difficult to establish. In whatever manner convergence is accommodated in the lithospheric mantle, the consequence for the crust is shortening and thickening, and it is the mechanical consequence and rheologic dependence of that process that we investigate here. One potential mode of convergence in collision zones is detachment and subduction of the lithospheric mantle such that the crust remains at the surface and is shortened above the converging plates (Mattauer 1986; Willett et al. 1993) (figure 7.12b). This mode of convergence leads to rapid accretion of continental crust into an orogen, which develops a deforming wedge geometry in response to the opposing basal tractions applied by the converging mantle plates. This is the problem addressed by the orogenic wedge models of the preceding section and the model of figure 7.11. The model of figure 7.11 represents an early stage of development of an orogen under these conditions.

Continued convergence and crustal thickening with a temperature-dependent rheology leads to weakening of the lower crust and a decrease in the surface slope. This process is shown in the model of figure 7.14. This model differs from the model of figure 7.11 in that temperature is time dependent in the latter model; otherwise, rheologic parameters are the same, and figure 7.11 is a representation of the early development of this model. As in figure 7.13, the subducting slab, in this case the lithospheric mantle, has a strong thermal effect, carrying low-temperature material into the deeper mantle reflecting the Pe of 50. The cooling of the crust due to this process is limited to the region near the detachment point. With the large convergence applied in this model, the crustal orogen has grown so large that much of its thermal field is not influenced by the downgoing slab. Convergence in this model is very large with a nondimensional time of more than 100. For an accreted crustal layer of 30 km, this corresponds to greater than 3,000 km of convergence. The temperature in the crust is dominated by heating due to thickening, including thickening of the more radioactive crust. The result of this heating is weakening of the lower crust to the point that the surface of the orogenic "wedge" becomes nearly flat above this weak lower crust (figure 7.14b,c). The elevation of the surface is shown with vertical exaggeration in figure 14c to show how flat the plateau becomes. Note that once the plateau begins forming (figure 7.14a), the maximum crustal thickness does not increase; the orogen grows outward by widening of the plateau, while maintaining a near-constant crustal thickness.

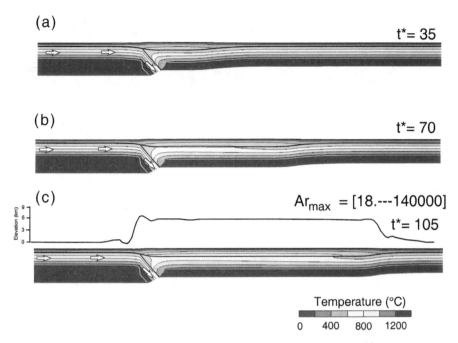

Figure 7.14 Finite-element model of coupled temperature and deformation of lithosphere for model with mantle subduction and crustal accretion as described in figure 7.12b. Non-dimensional viscosity of the lower crust decreases with temperature so that maximum *Ar* increases from 18 at time 0 to 140,000 at nondimensional time of 105. Dimensional results shown for initial continental crustal thickness of 30 km and a subduction rate of 50 km/Ma. A crustal plateau with an elevation just under 6 km has developed in (a) and widens with additional convergence. Topography is shown vertically exaggerated in (c). See text for additional details.

The viscosity of the lower crust decreases markedly during the course of the model run. Reflecting this viscosity, *Ar* varies through the crust, but has an initial maximum value of 18.0 in this model. This maximum value occurs at the base of the crustal layer where temperature is a maximum and viscosity has its minimum value. As temperature increases, the maximum value of *Ar* reaches a final value of 140,000, reflecting nearly four orders of magnitude decrease in the lower crustal viscosity.

The thickness of the crustal layer is more than doubled in this model. It should be noted that although we refer to this layer as the "crust," it need not represent the entire crust; it could represent a fraction of the continental crust with the remainder being subducted. The base of the accreted crustal layer is determined by the rheologic stratification of the crust. If the model results are dimensionalized by assuming a crustal thickness of 30 km, and a convergence velocity of 50 km/m.y., the total crustal thickness under the plateau is approximately 75 km, and the elevation of the compensated plateau is just under 6 km. These numbers are close to those observed for the Tibetan plateau (Owens and Zandt 1997), but as discussed in the following text they are a direct function of the assumed crustal rheology.

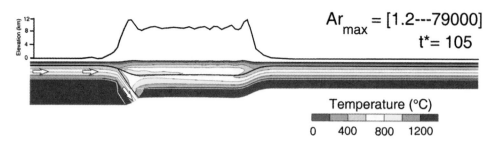

Figure 7.15 Finite-element model of coupled temperature and deformation of lithosphere for model with mantle subduction and crustal accretion as described in figure 7.12b. Non-dimensional viscosity of the lower crust increases with temperature so that maximum *Ar* increases from 1.2 at time 0 to 79,000 at nondimensional time of 105. Dimensional results shown for initial continental crustal thickness of 30 km and a subduction rate of 50 km/Ma. A crustal plateau with an elevation near 9 km has developed. Topography is shown vertically exaggerated. See text for additional details.

The topographic highs at the edge of the plateau are the result of flexural compensation. The edge of the plateau is supported by the elastic plate and thus is held at a slightly elevated height. The topographic lows on either side of the plateau are the flexural effect of the load of the plateau.

The stable thickness of the crust under the plateau is a direct consequence of the rheology of the crust. A stronger crust implies a thicker equilibrium crust under the plateau. This is demonstrated in the model of figure 7.15 in which the minimum viscosity of the crust is increased by an order of magnitude. The initial, maximum *Ar* in this model is 1.2. Other parameters, including the plastic yield stress, *Pe*, and the kinematic model for the mantle, remain the same. As in the previous model (figure 7.14), the crust thickens to a stable, near-constant value, but in this case the thickness is more than four times the original crustal thickness. For an initial crustal thickness of 25 km, this produces nearly 10 km of elevation. This is clearly more elevation than is observed in any modern orogenic systems, but it demonstrates the sensitivity of the crustal structure to the crustal rheology. If this model is regarded as being outside the geologically acceptable range of crustal thickness, it thus excludes the rheologic parameters that produced this result.

One feature that is observed in figure 7.15, but not figure 7.14, is the topographic undulations of the surface of the plateau. These are the result of strain localization in the brittle, upper layer of the crust. The plastic deformation localizes into discrete shear zones (e.g., figure 7.5) and these produce variable uplift.

Mantle Shortening Model

The primary alternative to the kinematic model of mantle subduction is uniform mantle shortening, in which a region of the lithospheric mantle is shortened and thickened (figure 7.12c). In this case, the width of the crustal shortening is largely determined by the width of the deforming mantle region. In the model considered here, the width of deforming mantle is fixed at L (figure 7.12c). Within this region there is a constant rate of pure-shear shortening and thickening. The thickening zone width, L, is taken to be constant with time. There is a transition zone on either side over which the strain rate increases linearly from zero. Using this kinematic model for the thermal calculation and as an imposed boundary condition at the base of the crust gives the model result of figure 7.16. The mantle strain field produces symmetric deformation and thermal fields. With constant convergence velocity and a constant width to the mantle shortening zone, a deep mantle root develops and grows monotonically. This mantle root has an associated thermal anomaly that, in contrast to the subduction models, is symmetric. The constant width of the mantle shortening zone dictates the width of the crustal thickening in the early development of the orogen. The crust thickens uniformly over the entire width, L (figure 7.16a,b). The upper surface thus rises uniformly, rather than building outward from its initiation at the point of subduction as in the previous

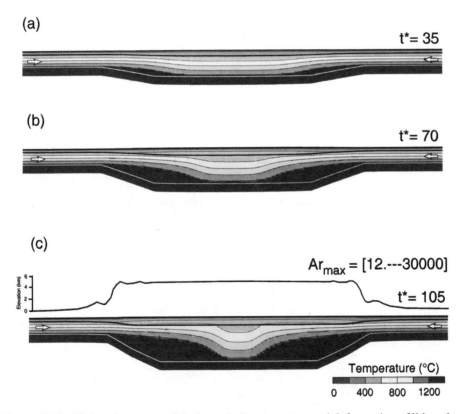

Figure 7.16 Finite-element model of coupled temperature and deformation of lithosphere for model with uniform mantle lithosphere shortening as described in figure 7.12c. Nondimensional viscosity of the lower crust increases with temperature so that maximum *Ar* increases from 12 at time 0 to 30,000 at nondimensional time of 105. Dimensional results shown for initial continental crustal thickness of 30 km and a subduction rate of 50 km/ Ma. A crustal plateau with an elevation near 5 km has developed. Topography is shown vertically exaggerated in (c). See text for additional details.

models (figures 7.14 and 7.15). However, the deformation and uplift pattern changes in the later stages of development. As the crust thickens and warms, eventually it reaches a maximum sustainable thickness, at which point it begins to grow outward beyond the mantle-thickening zone (figure 7.16c). This does not imply extension or outward flow of the preexisting plateau crust; the crust simply thickens by accretion into a bounding orogenic wedge that propagates out over the mantle that is moving in toward the thickening region. At termination of the model, the width of the plateau (figure 7.16c) is thus wider than the mantle-thickening zone, and were convergence to continue, the crustal orogen would continue to grow outward, rather than thicken. The decrease in crustal viscosity through the duration of the model is reflected in an increase in the maximum value of *Ar*, which increases from 12 to 30,000.

The topography in this model is nearly indistinguishable from the previous models. There is a high, flat plateau in the interior of the orogen with flexurally induced peaks at each end. The crust thickens by a factor of 2.6, so that a 30-km-thick crustal layer would thicken to 77 km and produce just under 6 km of relief. Elevation decreases rapidly away from the plateau across the orogenic wedge that bounds the plateau on each side. The contrast in mantle thermal structure between this model and the mantle-subduction models has little effect on the crustal deformation. Although there is local cooling near the subducting mantle slab (figures 7.14 and 7.15), most of the thermal structure of the crust is dictated by the crustal thickness and internal heat production. Once the crust heats enough that a plateau has formed, it effectively detaches from the underlying mantle, thereby decoupling the deformation in the crust from the deformation of the underlying mantle. The crustal deformation is thus largely the result of the gravitational forces associated with the thickened crust and the crust-mantle coupling outside of the plateau where the crust is thinner, cooler, and more strongly coupled to the underlying mantle.

The crustal structure and thickness depends on the strength of the crust. As with the mantle-subduction models, a stronger crust produces a thicker stable crustal thickness. This is demonstrated with the model of figure 7.17 in which the rheologic parameters are changed such that the crustal viscosity is higher than in the model of figure 7.16. Thermal parameters were changed to achieve this. The heat production in the crust was reduced by half and the Peclet number was decreased from 74 (figure 7.16) to 48 (figure. 7.17). Only the final configuration after a nondimensional convergence time of 105 is shown in figure 7.17. With the stronger rheology, the crustal thickness has increased to a thicker stable value under the plateau. In this case, the crustal layer has thickened by a factor of 3.5; for a 30-km-thick crustal layer, this corresponds to a crustal thickness of 105 km and a mean plateau elevation of greater than 9 km. At least by comparison with the Tibetan plateau, these values are too high, demonstrating the unreasonableness of these assumed rheologic parameters. The initial maximum *Ar* in this model was 0.4; it increased to a value of 23,000 at the model termination. This change of just over an order of magnitude in the initial maximum viscosity was thus sufficient to change an acceptable model into an unacceptable one.

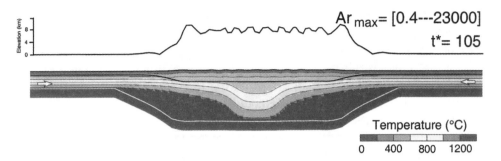

Figure 7.17 Finite-element model of coupled temperature and deformation of lithosphere for model with uniform mantle lithosphere shortening as described in figure 7.12c. Non-dimensional viscosity of the lower crust increases with temperature so that maximum *Ar* increases from 0.4 at time 0 to 23,000 at nondimensional time of 105. Dimensional results shown for initial continental crustal thickness of 30 km and a subduction rate of 50 km/Ma. A crustal plateau with an elevation greater than 8 km has developed. Topography is shown vertically exaggerated. Topographic undulations are due to plastic localization of shear. See text for additional details.

Discussion

The models in this section describe a specific mechanism for the formation of orogenic plateaus, a mechanism that is directly dependent on the rheology of the crust. In this model, the development of an orogenic plateau is a direct consequence of crustal thickening by convergence and the subsequent heating of the lower crust. Crustal thickening by convergence, shortening, and the formation of an orogenic crustal wedge inevitably leads to heating of the lower crust. This heating decreases the crustal viscosity to the point that topographic relief in the form of a regional surface slope cannot be maintained and a flat plateau of thick crust develops. This is a simple mechanism that predicts the formation of orogenic plateaus with only temperature-dependent viscosity and structural thickening as necessary conditions. It also has the characteristic of strong feedback between crustal thickening, temperature, and a maximum crustal thickness thus providing a mechanism for maintaining stable orogenic plateaus.

Flow of the lower crust exists in all models that have the low viscosity necessary for formation of a plateau. This takes the form of a channel flow in which material extrudes away from the highest topography through a depth-limited channel at the base of the crustal layer. This process has been described by Bird (1991), Royden (1996), and Beaumont et al. (2001) and was found to be significant in all these models. Beaumont et al. (2001) demonstrated that this outward channel flow is even more important if lower crustal temperatures are high enough to partially melt the middle or lower crust. Outward flow of the lower crust can produce surface uplift as observed in the eastern Tibetan plateau (Royden et al. 1997) or can drive the deep exhumation of the Himalaya (Beaumont et al. 2001).

It is notable that an orogenic plateau represents the mechanical limit to crustal thickening. This differs in concept from the mechanical limit discussed by Molnar and Lyon-Caen (1988) by which the height of a mountain range was limited by the forces imposed by the bounding plates. In the model presented here, the topography is limited not by the imposed forces but by the strength of the crust. Once a threshold crustal thickness is reached, any additional contraction results in lower crustal flow without additional crustal thickening. Thus, the maximum height attained by a mountain range is modulated by the crustal strength, thickness, and thermal state of the crust.

In nearly all models, after a significant level of convergence, the crust attains a near-constant maximum thickness that, through isostatic compensation, determines the elevation of the plateau. That maximum thickness is determined by the effective viscosity of the lower crust, which depends on the temperature, rheologic parameters, and strain rate. There is a clear relationship between the maximum crustal thickness and these physical parameters as characterized by the Peclet and Argand numbers (figure 7.18). The Argand number is a function of space and time, so the initial maximum Ar is used to characterize the crustal rheology. This Ar corresponds to the highest temperature, which typically occurs at the base of the crust. The maximum Ar increases with time as the crustal temperature increases, but interestingly, it was the initial value that was found to provide the best predictor for the stable crustal thickness. In fact, it provides a very good predictor of the final crustal thickness under the plateau as indicated by figure 7.18b. The initial maximum Ar is also representative of the lower crustal viscosity in the regions adjacent to the plateau. It is the viscosity of this crust that limits the height of the plateau by restricting the lower crustal flow from beneath the region of high elevation. If the surrounding regions are cold and strong (low initial Ar), this flow is inhibited and a higher plateau can be maintained. A hotter, weaker crust permits lower-crustal flow to reduce plateau height, provided that flow can be maintained to sufficient distance into the surrounding low-elevation regions. The viscosity of the lower crust beneath the plateau is important in limiting the surface and moho topography within the plateau, but once it drops below the threshold for plateau formation, it affects primarily the low-wavelength features on the plateau surface.

The rate of heating of the crust also plays an important role in plateau development as indicated by the dependence of the plateau crustal thickness on the Peclet number, Pe (figure 7.18a). At high Pe, thermal diffusion is so slow that the crust does not equilibrate, thereby retaining its initial, relatively low temperatures, even following significant thickening. In fact, thickening can be extreme at Pe over about 500 with the crust thickening to several times its original thickness, while retaining the temperatures and viscosity of its initial state. In contrast, at low Pe, the crust quickly comes into equilibrium with its boundary conditions and structure, relaxing the cooling effects of crustal thickening. Under these conditions, the crustal temperatures increase, viscosity decreases, and the crust attains an equilibrium thickness that is thin in contrast to the high Pe case.

The timescale of thermal equilibration also has implications for transient deformational behavior that appears in some models as a type of gravitational collapse (Dewey 1988). We use the term gravitational collapse to describe transient

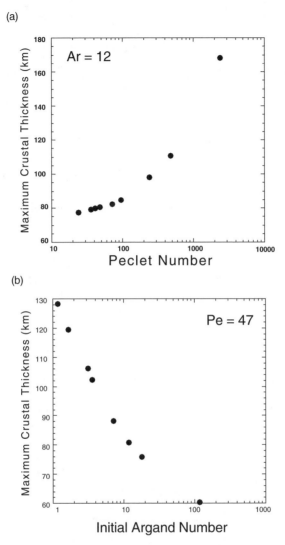

Figure 7.18 Dependence of plateau crustal thickness on model parameters: (a) Peclet number and (b) initial, maximum Argand number. Individual model results shown as points. *Pe* and initial maximum *Ar* are both strong predictors of the equilibrium crustal thickness for reasons discussed in the text.

behavior in which the crust initially thickens but subsequently thins in response to gravitational forces, not to far-field stresses. This process occurs in many models, but only over a limited range of physical parameters. The transient nature of the deformation is linked to the thermal equilibration rate. There are two competing thermal processes in this situation. Crustal thickening leads to cooling of the crust. The cooling is actually only in a spatial (Eulerian) reference frame. In most cases material points in the crust warm during thickening, but the temperature at any specific depth below the surface will decrease with time. Thermal diffusion leads to heating in both material and spatial reference frames. These two processes are thus in competition in the spatial reference frame, and temperature at a specific point in depth may either increase or decrease depending on the relative rates of thickening and diffusion. In models that exhibit a phase of gravitational collapse, the initial stage of crustal thickening occurs without full thermal equilibration, so that the cooling due to thickening dominates the temperature. This leads to a relatively cool lower crust with a correspondingly high viscosity and development of a thick crust. However, with time, thermal diffusion heats the crust and viscosity decreases. If this occurs at a significant rate over the duration of the model run, the viscosity will decrease to the point that the crust cannot support the stresses associated with the elevated thickness, and the crust will begin to thin by viscous flow. This flow results in the thickened crust spreading to a near-uniform thickness, the value of which is determined by the strength of the surrounding crust, as discussed previously.

The timescale for thermal equilibration is characterized by the Peclet number for the system, and it is not surprising that gravitational collapse occurs only for a limited range of *Pe* as illustrated schematically in figure 7.19. In figure 7.19, stable and unstable fields are shown for conditions of crustal thickness and crustal viscosity (*Ar*). Crust of a given thickness and viscosity plots in one of these fields. Convergence and thickening moves a point to the right; heating by thermal diffusion moves a point upward. If *Pe* is very small, thermal diffusion dominates and the crustal temperature increases rapidly with thickening; the consequent viscosity (*Ar*) is always in equilibrium with the crustal thickness. There is thus no potential for gravitational collapse. If *Pe* is very large, the temperature field is dominated

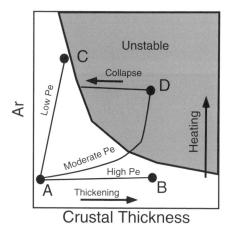

Figure 7.19 Schematic explanation for *Pe* and *Ar* control on gravitational collapse of plateau. Unstably thick crust for a given *Ar* is indicated by shaded region. Crustal shortening leads to crustal thickening or motion to the right in figure. At high *Pe*, there is little change in temperature or *Ar*, so crust remains stable (path A–B). At low *Pe*, temperature and *Ar* remain in equilibrium with thickness, so crust remains stable (path A–C). At intermediate values of *Pe*, crust can thicken without thermal equilibration, but subsequent heating can make the crust unstable, leading to collapse (path A–D).

by deformation and crustal thickening; diffusion never catches up to thickening, at least not over the timescale of the orogenic evolution. Temperature remains low and out of equilibrium, leading to high viscosity and low Ar, and in Ar-thickness space (figure 7.19) the crustal state follows a near-horizontal path, remaining in the stable field. Eventually thermal diffusion will become significant and this orogen might collapse, but the timescale of this process could be prohibitively long. The collapse case occurs for moderate values of Pe in which the initial temperature field is dominated by thickening, moving a crustal state from A to the right in figure 7.19, but with thermal diffusion leading to an increase in Ar and upward motion in figure 7.19, bringing the crustal mechanical state into the unstable field. Collapse thins the crust, bringing it back into a stable regime.

This mechanism and analysis of gravitational collapse is in contrast to the study of England and Houseman (1989), in which they concluded that thermal weakening of thickened crust leads to greater contraction, not extension. This difference is a consequence of different modeling approaches. England and Houseman (1989) used a thin-sheet model in which only depth-averaged velocity is used in the equation of motion. This model formulation precludes extension of the upper crust in a mode in which it is detached from the underlying lithosphere. It also precludes any effects of lower-crustal extrusion. To produce surface extension in a thin-sheet model it is necessary to extend the entire lithosphere, a much more difficult proposition than the shallow, upper-crustal extension observed in the models of this chapter. This comparison is an interesting demonstration of the importance of model formulation as most of the modes of deformation observed in plane-strain or fully three-dimensional models are not predicted by thin-sheet models, simply because of the lack of generality implicit in the use of depth-averaged velocity and rheology.

Another important conclusion derived from these models is that the crust and mantle are effectively decoupled by lower-crustal flow, so that the surface deformation is not sensitive to the mantle kinematics or thermal conditions. All three modes of mantle deformation investigated here (figure 7.12) produced plateaus with similar crustal thickness structure. The deformation history, crustal thickness, and crustal structure depend strongly on the rheologic properties of the crust but only weakly on the motion or deformation of the underlying mantle. This suggests that it is difficult to resolve mantle deformation from surface or crustal-scale observations.

Summary and Conclusions

The crustal deformation associated with orogenesis is fundamentally linked to crustal rheology thereby demonstrating a need to understand crustal rheology but also providing an opportunity to infer information regarding rheology from crustal-scale observations. To do this requires knowledge of the driving forces and boundary conditions in an orogenic system and a model to link boundary conditions, physical parameters, and observables such as deformation rates, strain, or topography. Finite-element methods provide such a tool, and one model with two applications has been presented in this chapter.

The Lagrangian-Eulerian finite-element model presented here provides a useful method for modeling large strain with a variety of constitutive behavior, including both viscous and plastic materials, allowing investigation of the form of the constitutive law as well as the physical parameters and state-variable dependencies of those parameters. These relationships are demonstrated through the two geodynamic processes considered here.

In the first case study, the formation of orogenic wedges of different constitutive behavior was investigated through a suite of models with viscous or plastic rheology. This provided a test of the finite-element model as there are analytic solutions and published analog models for the simple cases of deformation of a wedge with Coulomb plastic or linear viscous constitutive laws. The finite-element model compared favorably with the predictions from other models, although the continuum finite-element model cannot produce the discrete faults observed in the plastic, sandbox analog. Otherwise, the geometry and deformation patterns were reproduced well, including the geologically important phenomenon of surface extension of a contracting, viscous wedge. The finite-element models also addressed the more complicated situation of a combined viscous-plastic crustal rheology and isostatically compensated topography. This led to the conclusion that a Coulomb plastic rheology prevented viscous extension of the orogenic wedge.

The second application considered the formation of orogenic plateaus as a response to contraction, crustal thickening, and thermal weakening of the lower crust. This is viewed as a natural consequence of growth of an orogenic wedge with temperature-dependent crustal viscosity. With a temperature-dependent viscosity, lower crustal heating and weakening eventually reach a point where high-frequency surface topography cannot be maintained and a high-elevation, low-slope "plateau" results. The series of numerical models presented here demonstrated the importance of internal heating and thermal diffusion in determining the timescale of plateau evolution. The limiting crustal thickness and hence elevation of an orogen was found to be a function of the crustal viscosity structure characterized by a maximum Argand number and the Peclet number that characterizes the thermal equilibration rate. Crustal thickness was found to be insensitive to the kinematic model of convergence in the upper mantle, although internal crustal structure was distinctive. In particular, models that parameterized lithospheric mantle shortening by pure-shear, distributed shortening, mantle subduction, or ablative mantle subduction were found to produce crustal plateaus of similar topographic character. We conclude that it is very difficult to use crustal-thickness structure or deformation patterns at the surface of an orogenic plateau to infer mantle kinematics.

These examples are a demonstration of a powerful methodology using numerical models of geologic processes to calibrate rheologic parameters at the crustal scale. The models in this chapter are simple and reflect our state of knowledge of the processes and effective rheology of the crust at this scale, but as our understanding of the tectonic processes and the rheology of the lithosphere increases, so too will the sophistication of mechanical models and their application.

References

Bathe, K.-J. 1996. *Finite Element Procedures*. New York: Prentice Hall.

Barr, T. D. and F. A. Dahlen. 1989. Brittle frictional mountain building: 2. Thermal structure and heat budget. *J. Geophys. Res.* 94:3923–3947.

Batt, G. E. and J. Braun. 1997. On the thermo-mechanical evolution of compressional orogens. *Geophys. J. Int.* 128:364–382.

Beaumont, C., P. Fullsack, and J. Hamilton. 1992. Erosional control of active compressional orogens. In K. R. McClay, ed., *Thrust Tectonics*, pp. 2–18. London: Chapman and Hall.

Beaumont, C., P. Fullsack, and J. Hamilton. 1995. Styles of crustal deformation caused by subduction of the underlying mantle. *Tectonophysics* 232:119–132.

Beaumont, C., S. Ellis, J. Hamilton, and P. Fullsack. 1996a. Mechanical model for subduction-collision tectonics of Alpine-type compressional orogens. *Geology* 24:675–678.

Beaumont, C., R. A. Jamieson, M. H. Nguyen, and B. Lee. 2001. Himalayan tectonics explained by extrusion of a low-viscosity crustal channel coupled to focused surface denudation. *Nature* 414:738–742.

Beaumont, C., P.J.J. Kamp, J. Hamilton, and P. Fullsack. 1996b. The continental collision zone, South Island New Zealand: Comparison of geodynamic models and observations. *J. Geophys. Res.* 101:3333–3359.

Beaumont, C., J. A. Munoz, J. Hamilton, and P. Fullsack. 2000. Factors controlling the Alpine evolution of the Central Pyrenees inferred from a comparison of observations and geodynamical models. *J. Geophys. Res.* 105:8121–8145.

Beck, S. L., G. Zandt, S. C. Myers, T. C. Wallace, P. G. Silver, and L. Drake. 1996. Crustal-thickness variations in the central Andes. *Geology* 24:407–410.

Bird, P. 1989. New finite element techniques for modeling deformation histories of continents with stratified temperature-dependent rheology. *J. Geophys. Res.* 94:3967–3990.

Bird, P. 1991. Lateral extrusion of lower crust from under high topography in the isostatic limit. *J. Geophys. Res.* 96:10275–10286.

Bird, P. 1996. Computer simulations of Alaskan neotectonics. *Tectonics* 15:225–236.

Bird, P. and J. Baumgartner. 1984. Fault friction, regional stress, and crust-mantle coupling in southern California from finite element models. *J. Geophys. Res.* 89:1932–1944.

Bird, P. and X. Kong. 1994. Computer simulations of California tectonics confirm very low strength of major faults. *Bull. Geol. Soc. Am.* 106:159–174.

Bird, P. and K. Piper. 1980. Plane-stress finite-element models of tectonic flow in southern California. *Phys. Earth Planet Inter.* 21:158–175.

Braun, J. 1993. Three dimensional numerical modeling of compressional orogens: Thrust geometry and oblique convergence. *Geology* 21:153–156.

Buck, W.R. and D. Sokoutis. 1994. Analogue model of gravitational collapse and surface extension during continental convergence. *Nature* 369:737–740.

Chapple, W. M. 1978. Mechanics of thin-skinned fold-and-thrust belts. *Geol. Soc. Am. Bull.* 89:1189–1198.

Crespi, J. M., Y.-C. Chan, and M. S. Swaim. 1996. Synorogenic extension and exhumation of the Taiwan hinterland. *Geology* 24:247–250.

Dahlen, F. A. 1984. Noncohesive critical Coulomb wedges: An exact solution. *J. Geophys. Res.* 89:10125–10133.

Dahlen, F. A. 1990. Critical taper model of fold-and-thrust belts and accretionary wedges. *Annu. Rev. Earth Planet Sci.* 18:55–99.

Dahlen, F. A. and T. D. Barr. 1989. Brittle frictional mountain building: 1. Deformation and mechanical energy balance. *J. Geophys. Res.* 94:3906–3922.

Dahlen, F. A., J. Suppe, and D. Davis. 1984. Mechanics of fold-and-thrust belts and accretionary wedges: cohesive Coulomb theory. *J. Geophys. Res.* 89:10125–10133.

Davis, D., J. Suppe, and F. A. Dahlen. 1983. Mechanics of fold-and-thrust belts and accretionary wedges. *J. Geophys. Res.* 88:1153–1172.

Dewey, J. F. 1988. Extensional collapse of orogens. *Tectonics* 7:1123–1139.

Dewey, J. F. and J. M. Bird. 1970. Mountain belts and the new global tectonics. *J. Geophys Res.* 75:2625–2647.

Dewey, J. F., R. M. Shackleton, C. Chengfa, and S. Yiyin. 1988. The tectonic evolution of the Tibetan Plateau. *Philos. Trans. R. Soc. Lond. A. Math. Phys. Sci.* 327:379–413.

Dunbar, J. A. and D. S. Sawyer. 1989. How pre-existing weaknesses control the style of continental breakup. *J. Geophys. Res.* 94:7278–7292.

Ellis, S., P. Fullsack, and C. Beaumont. 1995. Oblique convergence of the crust driven by basal forcing: Implications for length-scales of deformation and strain partitioning in orogens. *Geophys. J. Int.* 120:24–44.

Ellis, S., C. Beaumont, R. A. Jamieson, and G. Quinlan. 1998. Continental collision including a weak zone: The vise model and its application to the Newfoundland Appalachians. *Can. J. Earth Sci.* 1323–1346.

England, P. and G. Houseman. 1986. Finite strain calculations of continental deformation: 2. Comparison with the India-Asia collision. *J. Geophys. Res.* 91:3664–3676.

England, P. and G. Houseman. 1989. Extension during continental convergence, with application to the Tibetan Plateau. *J. Geophys. Res.* 94:17561–17579.

England, P. and D. P. McKenzie. 1982. A thin viscous sheet model for continental deformation. *Geophys. J. R. Astron. Soc.* 70:295–321.

England, P. and D. P. McKenzie. 1983. Correction to: A thin viscous sheet model for continental deformation. *Geophys. J. R. Astron. Soc.* 73:523–532.

England, P., G. Houseman, and L. Sonder. 1985. Length scales for continental deformation in convergent, divergent and strike-slip environments: analytical and approximate solutions for a thin viscous sheet model. *J. Geophys. Res.* 90:4797–4810.

Fullsack, P. 1995. An arbitrary Lagrangian-Eulerian formulation for creeping flows and its applications in tectonic models. *Geophys. J. R. Astron. Soc.* 120:1–23.

Houseman, G. and P. England. 1986. Finite strain calculations of continental deformation: 1. Method, and general results for convergent zones. *J. Geophys. Res.* 91:3651–3663.

Isacks, B. L. 1988. Uplift of the central Andean plateau and bending of the Bolivian Orocline. *J. Geophys. Res.* 93:3211–3231.

Karig, D. E. and G. F. Sharman. 1975. Subduction and accretion in trenches. *Bull. Geol. Soc. Am.* 86:377–389.

Kohlstedt, D. L., B. Evans, and S. J. Mackwell. 1995. Strength of the lithosphere: Constraints imposed by laboratory experiments. *J. Geophys. Res.* 100:17587–17602.

Kronenberg, A. K. and J. Tullis. 1984. Flow strengths of quartz aggregates: Grain size and pressure effects due to hydrolytic weakening. *J. Geophys. Res.* 89:4281–4297.

Le-Pichon, X., M. Fournier, and L. Jolivet. 1982. Kinematics, topography, shortening, and extrusion in the India-Eurasia collision. *Tectonics* 11:1085–1098.

Malavieille, J. 1984. Modelisation experimentale des chevauchements imbriques: Application aux chaines de montagnes. *Bull. Soc. Geol. France* 26:129–138.

Malavieille, J. 1993. Late orogenic extension in mountain belts: Insights from the basin and range and the late Paleozoic Variscan belt. *Tectonics* 12:1115–1130.

Malvern. 1969. *An Introduction to the Mechanics of a Continuous Medium.* New York: Prentice-Hall.

Mattauer, M. 1986. Intracontinental subduction, crust-mantle decollement and crustal-stacking wedge in the Himalayas and other collision belts. In M. P. Coward and C. Ries-Alison, eds., *Collision Tectonics.* Geological Society Special Publication 19, pp. 37–50. London: Geological Society of London.

Medvedev, S. 2002. Mechanics of viscous wedges: Modeling by analytical and numerical approaches. *J. Geophys. Res. B, Solid Earth Planets* 107:15.

Molnar, P. and H. Lyon-Caen. 1988. Some simple physical aspects of the support, structure, and evolution of mountain belts. Processes in continental lithospheric deformation, Special Paper 218, pp. 179–207. Boulder, CO: Geological Society of America.

Molnar, P. and P. Tapponnier. 1975. Cenozoic tectonics of Asia: Effects of a continental collision. *Science* 189:419–426.

Nelson, K. D., W. Zhao, L. D. Brown, J. Kuo, J. Che, X. Liu, S. L. Klemperer, Y. Makovsky, R. Meissner, J. Mechie, R. Kind, F. Wenzel, J. Ni, J. Nabelek, L. Chen, H. Tan, W. Wei, A. G. Jones, J. Booker, M. Unsworth, W.S.F. Kidd, M. Hauck, D. Alsdorf, A. Ross, M. Cogan, C. Wu, E. A. Sandvol, and M. Edwards. 1996. Partially molten middle crust beneath southern Tibet: Synthesis of Project INDEPTH results. *Science* 274:1684–1688.

Owens, T. J. and G. Zandt. 1997. Implications of crustal property variations for models of Tibetan Plateau evolution. *Nature* 387:37–43.

Pfiffner, O. A. 1992. Tectonic evolution of Europe: Alpine orogeny. In D. Blundell, R. Freeman, and S. Müller, eds., *A Continent Revealed: The European Geotraverse*, pp. 180–190. Cambridge, UK: Cambridge University Press.

Platt, J. P. 1986. Dynamics of orogenic wedges and the uplift of high-pressure metamorphic rocks. *Geol. Soc. Am. Bull.* 79:1037–1053.

Pope, D. C. and S. D. Willet. 1998. Thermal-mechanical model for crustal thickening in the central Andes driven by ablative subduction. *Geology (Boulder)* 26:511–514.

Pysklywec, R., C. Beaumont, and P. Fullsack. 2000. Modeling the behavior of the continental mantle lithosphere during plate convergence. *Geology* 28:655–658.

Royden, L. 1996. Coupling and decoupling of crust and mantle in convergent orogens: Implications for strain partitioning in the crust. *J. Geophys. Res.* 101:17679–17705.

Royden, L., B. C. Burchfiel, R. W. King, E. Wang, Z. Chen, F. Shen, and Y. Liu. 1997. Surface deformation and lower crustal flow in eastern Tibet. *Science* 276:788–790.

Schmid, S. M., O. A. Pfiffner, N. Froitzheim, G. Schönborn, and E. Kissling. 1996. Geophysical-geological transect and tectonic evolution of the Swiss-Italian Alps. *Tectonics* 15:1036–1064.

Seely, D. R. 1979. The evolution of structural highs bordering major forearc basins. In J. S. Watkins, L. Montadert, and P. W. Dickerson, eds., *Geological and Geophysical Investigations of Continental Slopes and Rises*. American Association of Petroleum Geology Memoir 29, pp. 245–260.

Sheffels, B. M. 1995. Mountain building in the central Andes: An assessment of the contribution of crustal shortening. *Int. Geol. Rev.* 37:128–153.

Sonder. L. J., P. C. England, and G. A.. Houseman. 1986. Continuum calculations of continental deformation in transcurrent environments. *J. Geophys. Res.* 91:4797–4810.

Stockmal, G. S. 1983. Modelling of large scale accretionary wedge deformation. *J. Geophys. Res.* 88:8271–8287.

Suppe, J. 1981. Mechanics of mountain-building and metamorphism in Taiwan. *Mem. Geol. Soc. China* 4:67–89.

Tao, W. C. and R. J. O'Connell. 1992. Ablative subduction: A two-sided alternative to the conventional subduction model. *J. Geophys. Res.* 97:8877–8904.

Villotte, J. P., Daignieres, M., and R. Madariaga. 1982. Numerical modelling of interplate deformation: Simple mechanical models of continental collision. *J. Geophys. Res.* 87:10709–10728.

Villotte, J. P., M. Daignieres, and R. Madariaga. 1986. Numerical Study of continental collision: Influence of buoyancy forces and an initial stiff inclusion. *Geophys. J. R. Astron. Soc.* 84:279–310.

Wang, W.-H. and D. M. Davis. 1996. Sandbox model simulation of forearc evolution and non-critical wedges. *J. Geophys. Res.* 101:11329–11339.

Wdowinski, S. and Y. Bock. 1994. The evolution of deformation and topography of high elevated plateaus: 1, Model, numerical analysis, and general results. *J. Geophys. Res.* 99:7103–7119.

Wdowinski, S., R. J. O'Connell, and P. England. 1989. A continuum model of continental deformation above subduction zones: Application to the Andes and the Aegean. *J. of Geophys. Res. B, Solid Earth Plants* 94:10331–10346.

Willett, S. D. 1992. Dynamic and kinematic growth and change of a Coulomb wedge. In K. R. McClay, ed., *Thrust Tectonics*, pp. 19–31. London: Chapman and Hall.

Willett, S. D. 1999a. Rheological dependence of extension in wedge models of convergent orogens. *Tectonophysics* 305:419–435.

Willett, S. D. 1999b. Orogeny and orography: The effects of erosion on the structure of mountain belts. *J. Geophys. Res.* 104:28957–28981.

Willett, S. D. and C. Beaumont. 1994. Subduction of Asian lithospheric mantle beneath Tibet inferred from models of continental collision. *Nature* 369:642–645.

Willett, S. D., C. Beaumont, and P. Fullsack. 1993. Mechanical model for the tectonics of doubly vergent compressional orogens. *Geology* 21:371–374.

Zhao, W.-L., D. M. Davis, F. A. Dahlen, and J. Suppe. 1986. Origin of convex accretionary wedges: Evidence from Barbados. *J. Geophys. Res.* 91:10246–10258.

Zienkiewicz, O. C. 1977. *The Finite Element Method*. New York: McGraw-Hill.

Zienkiewicz, O. C. and P. N. Godbole. 1975. Penalty function approach to problems of plastic flows of metals with large surface deformations. *J. Strain Anal.* 10:180–183.

CHAPTER EIGHT

Structure of Large-Displacement, Strike-Slip Fault Zones in the Brittle Continental Crust

F. M. Chester, J. S. Chester, D. L. Kirschner,
S. E. Schulz, and J. P Evans

Introduction

Characterizing the structure of large-displacement, plate-boundary fault zones is central to the MARGINS initiative to understand the mechanisms that allow continental lithosphere to deform by weak tectonic forces, strain partitioning, and movement of fluids during margin formation. At many boundaries, plate motions primarily are accommodated along large fault zones that achieve significant displacement over time. In addition, fluid migration through the crust often is intimately linked to the fluid-flow properties of these zones. At the crustal scale, faults may be idealized as simple discontinuities, or surfaces, between relatively rigid blocks along which shear displacement has occurred. Considerable work has elucidated the geometry of faults and the shear-displacement distribution along faults, as well as scaling relations and the manner in which faults grow and link with strain (Cowie et al. 1996; Davison 1994). Nonetheless, continued effort to better define the structural, physical, and chemical properties of fault zones is necessary to understand the mechanical, fluid-flow, and geophysical properties of the lithosphere.

Our current understanding of the mesoscopic structure of large-displacement, strike-slip fault zones primarily results from study of uplifted and exhumed faults (Little 1995; Chester and Logan 1986; Flinn 1977) and faults exposed in mines and drill core (Ohtani et al. 2000; Wallace and Morris 1986). Early petrographic study of mylonites and cataclasites initiated considerable research into the classification and origin of fault rocks (for review, see Snoke and Tullis 1998). The most commonly used fault-rock classification schemes are based on particle size, fabric, and cohesion. For example, in Sibson's classification (1977), mylonites and cataclasites both are distinguished by primary cohesion and tectonic grain-size reduction, but mylonites have a well developed foliation with some evidence of syntectonic recovery and recrystallization, whereas cataclasites have random fabrics and microstructures indicative of cataclasis. The breccia-gouge series is distinguished by lack of primary cohesion and cataclastic microstructures (Snoke and Tullis 1998; Sibson 1977). The general progression from breccia-gouge to cataclasites to mylonites to blastomylonites (mylonites with extensive recovery and recrystallization) reflects differences in the relative contribution of different deformation and recovery mechanisms with increasing depth of formation within the crust (Reed 1964; Christie 1960). The importance of metamorphic reactions coupled with deformation processes is reflected in the fact that mylonites often are viewed as transitional between the more brittle fault rocks (breccia and gouge) and classic metamorphic rocks (schist and gneiss) (Waters and Campbell 1935). Current synoptic models of fault zones over the entire depth range of the crust describe systematic variations in mechanical properties expected with depth on the basis of our understanding of faulting processes and products of faulting (Kohlstedt et al. 1995; Scholz 1990; Sibson 1977). A particularly important change in fault-zone structure and mechanical properties occurs at upper- to middle-crustal depths. At these depths there is a change in the dominant deformation process from brittle, cataclastic mechanisms at lower pressures in the upper crust to intracrystalline plastic mechanisms at higher temperatures in the middle and lower crust (Schmid and Handy 1991; Sibson 1977).

Deformation in brittle fault zones is heterogeneous at the mesoscopic scale. Individual fault zones may display fractured zones, brecciated zones, protocataclasites, cataclasites, gouge, and ultracataclasite. Degree of cataclastic grain-size reduction reflects magnitude of strain, in part, so the typical occurrence of gouge and ultracataclasite as relatively narrow layers or seams within much broader zones of brittle deformation is taken as evidence that shear was localized (Reed 1964). Localization of shear is consistent with the common observation that host-rock lithology is continuous up to, and discontinuous across, gouge layers (Chester and Logan 1986; Flinn 1977).

Brittle faults are geometrically complex, ranging from individual curviplanar and discrete surfaces to branching and braided surfaces within broad zones of deformation (Wallace and Morris 1986). Faults often consist of multiple segments that are linked in a variety of ways (Davison 1994). Although the thickness of gouge and ultracataclasite layers and total thickness of fault zones generally appear to increase with displacement, the distribution of deformation within a fault zone may be highly variable. Fault zones may contain a single gouge layer near the center or along one side of the zone (Flinn 1977; Wallace and Morris 1986). In addition, fault zones may contain two or more gouge layers that are subparallel and located along the boundaries of the zone (Sibson 1986) or are woven throughout the zone (Rutter et al. 1986; Wallace and Morris 1986).

Studies of brittle fault zones document a considerable variation in structure with tectonic setting and host lithology. Dip-slip faults tend to be asymmetric in that they juxtapose wall rocks with markedly different burial and deformation history. Often the uplifted side contains fault rocks that were formed at deep crustal levels and repeatedly overprinted by faulting at shallower levels during upward transport (Sibson 1977). In contrast, wall rocks of strike-slip faults are more likely to have common burial and deformation histories, though these faults also may experience overprinting by faulting at different depths if faulting is concurrent with burial or exhumation. At the same burial conditions, the structure of faults and processes of deformation are markedly different in unconsolidated, poorly lithified sediments as compared with well lithified rock. In addition, physical properties of fault rocks relative to host rocks may be considerably different for faulting in highly porous sedimentary rocks and crystalline igneous and metamorphic rocks (Goodwin et al. 1999). The focus of this chapter is the structure of large-displacement, strike-slip faults in relatively low-porosity, well lithified siliciclastic, igneous, and metamorphic rocks typical of continental crust. In addition, we focus on faulting in the several-kilometer depth range over which the transition from incohesive breccia-gouge to cohesive cataclasites occurs.

Many brittle fault zones in lithified rocks can be divided into two structural domains (Chester et al. 1993; figure 8.1); a fault core (also referred to as "breccia zone" by Robertson [1983] and "sheared zone" by Wallace and Morris [1986]), and a surrounding damage zone (also referred to as "zone of fractured rock" by Wallace and Morris [1986] and "transition zone" by Bruhn et al. [1994]). In this model the entire fault zone is defined as the volume of rock about a fault that exhibits a deformation intensity that exceeds background levels. In the simplest case of a single fault with a symmetric damage zone, the fault core forms a tabular zone of highly deformed rock. The core displays highly deformed and reoriented host-rock structures, cataclastic foliations, severely comminuted and altered rocks such as gouge and ultracataclasite, and synfaulting mineralization. Almost all shear displacement across a fault occurs in the fault core, and thus at the macroscopic scale the fault core represents the "fault surface" (Chester and Chester 1998). The damage zone is a much thicker zone of fractured and faulted rock that can display folded strata, subsidiary faults, fractures, veins, solution seams, comminuted grains, microfractures, localized alteration zones, and mineralization. In general, intensity of damage increases toward the fault core, thickness of the damage zone

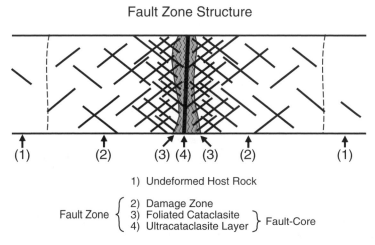

1) Undeformed Host Rock

Fault Zone { 2) Damage Zone
3) Foliated Cataclasite } Fault-Core
4) Ultracataclasite Layer

Figure 8.1 Fault-core damage-zone model of a large-displacement, strike-slip fault zone in well lithified sedimentary and crystalline rock. Figure is schematic and not to scale. After Chester et al. (1993).

varies laterally, and the transition from undeformed host rock to damaged rock is gradational (Chester and Logan 1986).

The detailed structure of brittle fault zones not only provides evidence of localization of shear to the fault core but also evidence for extreme localization of slip within the fault core (Chester et al. 1993; Chester and Chester 1998). Localization of slip is recorded by presence of discrete layers of gouge or ultracataclasite and mesoscopic slip surfaces. Cataclastic foliations within a fault core records distributed shear. Cataclasites often display a progressive grain-size reduction and rotation of cataclastic foliations parallel to the fault with proximity to the ultracataclasite layer, consistent with an increase in shear toward the ultracataclasite. The extremely small and uniform particle size of ultracataclasite and sharp contact with surrounding cataclasites are taken as evidence of an abrupt change in magnitude of shear and may reflect a change in the process of comminution, i.e., process of grinding and crushing solid rock into finer particles (Chester and Chester 1998). As such, cores of brittle fault zones record valuable information about mode of failure, processes of localization, and extent of mixing of host-rock lithologies across a fault (figure 8.2).

In this chapter we provide a detailed characterization of the structure of two large-displacement, strike-slip fault zones—the Punchbowl and North Branch San

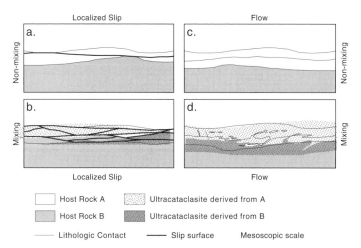

Figure 8.2 Four end-member modes of failure in ultracataclasite as a function of slip localization and degree of mixing. (a) Slip on a single, stationary, through-going slip surface. Results in juxtaposition of contrasting host rock and ultracataclasite across the slip surface with little mixing. (b) Contemporaneous or alternating slip on multiple, anastomosing, through-going slip surfaces. Fault slices of host rock and ultracataclasite mixed throughout the layer. For the f closely spaced slip surfaces and large slip, the layer may appear homogeneous. (c) Distributed deformation within ultracataclasite by streamline or laminar flow. Contrasting ultracataclasite is in contact but not mixed. (d) Distributed deformation by turbulent or nonlaminar flow. Contrasting ultracataclasite is mixed by processes such as fluidization and injection. As in (b), mixing could homogenize the ultracataclasite. Modified from Chester and Chester (1998).

Gabriel faults. Both faults are inactive traces of the San Andreas system in southern California. The San Andreas fault system is an integral component of the rifted margin between the Pacific and North American plates in California and Mexico. The San Andreas fault is one of the best studied faults in the world, and it is central to the long-standing debate about the strength of faults that rupture the lithosphere (Brune et al. 1969; Hickman 1991; Scholz and Hanks 2003). The Punchbowl and San Gabriel faults have been exhumed to reveal the products of faulting well lithified sedimentary and crystalline rocks under brittle conditions characteristic of the upper continental crust (Chester and Logan 1986; Chester et al. 1993). In the following text, we present a synthesis of new and previously published work to characterize and compare a fault zone containing a single, centrally located fault core and a fault zone containing paired fault cores. In addition, we document, through particle-size analysis and detailed mapping, comminution processes and distribution of slip in the fault cores. Lastly, we address the origin of fault-zone structure and relation of structure to mechanical and fluid-flow properties.

Geology of the Punchbowl and North Branch San Gabriel Faults

The San Gabriel fault is one of the oldest onshore components of the modern San Andreas transform system, and during the Miocene (12–5 Ma) was the main fault of this system in the central Transverse Ranges of southern California (Powell et al. 1993). The San Gabriel fault splays near Big Tujunga Creek in the western San Gabriel Mountains to form the South Branch and North Branch San Gabriel faults (figure 8.3). The North Branch cuts post-Paleocene sedimentary rock and Proterozoic, Jurassic, and Cretaceous crystalline rock of the San Gabriel basement complex (Ehlig 1981). The juxtaposition of igneous and metamorphic assemblages along the near-vertical North Branch San Gabriel and fabric of nearby subsidiary faults (Chester et al. 1993) are consistent with approximately 21 km of right-lateral strike slip. Following the late Miocene, most transform displacement in the central Transverse Ranges occurred 30 km northeast on the Punchbowl fault and historically active trace of the San Andreas fault (Powell et al. 1993).

The Punchbowl fault accommodated approximately 44 km of right-lateral separation in the San Gabriel Mountains during the Miocene and Pliocene as shown by offset of older structures, including the San Francisquito and Fenner faults (Dibblee 1968). In Devil's Punchbowl Los Angeles County Park, the Punchbowl fault zone dips steeply to the southwest. Subsidiary fault and fold fabrics in the Punchbowl Park indicate right-lateral, oblique-reverse kinematics with a slip vector plunging approximately 30° to the southeast (Chester and Logan 1987).

Uplift and erosion of the San Gabriel Mountains have exhumed the Punchbowl and North Branch San Gabriel faults, resulting in excellent exposures of the products of faulting from 2 to 5 km depth. Uplift of the San Gabriel Mountains since Pliocene is largely the result of dip-slip motion on the north-dipping Sierra

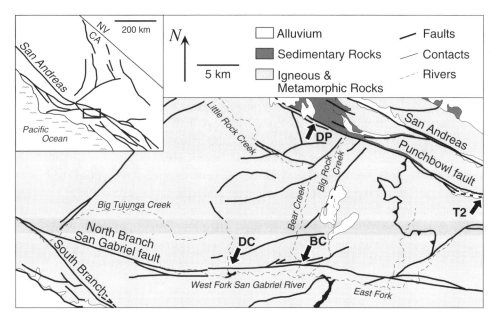

Figure 8.3 Map of the Punchbowl and North Branch San Gabriel faults in the central San Gabriel Mountains, southern California. Arrows identify the location of sites of detailed study along the Punchbowl fault at the Devil's Punchbowl Los Angeles County Park (DP) and at the saddle between Wright Mountain and Pine Mountain, T2, and along the North Branch San Gabriel fault at Devil's Canyon, D.C., and Bear Creek, B.C.

Madre-Cucamonga thrust system and regional arching of the Transverse Ranges (Oakeshott 1971; Morton and Matti 1987). Estimates of depth of faulting for the Punchbowl and North Branch San Gabriel faults is based on post-Pliocene uplift and erosion rates for the San Gabriel Mountains (Oakeshott 1971; Morton and Matti 1987), thickness of the sedimentary sequence in Devil's Punchbowl basin cut by the Punchbowl fault (Chester and Logan 1986), and mineral assemblages and microstructures of fault rocks (Anderson et al. 1983; Chester and Logan 1986; Evans and Chester 1995). Alteration mineralogy is consistent with a slightly greater depth of faulting recorded along the North Branch San Gabriel fault than that along the Punchbowl fault (Chester et al. 1993).

At the two locations of the North Branch San Gabriel fault studied in detail (Bear Creek and Devil's Canyon; figure 8.3), the fault zone consists of a single fault core located near the center of the damage zone (Chester et al. 1993). At both locations, the fault core contains a layer of ultracataclasite that is several centimeters to several decimeters thick. Distinctly different host-rock lithologies are juxtaposed across the ultracataclasite layer. At Devil's Canyon, a banded hornblende- and biotite-rich gneiss (Precambrian) occur on the north side, and a hornblende diorite (Precambrian) cut by a complex of granitic dikes (Cretaceous) occurs on the south side. At Bear Creek, a granite (Cretaceous) and granodiorite (Permo-Triassic) on the north are juxtaposed against a quartz monzogranite (Cretaceous) and diorite gneiss (Precambrian) on the south. At both locations, the damage zone and neighboring host rock are derived from the same protolith, and the only major discontinuity in protolith type occurs across the ultracataclasite layer of the fault cores. This relation indicates that the majority of the shear displacement on the fault occurred in the ultracataclasite layers.

At the macroscopic scale, the Punchbowl fault zone consists locally of two semiparallel faults bounding a sliver of damaged host rock tens to several hundreds of meters thick. In Devil's Punchbowl Park (figure 8.3), the northern fault is defined by a fault core with a single layer of ultracataclasite several centimeters to 1 m thick (Chester and Logan 1986). Arkosic sedimentary rocks of the Punchbowl Formation (Miocene-Pliocene) are in fault contact with an assortment of igneous and metamorphic rocks of the San Gabriel basement complex (Precambrian gneiss and Cretaceous quartz diorite, quartz monzonite, and monzogranite) across the ultracataclasite layer. The southern fault is segmented, poorly developed, and often does not juxtapose different host-rock lithologies. These and regional relations led Chester and Chester (1998) to conclude that nearly all of the 44 km of displacement across the Punchbowl fault zone occurred in the fault core of the northern fault. To the southeast of the Devil's Punchbowl area, the Punchbowl fault zone cuts Paleocene Pelona schist, an interlayered greenschist facies metagraywacke, metabasalt, and metachert (Ehlig 1981). At Wright and Pine Mountains, the fault zone contains two faults that are 153 m apart (Schulz and Evans 1998, 2000). The magnitude of displacement on each fault is unknown.

Fault-Zone Structure

North Branch San Gabriel Fault Zone: Damage About a Single Fault Core

Variation in intensity of damage across the North Branch San Gabriel fault zone is illustrated by the distribution of mesoscopic fractures, veins, subsidiary faults, and microfractures, as a function of distance from the central ultracataclasite layer of the fault core (figure 8.4). Although locally variable, each fabric element displays an overall decrease in abundance with increased distance from the ultracataclasite layer. Damage decreases to values representative of regional deformation at a distance of approximately 100 m from the ultracataclasite layer. Local increases in abundance and variability in damage at large distances are particularly evident in the mesoscopic data. In part, variations reflect local damage associated with large subsidiary faults. Overall, the damage zone displays a nested structure consisting of larger subsidiary faults clustered around the fault core, each having an associated halo of smaller faults and fractures. These data are consistent with a gradational and irregular boundary between the damage zone and surrounding host rock at the macroscopic scale.

Along the North Branch, the fault core is decimeters to meters thick and displays foliated and nonfoliated cataclasites and ultracataclasite. The boundary between the fault core and bounding damage zone is gradational and irregular at

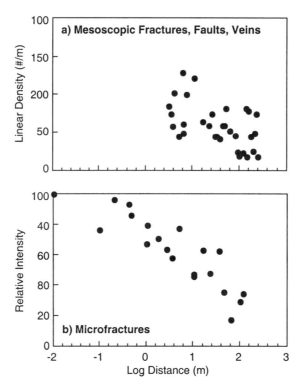

Figure 8.4 Variation in damage intensity across a fault zone containing a single fault core, North Branch San Gabriel fault. Distance is measured normal to the fault from the ultracataclasite-cataclasite contact. (a) Linear density of mesoscopic fractures, faults, and veins versus log distance. Linear density is determined by counting the number of intercepts along two orthogonal lines. (b) Relative intensity of microfractures versus log distance. Intensity determined with an optical microscope by point-counting 100 grains and characterizing the number of fractures in each grain. Microfracture count normalized by the value of the sample with the greatest number of fractures.

the mesoscopic scale. Primary igneous and metamorphic textures of the host rock are partly destroyed through extreme cataclasis and mineralization within the fault core. Mesoscopic cataclastic foliations are developed best in gneissic host rock containing melanocratic layers and near the ultracataclasite. The cataclastic foliations are oriented subparallel to the fault consistent with a concentration of right-lateral shear to the fault core.

At Devil's Canyon, the relative distribution of shear strain in the fault zone is qualitatively recorded by reorientation of preexisting fabric elements of the host rock. The host rock is a hornblende diorite cut by numerous granitic dikes having an approximately random orientation distribution. Dikes are fractured within the damage zone, but are significantly reoriented only within several meters of the ultracataclasite layer (figure 8.5a). The orientation of rotated dikes near the ultra-cataclasite layer is consistent with distributed right-lateral shear of passive markers. At this location, dike fabric may be used to demarcate the zone of high shear strain, i.e., core of the fault.

The fault core also may be identified on the basis of several microscopic fabric elements. In the North Branch San Gabriel fault zone at Bear Creek, the volume percent of mineralized particles and comminuted particles less than 10 μm in diameter increase dramatically within several meters of the ultracataclasite (figure 8.5b, c; Chester et al. 1993). Volatile concentration (primarily H_2O and CO_2) in samples collected from the fault zone, as estimated by loss on ignition, increases markedly at a distance of several meters from the ultracataclasite (figure 8.5d). The increase reflects the increased abundance of synfaulting alteration products of clay and zeolite within the fault core relative to that in the surrounding damage zone (Evans and Chester 1995).

At many locations, the cataclasite in the fault core displays a composite planar fabric. Typically, the composite fabric consists of a crude layering developed through distributed cataclastic flow and offset along synthetic and antithetic shears. Layering is defined by compositional banding, crude shape fabrics, and preferred crystallographic orientation of phyllosilicate minerals; it is geometrically similar to the S-foliation and P-foliation in ductile- and brittle-shear zones, respectively (e.g., Rutter et al. 1986). Shears are preferentially parallel to subparallel to the fault and geometrically similar to C and C′ shears and to R and Y shears in ductile- and brittle-shear zones, respectively (e.g. Rutter et al. 1986). The fabric of shears in the fault core is different from that of subsidiary faults in the bounding damage zone. Synthetic and antithetic subsidiary faults associated with the North Branch San Gabriel fault constitute a quasi-conjugate geometry with the bisector of the conjugates (i.e., the direction of contraction) oriented at high angles to the master fault.

Punchbowl Fault Zone:
Damage About Paired Fault Cores

Distribution of damage within a fault zone about paired fault cores is illustrated by data from a traverse across the Punchbowl fault zone at the saddle between Wright Mountain and Pine Mountain (traverse 2 of Schulz and Evans 2000). The linear density of mesoscopic fractures and subsidiary faults increases from the outer boundaries of the damage zone inward toward each fault core, similar to that documented along the North Branch San Gabriel fault and elsewhere along the Punchbowl fault (e.g., Chester and Logan 1986). However, throughout the sliver of rock bounded by the two fault cores, the intensity of damage is equal to or greater than that observed in the damaged host rock outside the sliver (figure 8.6a).

The spatial distribution of microscopic fractures, shears, and veins is similar to the mesoscopic data in that damage is high throughout the sliver. Microscopic data also show locally high damage adjacent to the fault cores (figure 8.6b). Abundance of veins and volatile concentration also increase near the two fault cores (figure 8.6c, d).

Particle-Size Distributions

Herein, microscopic particles are defined on the basis of mineralogy and crystallography, i.e., particle boundaries include grain boundaries and microfractures separating different minerals or the same mineral but with different crystallographic orientation. By this definition, the undeformed host rocks have particle-

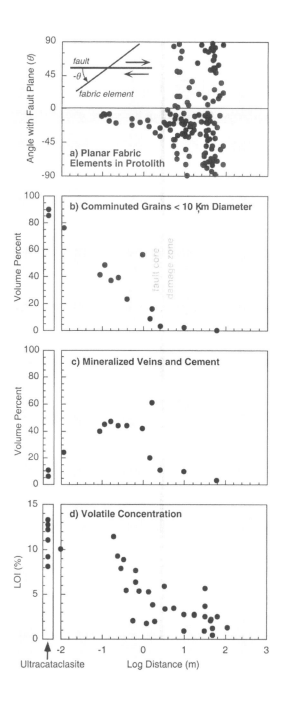

size distributions that reflect igneous, metamorphic, and sedimentary processes of formation. Cataclasis leads to grain-size reduction and modifies the particle-size distribution. Samples of ultracataclasite, cataclasite, protocataclasite, and undeformed host rock were collected along traverses across the Punchbowl and North Branch San Gabriel faults at three different locations. Along the Cruthers Creek traverse (northwest of Devil's Punchbowl Park), where the southern core of the Punchbowl fault cuts monzogranite, syntectonic alteration was insignificant and ultracataclasite of the fault is mineralogically equivalent to surrounding host rock. Along the two traverses across the North Branch San Gabriel fault, at Devil's Canyon and Bear Creek, low-grade, syntectonic hydration reactions of amphiboles and micas to clays, and feldspar to laumontite, were significant (Evans and Chester 1995).

Particle-size distributions of all samples are similar if grouped according to structural position (figure 8.7). Undeformed host rock displays a narrow range of particle sizes. Cataclastic rocks have a lower density of large particles and higher density of small particles relative to host rock. Protocataclasites from the damage zone and cataclasites from the fault core have a fractal particle-size distribution with a fractal dimension of 1.6 to 1.8 in a planar section. If fault-rock fabric is isotropic, then the fractal dimension for a plane section would be increased by one for a three-dimensional volume (e.g., Sammis et al. 1987). Ultracataclasite from the core of each fault is characterized by a fractal particle-size distribution with a fractal dimension greater than 2 in a two-dimensional cross section, which is significantly different from the neighboring cataclastic rocks (figure 8.7). The size distribution for ultracataclasites records a substantial increase in density of small particles and a relative decrease in density of large particles relative to cataclasites.

Similarity in particle-size distribution within each structural domain, regard-

Figure 8.5 Variation in structural features across a fault zone containing a single fault core, North Branch San Gabriel fault. Distance is measured normal to the fault from the ultracataclasite-cataclasite contact. Columns on the left side of the plots in (b), (c), and (d) show measurements from within the ultracataclasite layer. All features show a marked change at several meters distance and demarcate the fault-core boundary. (a) Orientation of planar fabric elements present in the protolith as a function of log distance at Devil's Canyon. Fabric elements are granitic dikes that predate brittle faulting. Orientation is represented by the acute angle between the dike and fault surface viewed in a plane perpendicular to the fault surface and parallel to the slip direction (horizontal); angle convention is shown in figure. (b) Volume of fault rocks composed of fractured grains less than 10 μm diameter as a function of log distance at Bear Creek. Volume determined with an optical microscope by 100 point counts. (c) Volume of fault rocks composed of veins and other mineralization as a function of log distance at Bear Creek. Volume determined with an optical microscope by 100 point counts. (d) Volatile concentration as determined from loss on ignition (LOI) of fault rocks as a function of log distance at Bear Creek.

less of lithology and degree of mineral alteration, suggests that comminution processes in each structural domain were independent of these parameters. Distinctly different particle-size distributions of ultracataclasite and neighboring cataclasite, and the sharp contact between the two, likely reflect different comminution processes, or a marked change in degree of comminution by the same process, across the contact (Sammis et al. 1987; Blenkinsop 1991).

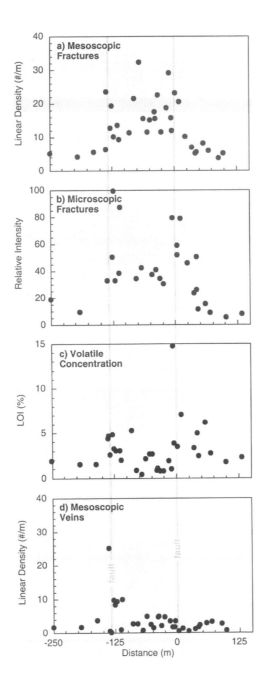

Fault Cores

Punchbowl Fault

The ultracataclasite layer in the northern fault core of the Punchbowl fault zone has been mapped at scales of 1:20 to 1:1 at four different locations in Devil's Punchbowl Park (figures 8.8 and 8.9). At every location mapped, a single, continuous ultracataclasite layer juxtaposes cataclastic basement on the south and cataclastic sandstones of the Punchbowl Formation on the north. Fault-rocks derived from the Punchbowl Formation and crystalline basement are never found on the same side of the ultracataclasite layer. As shown in figure 8.9, fault-parallel lithologic layering is developed locally in the fault core adjacent to the ultracataclasite layer as a result of distributed cataclasis and subsidiary faulting. Similar to that documented through fabric analysis of the entire damage zone of the Punchbowl fault (Chester and Logan 1987), synthetic and antithetic subsidiary faults near the ultracataclasite define a quasi-conjugate geometry with the bisector of the conjugates (axis of shortening) oriented at high angles to the layer.

The contact boundaries between the ultracataclasite layer and surrounding cataclastic rocks are irregular and nonparallel (figure 8.8). At the mesoscopic scale, the boundaries display a roughness (ratio of the amplitude to the wavelength of geometric irregularities) of approximately 10^{-1} (Chester and Chester 1998). A single Prominent Fracture Surface (PFS) cuts the ultracataclasite in all mapped exposures (Chester and Chester 1998). The PFS is continuous across each mapped exposure, is located within the ultracataclasite or along the boundary of the layer, and cuts across the layer (boundary to boundary) in some locations. The PFS displays a mesocopic roughness of approximately 10^{-3}, and is much more planar than the boundaries of the ultracataclasite layer. All mesoscopic-scale structures truncate against the PFS (figures 8.8 and 8.9).

Different ultracataclasite units have been identified within the layer on the basis of grain size, color, cohesion, fracture fabric, veins, and porphyroclast lithology (Chester and Chester 1998). At all locations, two basic types of ultraca-

Figure 8.6 Variation in structural and chemical features across a fault zone containing paired fault cores at location T2, Punchbowl fault. Distance is measured from the northernmost fault core, and negative values are to the south. (a) Density of mesoscopic fractures and faults as a function of distance. Linear density is determined by counting the number of intercepts along two orthogonal lines. (b) Relative intensity of microscopic fractures, shears, and veins, versus distance. Intensity determined with an optical microscope by counting the number of intercepts along two orthogonal lines at 4X. The intensity is normalized by the value of the sample with the greatest count. (c) Volatile concentration as determined from loss on ignition (LOI) of fault rocks as a function of distance. (d) Density of mesoscopic veins as a function of distance. Linear density is determined by counting the number of intercepts along two orthogonal lines. Modified from Schulz and Evans (1998, 2000).

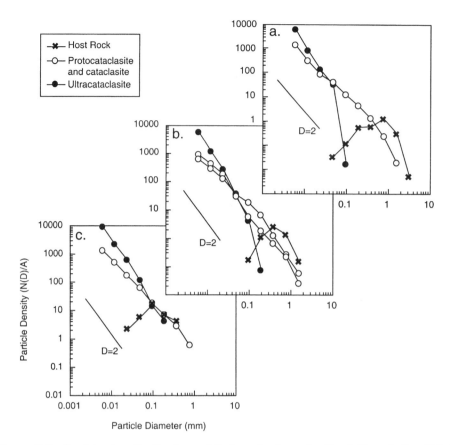

Figure 8.7 Particle-size distributions of fault rocks from the North Branch San Gabriel fault at (a) Bear Creek, (b) Devil's Canyon, and (c) from the southernmost fault core of the Punchbowl fault zone at Cruthers Canyon. Samples are grouped as undeformed host rock, protocataclasite and cataclasite from the damage zone, and ultracataclasite from the fault core. Slope for a fractal particle-size distribution with two-dimensional fractal dimension of 2 is shown for reference. Particle-size distributions determined from digital images of samples at magnifications of $\times 15$, $\times 30$, $\times 55$, $\times 130$, $\times 325$, $\times 814$, $\times 1630$. Particles were sorted into four classes, where each class differs in mean diameter by a factor of 2. Representative areas that do not contain particles larger than the largest class at the image magnification were measured, which imposes constraints on the determination of particle distributions as discussed by Sammis et al. (1987). Because the magnification varied by a factor of 3 or less, several independent measurements of particle density were measured for all but the largest and smallest particle-size classes. Particle density given as the number of particles in a class size, N(D), divided by area of image.

taclasite comprise the layer as illustrated in figure 8.8. An olive-black ultracataclasite (hereafter referred to as black ultracataclasite) always is found in contact with cataclastic basement on the south side of the layer. A dark yellowish-brown ultracataclasite (hereafter referred to as brown ultracataclasite) always is found in contact with the cataclastic sandstone on the north side. The contact between the black and brown ultracataclasites is sharp, continuous, and either coincides with

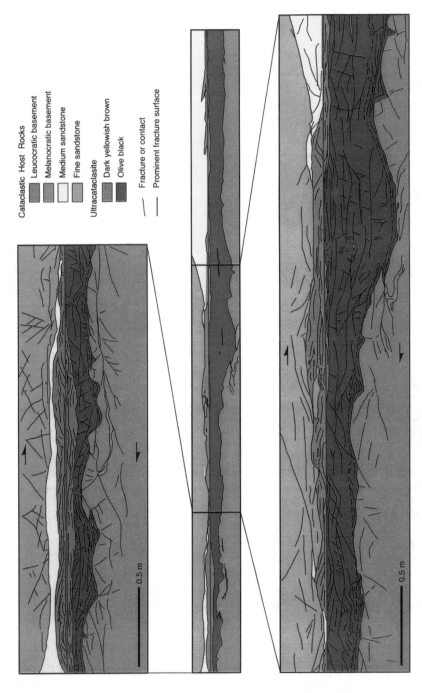

Figure 8.8 Map of a portion of the ultracataclasite layer in the fault core of the Punchbowl fault, Devil's Punchbowl Park. The orientation of the map projection is approximately perpendicular to the layer and parallel to the slip direction. The central panel shows the overall geometry of the section mapped, and the upper and lower panels show greater detail. The thick red line through the center of the layer is the prominent fracture surface (PFS). The thin black lines represent the location of contacts and fractures. The ultracataclasite layer has irregular boundaries and variable thickness, and consists of two basic types of ultracataclasite juxtaposed along a continuous contact that is nearly coincident with the PFS (see text). Modified from Chester and Chester (1998).

Cataclastic Host Rocks
- Leucocratic basement
- Melanocratic basement
- Medium sandstone
- Fine sandstone

Ultracataclasite
- Dark yellowish brown
- Olive black

— Fracture or contact
— Prominent fracture surface

0.5 m

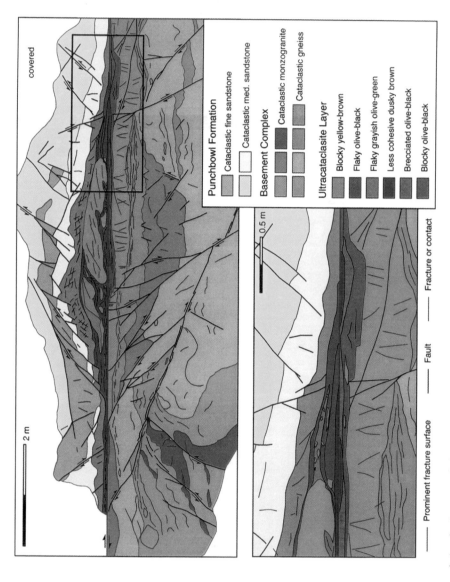

Figure 8.9 Map of the fault core of the Punchbowl fault, Devil's Punchbowl Park. The orientation of the map projection is approximately perpendicular to the layer and parallel to the slip direction. Compositional layering in the fault core is subparallel to the fault from the combined effect of subsidiary faulting and cataclastic flow. A sliver of basement within the ultracataclasite layer, but on the opposite side of the PFS as the basement host rock, is evidence for branching of the PFS (see text). Offset on subsidiary faults contributes to roughness of the ultracataclasite-cataclasite contact. Detail of structure in vicinity of the basement sliver shows the distribution of, and truncation relations for, the different ultracataclasite units.

the PFS or is located near the PFS. The contact has mesoscopic roughness intermediate between that of the PFS and outer boundaries of the ultracataclasite layer. The relatively few pebble- to cobble-size porphyroclasts that are present in the ultracataclasite tend to occur in clusters (figure 8.8). Porphyroclasts of cataclastic sandstone are confined to the brown ultracataclasite and porphyroclasts of cataclastic basement are confined to the black ultracataclasite.

We infer that the brown ultracataclasite was derived primarily from the Punchbowl Formation sandstone because (1) it is always in contact with the cataclastic sandstone, (2) it contains only porphyroclasts of sandstone, (3) it is very similar in appearance to ultracataclasites in subsidiary faults of the Punchbowl Formation, and (4) porphyroclasts of sandstone are restricted to the brown ultracataclasite (Chester and Chester 1998). On the basis of similar observations and reasoning, we infer that the black ultracataclasite primarily originated from comminution of crystalline basement.

Most of the ultracataclasite is cohesive due to syn-faulting cementation and fracture sealing recorded by veins and vein fragments. A centimeter-thick accumulation of less-cohesive ultracataclasite occurs locally along the PFS. This less cohesive ultracataclasite generally lacks veins and vein fragments. We infer that the PFS was a through-going, mesoscopic slip surface within the ultracataclasite because (1) the PFS is present and continuous in all exposures of the ultracataclasite; (2) all contacts, layering, and fracture surfaces in the ultracataclasite either merge with or truncate against the PFS; (3) the PFS forms a contact between different fault rocks; (4) the PFS displays a smaller roughness in the slip-parallel than in the slip-perpendicular exposures, consistent with mesoscopic-scale corrugations of the surface aligned with the inferred direction of slip on the Punchbowl fault; and (5) the PFS is spatially associated with the thin layer of less cohesive ultracataclasite (Chester and Chester 1998).

On the basis of geometry, location, cohesion, and porphyroclast content, we infer that the less cohesive ultracataclasite was produced by attrition wear during sliding on the PFS after the last major episode of cementation and veining. If true, the less cohesive ultracataclasite was derived primarily by reworking the older, cohesive ultracataclasite, because it and the PFS are almost wholly contained within the ultracataclasite layer. Only along the sections of the PFS that coincided with the contact between the ultracataclasite and surrounding cataclasite could new ultracataclasite material be produced by attrition of the cataclasite.

Chester and Chester (1998) conclude that the Punchbowl Formation and crystalline basement were juxtaposed at the mapped location for at least the last two to ten kilometers of Punchbowl fault displacement on the basis of regional geology and outcrop relations in Devil's Punchbowl Park. We infer that the contact between the brown and black ultracataclasite units, which were derived from the Punchbowl Formation and crystalline basement, respectively, was originally formed by slip on the PFS. Formation of the contact by slip on the PFS is consistent with observations that the PFS does not cut across and offset the contact, and the PFS either coincides with, or is separated from, the contact by the less cohesive, brown ultracataclasite. After the contact was formed, continued slip on the PFS resulted in the formation of a narrow layer of reworked, less cohesive ultracataclasite.

The structure of the Punchbowl ultracataclasite appears consistent with the end-member mode of failure of little mixing and localized slip as shown in figure 8.2a. The fact that brown ultracataclasite and associated cataclastic sandstone, and black ultracataclasite and associated cataclastic basement, are confined to opposite sides of the PFS is evidence that slip occurred on the PFS with little branching or off-PFS shearing during the final phase of Punchbowl faulting. Apparently the change in location of the PFS relative to bounding rock was minor and occurred by migration via local reworking and accumulation of the less cohesive ultracataclasite.

Although most exposures indicate that the PFS was a fairly stable feature during the final stages of fault displacement, there is structural evidence for a branch event at one of the mapped locations (figure 8.9). The branch event involved a relocation of the PFS and isolation of a 0.5 × 2 m sliver of cataclasitic-granite and -gneiss within the ultracataclasite layer on the northern (Punchbowl Formation) side of the PFS. The older, blocky, yellow-brown ultracataclasite and abandoned segment of the PFS is preserved between the sliver and the cataclastic sandstone. After branching, the PFS was reestablished and the sliver was displaced with the sandstone host rock (figure 8.9a). Displacement of the sliver was accommodated in the bounding fault core and damage zone by offset on a network of antithetic and synthetic subsidiary faults. Trails of younger, flaky ultracataclasite containing many porphyroclasts from the basement sliver record localized slip and attrition wear along the boundary of the sliver and are most likely associated with the branching event. In addition, a brecciated and cemented ultracataclasite occurs as thin lenticular deposits along the PFS opposite to the sliver (figure 8.9b). Consistent with all other mapped locations, the most recently formed PFS is the most planar feature present, and it truncates all other structures. Also similar, the youngest accumulation is a less cohesive ultracataclasite that occurs in thin lenses along and adjacent to the PFS. The cross-cutting relations are consistent with progressive accumulation of ultracataclasite, from oldest at the boundaries and youngest inward toward the PFS, except within the sliver generated by the branching event where a reversal in age progression would be expected.

North Branch San Gabriel fault

A 7-m-long section of the North Branch San Gabriel fault core was mapped at a scale of 1:5 (figure 8.10). The outcrop surface is approximately perpendicular to the fault, but oblique to the inferred slip direction. Deformation and alteration of host rock increase toward the ultracataclasite layer, and the orientation of cataclastic foliation and the kinematics of subsidiary faults near the ultracataclasite are consistent with distributed right-lateral shear. On the basis of porphyroclast lithology, and progressive increase in deformation and alteration from the host rock through the damage zone and fault core, it is clear that the neighboring host rock outside the fault zone sourced the cataclastic rocks in the fault zone. The

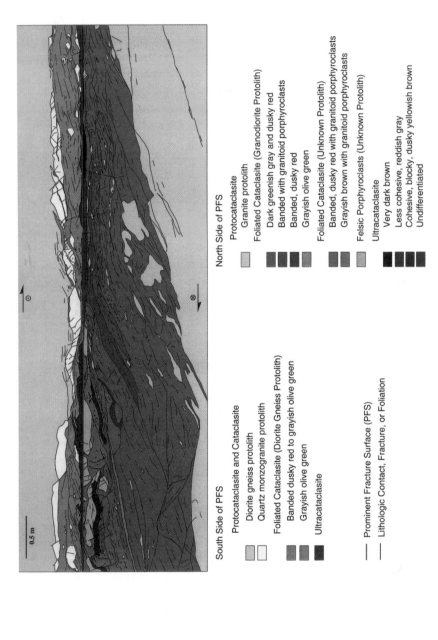

Figure 8.10 Map of the fault core of the North Branch San Gabriel fault, Bear Creek. Host rocks are granite and granodiorite on the north side, and quartz monzogranite and diorite gneiss on the south side, of the PFS. The foliated cataclasites in the fault core and on the south side of the ultracataclasite were derived from the diorite gneiss and quartz monzogranite, and those on the north side from the granodiorite. Several different types of ultracataclasite are distinguished on the basis of color, cohesion, fracture fabric, and porphyroclast occurrence. Most of the ultracataclasites occur in discontinuous layers along the PFS. A sliver of foliated cataclasite and ultracataclasite derived from an unknown protolith records one or more branching events (see text).

South Side of PFS

Protocataclasite and Cataclasite

- Diorite gneiss protolith
- Quartz monzogranite protolith

Foliated Cataclasite (Diorite Gneiss Protolith)

- Banded dusky red to grayish olive green
- Grayish olive green

Ultracataclasite

—— Prominent Fracture Surface (PFS)

—— Lithologic Contact, Fracture, or Foliation

North Side of PFS

Protocataclasite

- Granite protolith

Foliated Cataclasite (Granodiorite Protolith)

- Dark greenish gray and dusky red
- Banded with granitoid porphyroclasts
- Banded, dusky red
- Grayish olive green

Foliated Cataclasite (Unknown Protolith)

- Banded, dusky red with granitoid porphyroclasts
- Grayish brown with granitoid porphyroclasts

Felsic Porphyroclasts (Unknown Protolith)

Ultracataclasite

- Very dark brown
- Less cohesive, reddish gray
- Cohesive, blocky, dusky yellowish brown
- Undifferentiated

0.5 m

contact between the ultracataclasite and bounding foliated cataclasite is sharp at the mesoscopic scale, and the ultracataclasite layer constitutes a major discontinuity in protolith across the fault. Cataclastic foliations have been rotated into parallelism with the ultracataclasite at the contact, or they are truncated by the ultracataclasite (figure 8.10).

Although some of foliated cataclasites are very fine grained, the ultracataclasite is easily distinguished by a uniform micron-scale grain size and lack of mesoscopic porphyroclasts. A single PFS with roughness of approximately 10^{-3} is present within the ultracataclasite along the entire length of the outcrop. Contacts between the different ultracataclasites are subparallel to the overall layer and either merge with, or are cut by, the PFS (figure 8.10). Cross-cutting relations are compatible with the conclusion that the oldest ultracataclasite occurs along the boundary of the layer, and progressively younger ultracataclasite units occur toward the PFS. Each type of ultracataclasite is confined to one side of the PFS.

At the eastern end of the mapped exposure, a segmented layer of cohesive, blocky, dusky yellowish-brown ultracataclasite is present on the northern side of the PFS (figure 8.10). This segmented layer and the ultracataclasite layer along the PFS bound a sliver of grayish-brown, foliated cataclasite containing large granitoid porphyroclasts. The sliver contains rock types exotic to the outcrop and is interpreted to have formed by an ancient branch event. The segmented ultracataclasite layer likely reflects the portion of the ancient ultracataclasite layer and PFS that was abandoned by the branch event. Subsidiary faulting and distributed cataclasis after the branch event resulted in the present geometry. A few less obvious pods of ultracataclasite are isolated within the foliated cataclasites; these may record earlier branch events.

Some fault displacement was accommodated by shear in the foliated cataclasites of the fault core; however, the majority of fault displacement occurred within the ultracataclasite layer. Displacement was localized to the ultracataclasite layer early in the faulting history and to the PFS at least during the final stages of faulting. Overall, the North Branch San Gabriel fault structures are consistent with progressive production and accumulation of ultracataclasite along a single slip surface that was long lived and relatively stable, similar to the end-member case depicted in figure 8.2a of localized slip with little mixing.

Discussion

Fault-Zone Structure for Single and Paired Fault Cores

The variation in deformation intensity across the Punchbowl and North Branch San Gabriel fault zones, as defined by each individual fabric element, does not correlate exactly. However, we suggest that at first approximation, features of the fault zone may be grouped into those that define the fault core and those that define the damage zone (figure 8.11). Elements that define the fault core, such as fine grain size and mineral alteration, reflect high-shear strain, extreme commi-

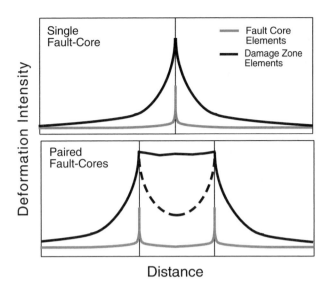

Figure 8.11 Relative intensity of deformation within a fault zone as a function of distance from principal fault surfaces (ultracataclasite) for two classes of fabric elements. One class of fabric element, which includes fractures and subsidiary faults, defines the damage zone. The other class, which includes comminuted particles, syntectonic mineralization, and cataclastic foliations, defines the fault core. (a) Symmetric deformation about a single fault surface as observed in the North Branch San Gabriel fault. Fabric elements of the damage zone decrease in density with log distance from the fault surface. Fabric elements of the fault core display high relative intensity only very near the fault surface. (b) Symmetric deformation about paired fault surfaces, as observed in the Punchbowl fault zone at location T2 (solid line) and as predicted by simple superposition of deformation associated with two individual fault surfaces (dashed line, see text).

nution, and enhanced fluid-rock reactions. Elements that define the damage zone, such as fractures and subsidiary faults, reflect stress cycling throughout displacement history of the fault.

The density of fabric elements in a damage zone bounding a single fault core, as in the North Branch San Gabriel fault, may be modeled as decreasing linearly with the logarithm of distance from the ultracataclasite layer, similar to that reported for other fault zones with smaller displacements (Anders and Wiltschko 1994; Scholz et al. 1993; Vermilye and Scholz 1998). Such a distribution describes a marked change in damage intensity with distance near the fault core, subtle changes in damage intensity with distance far from the fault core, and an outer boundary that is gradational with the undeformed host rock (figure 8.11a).

We use observations of the Punchbowl fault to test the hypothesis that the distribution of damage elements in a fault zone containing paired fault cores is predicted by a simple linear superposition of damage associated with each individual fault core (figure 8.11b). For the case of the Punchbowl fault at location T2 in the Pelona Schist, the distance between the paired fault cores is similar to the thickness of the damage zone of the North Branch San Gabriel fault (figures

8.4 and 8.6). Superposition of the damage distribution of the North Branch San Gabriel fault about each fault core of the Punchbowl fault predicts a decrease in damage with log distance from the paired fault cores outward toward the undeformed host rock, similar to that observed in the Punchbowl fault (figures 8.6 and 8.11). However, this prediction underestimates the observed damage in the region between the fault cores. We suggest that the high intensity of damage between paired fault cores reflects the mechanical interaction of the two faults. Mechanical analyses of paired faults and overlapping fault segments indicate local modification of stress states that can increase the likelihood of fracturing rock between two faults (e.g., Segall and Pollard 1980; Johnson and Fletcher 1994). Observations of small faults also indicate enhanced fracturing in regions between paired or overlapping fault segments (e.g., Segall and Pollard 1983; Martel 1990). Mechanical interaction between paired faults could provide an explanation for the very broad zones of damage observed along some large-displacement, strike-slip fault zones.

Fault Evolution and Origin of Damage Zones

Damage may accumulate during fault growth from process-zone deformation associated with propagation of fault tips (e.g., Cowie and Scholz 1992; Scholz et al. 1993), progressive deformation associated with stress cycling from displacement on nonplanar and segmented faults (e.g., Flinn 1977), and stress cycling from repeated passage of propagating slip events (e.g., earthquake ruptures, Rudnicki 1980). Experiments and field studies indicate that growth of brittle-fault zones occurs by a complex breakdown process at fault tips involving coalescence of fractures and shears to form through-going fault cores (e.g., Cowie and Scholz 1992; Reches and Lockner 1994). Concentration of shear stress at tips of faults produces arrays of cracks that extend, interact, and ultimately link. This growth process is associated with a reduction in strength from the intrinsic yield strength of the intact rock to residual frictional strength on the through-going fault core (Cowie and Scholz 1992). Scholz et al. (1993) suggested that a propagating fault tip would leave in its wake a zone of fractured rock about the fault core that is similar to the zone of damage observed along established faults. If true, the thickness of the damage zone should scale with the breakdown process zone.

Shear displacement on a nonplanar fault will lead to geometric mismatch, local stress concentration, and additional deformation of rock surrounding the fault (e.g., Flinn 1977; Chester and Chester 2000). Continued slip on irregular faults will produce stress cycling and damage in the host rock that ultimately may overshadow the relict damage from tip propagation. Fault surfaces display roughness at all scales (e.g., Scholz and Aviles 1986; Power and Tullis 1991), so outwardly expanding zones of damage might be expected as fault displacement accumulates because larger wavelength roughness elements will be juxtaposed with greater slip (e.g., Scholz 1987). In this case, thickness of a damage zone should scale with total fault displacement.

Inhomogeneous slip or propagating slip patches (e.g., earthquake ruptures) also will produce stress concentrations resulting in additional deformation adjacent to faults. Growth of an individual earthquake rupture involves a breakdown process at the rupture tip similar to the process of fault growth (e.g., Rudnicki 1980; Scholz et al. 1993; Swanson 1992). The structural signature from passage of a seismic rupture probably is a narrow zone or zones of concentrated shear demarcating the rupture surface within a broader zone of distributed fracturing. The cumulative effect of numerous ruptures on a large-displacement, seismic fault may be a damage zone characterized by a thickness that scales to the breakdown dimension of earthquake ruptures. The breakdown dimension for an earthquake rupture on an existing fault is probably less than that for formation of a new fault (Cowie and Scholz 1992). Thus, the thickness of a damage zone produced during seismic slip may be less than that produced during initial stages of fault formation and growth.

For small faults, damage associated with stress cycling from slip on rough surfaces or from repeated earthquake rupture is probably less important due to the small magnitude of total displacement. Observations of microfracture fabric and density distribution about small faults appear to compare well with expectations based on models of fault-tip stresses associated with fault growth (Vermilye and Scholz 1998). The studies of fault development in crystalline rock by shear along preexisting joints also show secondary fracturing concentrated in the tip regions and in areas of fault overlap and interaction (e.g., Segall and Pollard 1983; Martel 1990). These and numerous other studies of fault growth support the conclusion that faults lengthen through tip propagation and segment linkage, and that the process is somewhat scale invariant (e.g., Davison 1994; Ben-Zion and Sammis 2003). In many cases, fault segments are en echelon and linkage involves relatively intense fracturing and secondary faulting over a broad region. Damage associated with a large-displacement fault that is relict from development by segment growth and linkage likely would be heterogeneous in dimensions and intensity along strike. Portions of a fault zone developed through linkage may contain more distributed and intense damage than portions developed through activation of preexisting weaknesses and tip propagation.

Large-displacement faults described herein display a relatively ordered structure that is continuous along strike, albeit with some variability in relative intensity and dimensions. A relatively ordered damage zone is consistent with an origin by stress cycling and progressive accumulation of damage as slip accrues.

Slip on the Punchbowl and North Branch San Gabriel faults must have localized to a fault core early in the displacement history because host-rock lithology only changes across the ultracataclasite layer, cataclastic foliations largely are confined to the fault core, and foliations locally are truncated by the ultracataclasite. The basic features, damage zone, fault core, and localized slip surface, probably formed early in the faulting history, although the damage zone probably was disordered and the slip surfaces less stable during early stages. The structure of small, brittle, strike-slip faults in crystalline rock, formed at similar depths to the fault exposures studied herein, is consistent with the inference that slip is localized and with the suggestion that basic elements of fault zones are established early in displacement history (e.g., Evans et al. 2000).

Although localization to a slip surface occurred early, additional deformation in the damage zone and fault core of the Punchbowl and North Branch San Gabriel fault zones occurred throughout the displacement history. For example, the present structure of the ultracataclasite is most consistent with the branching events having occurred during intermediate or later stages of fault displacement history. Features produced during the branching event, such as the abandoned ultracataclasite contacts, are offset by subsidiary faults and distorted by local flow. Overall, the spatial relation between geometric irregularities of the ultracataclasite-cataclasite contact and subsidiary faults in the bounding cataclasite and damage zone records structural modification by fracturing, faulting, and flow during early, intermediate, and late stages of the displacement history.

The fabric of subsidiary faults in the damage zone of both faults also records deformation after fault formation. Quasi-conjugate subsidiary fault fabrics record a contraction direction at high angles to the principal faults. High-angle shortening is similar to that inferred from earthquake focal mechanisms for the modern San Andreas fault (e.g., Zoback and Beroza 1993; Savage 1994) and may be most compatible with slip on a preexisting fault rather than fault formation by tip propagation (e.g., Scholz et al. 1993; Chester and Chester 2000). Additional information on timing of deformation along the Punchbowl fault is provided by the study of microfractures in the sandstone of the Punchbowl Formation in Devil's Punchbowl Park (Wilson et al., unpublished manuscript). The fabric of healed (fluid inclusion planes), sealed (containing secondary mineralization), and open microfractures in the damage zone of the Punchbowl fault are consistent with the shortening direction inferred from the subsidiary fault fabric (Wilson et al., unpublished manuscript). Cross-cutting relations constrain the timing of fracturing during movement on the Punchbowl fault. The fabrics of healed (early phase), calcite-sealed, and open (latest phase) microfractures record fairly similar average stress for all phases of deformation and are consistent with progressive microfracture development from local stress cycling of the wall rock associated with movement along a weak, nonplanar fault.

Comminution Processes

Different types of fault rocks may reflect progressive comminution by a single process or different comminution processes that operate independently. Sibson (1986) identifies three types of cataclasites associated with different processes of cataclastic grain size reduction. Attrition breccias are highly comminuted rock produced by abrasive wear along sliding surfaces. Crush breccias record distributed fracturing and particle crushing due to stress cycling as might occur along faults. Implosion breccias have jigsaw-puzzle-type clast geometry and are thought to form by a sudden decrease in pressure as might occur during earthquake rupture propagation into a dilational jog (Sibson 1986). Both the crush and implosion breccias record substantial deformation, but may not necessarily record large-shear strain. In contrast, attrition breccias necessarily are associated with sites of shear. We interpret the different types of fault rocks in the Punchbowl and San Gabriel faults in the context of Sibson's (1986) attrition, crush, and implosion breccias.

Within the Punchbowl and North Branch San Gabriel fault zones, deformation in the damage zone is characterized by distributed fracture and particle-size reduction even though relatively little shear strain was achieved. The overall increase in intensity of fracturing toward the fault cores produced progressively finer particles. Observations are compatible with many episodes of healing, sealing, and fracturing in the damage zone and with rock repeatedly loaded to failure. Characteristics of the cataclastic rock in the damage zone are consistent with a crush-breccia classification (Sibson 1986).

The distribution of fracture and particle-size reduction recorded in crush breccias may be consistent with constrained comminution. Constrained comminution is the tensile fracturing of particles resulting from loading by neighboring particles where the greatest probability of fracture occurs if neighboring particles are similar in size. The constrained comminution model for fragmentation of a three-dimensional granular material (Sammis et al. 1987) predicts a fractal particle-size distribution with dimension of approximately 2.6, which is very similar to that observed for the cataclasites and protocataclasites bounding the ultracataclasite layers of the Punchbowl and San Gabriel fault zones (figure 8.7). Not only does constrained comminution produce a fractal particle-size distribution, it also results in a geometrical arrangement in which neighboring particles are different in size. This geometrical arrangement of particles is observed in the protocataclasites and cataclasites of the damage zone.

Similar to the findings from previous studies of fault rocks, we find the fractal dimension of the finer grained fault rocks tends to correlate with maximum fragment size, and fault rocks from regions of concentrated shear generally have a larger fractal dimension (Blenkinsop 1991; An and Sammis 1994). The particle-size distribution of the ultracataclasites is characterized by a fractal dimension larger than 3 (in three dimensions). The foliated cataclasites and ultracataclasites of the fault core consist of a fine-grained cataclastic matrix containing relatively large porphyroclasts of veins and host rock. The matrix supported porphyroclast texture is not the expected geometrical arrangement of particles for the constrained comminution process.

The particle-size distribution and particle arrangement of the foliated cataclasites and ultracataclasites could be produced by constrained comminution if an additional fragmentation process was active. An and Sammis (1994) suggest that finer grained fault rocks will result naturally from progressive deformation to high-shear strains if the probability of particle fracture in constrained comminution also depends on absolute particle size, such as would occur if there is a grinding limit. Given the large-shear strain in the fault cores, the particle-size distributions of the ultracataclasites could be the end product of progressive fragmentation by constrained comminution with a grinding limit. From such a model, one would expect a record of progressive comminution in the form of a continuous reduction in particle size toward the locus of highest shear strain. Such a progressive grain-size reduction is observed in the traverse from fractured host rock to protocata-clasite of the damage zone and through cataclasite and foliated cataclasite of the fault core. However, the contact with the ultracataclasite is sharp, forming a tex-turally discontinuous boundary. We suggest that an additional process contributes to the formation of the ultracataclasite and this distinct textural boundary.

Localization of Slip and Formation of Ultracataclasite

The PFS, a surface of localized sliding during late-stage faulting (Chester and Chester 1998), may have been a dominant slip surface throughout faulting history. If true, the generation and progressive accumulation of finely comminuted material from abrasive wear along such a slip surface could explain a number of important characteristics of the ultracataclasite layer including the (1) uniform and extremely fine grain size of ultracataclasite, (2) layered structure of ultracataclasite units and truncation of layers at the PFS, (3) general decrease in age of accumulation of the layers of ultracataclasite toward the PFS, and (4) sharp contact between the ultra-cataclasite and surrounding cataclastic rocks (Chester and Chester 1998).

Laboratory experiments of sliding friction between rock surfaces with and without simulated gouge demonstrate that comminution along localized sliding surfaces can produce extremely fine-grained wear products. Shear of granular material at high pressure to a shear strain of about one produces a fractal particle-size distribution similar to that produced by constrained comminution (e.g., Biegel et al. 1989). After additional shear, however, slip localizes to form discrete shears, and fractal dimension of the particle-size distribution is increased within the shears (Marone and Scholz 1989). Similarly, extremely fine particles are produced during sliding of rock-on-rock surfaces (e.g., Moody and Hundley-Goff 1980). Commi-nuted material produced by grinding and abrasive wear of micro-roughness fea-tures on sliding surfaces is an attrition breccia (Sibson 1986). We classify the ultracataclasite in the Punchbowl and San Gabriel faults as attrition breccias be-cause the ultracataclasites have a uniform texture throughout, generally lack sur-vivor porphyroclasts, have a particle-size distribution with a large fractal dimen-sion, and are spatially associated with slip surfaces. Specifically, we interpret that all the ultracataclasite was formed by a process of abrasive wear from sliding along a PFS.

For sliding along a slip surface, the wall rocks on opposing sides of the surface move parallel to the surface. However, where rock along the surface undergoes abrasive wear, wall rocks will converge gradually toward the sliding surface. At locations where the fine-grained product of abrasive wear accumulates along the surface, the wall rocks will diverge gradually from the sliding surface. Even for low rates of abrasive wear, significant accumulations would likely occur after kilometers of displacement. For a slip surface largely contained within an ultra-cataclasite layer, as is the PFS of the Punchbowl and North Branch San Gabriel faults, the preexisting ultracataclasite would be abraded by additional slip. Only in locations where the slip surface is against host rock would the abrasive wear product be derived directly from the cataclastic rocks outside the ultracataclasite layer.

If the slip surface is continuous and long lived, then the accumulation of ultracataclasite could be very ordered. Off-surface deformation of the wall rocks, such as by subsidiary faulting, folding, and cataclastic flow, may lead to local divergence of wall rock from the slip surface and provide a site for long-term accumulation of ultracataclasite. Similarly, wall-rock deformation can produce convergence of the wall rocks and long-term wear. In exposures mapped, the geometry of subsidiary faults adjacent to the ultracataclasite layers suggests that fault-parallel extension of the wall rock was common. Figure 8.12 is a schematic representation of progressive wear and accumulation that appears common in the mapped exposures of the Punchbowl and North Branch San Gabriel faults. In this example, movement on a subsidiary fault leads to local wear and accumulation along one side of the slip surface. Note that the sequential accumulation produces youngest of ultracataclasite inward toward the slip surface, as well as truncation of layering at the surface, as is observed in mapped exposures.

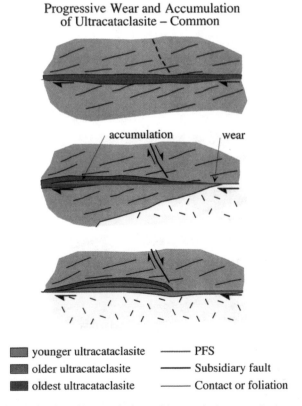

Progressive Wear and Accumulation
of Ultracataclasite – Common

accumulation   wear

■ younger ultracataclasite   ——— PFS
■ older ultracataclasite   ——— Subsidiary fault
■ oldest ultracataclasite   ——— Contact or foliation

Figure 8.12 Schematic showing evolution of layered ultracataclasite structure resulting from progressive wear and wear-product accumulation along a stable PFS. This process appears quite common in the ultracataclasite of the Punchbowl and North Branch San Gabriel faults.

Although not a common feature, incorporation of slivers of wall rock into the ultracataclasite layer by branching of the slip surface also must be considered in models of ultracataclasite evolution (figure 8.13). An observation from the mapped exposures is that the most recently active slip surface is planar and the abandoned slip surface is folded about the sliver. The sequence shown in the schematic implies that the branching fault formed along a curved trajectory, much like a side-wall rip-out (e.g., Swanson 1989), and then acquired a planar geometry through wall-rock deformation. It is equally plausible, however, that the wall-rock deformation occurred simultaneously or earlier to deform the original slip surface such that the branch propagated along a planar trajectory. An important consequence of branching is that the age progression of ultracataclasite accumulation may be disrupted through branching.

As suggested by the structure of fault cores studied herein, the PFS spans the ultracataclasite layers and can occur at the boundary of the surrounding cataclastic rock. Detailed inspection of these locations shows that the structures in the cata-clasite are truncated at the PFS and that the textural change from cataclasite to

Branching and Formation of
Exotic Sliver – Rare

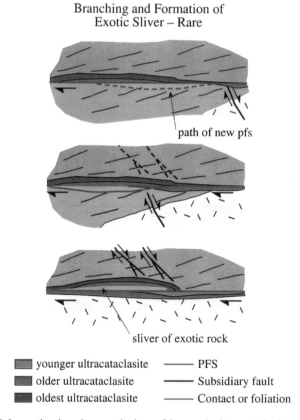

path of new pfs

sliver of exotic rock

▨ younger ultracataclasite	—— PFS
▨ older ultracataclasite	—— Subsidiary fault
▨ oldest ultracataclasite	—— Contact or foliation

Figure 8.13 Schematic showing evolution of layered ultracataclasite structure resulting from branching of the PFS to form an exotic sliver of rock within an ultracataclasite layer. Branching appears to have been fairly rare in the cores of the Punchbowl and North Branch San Gabriel faults.

ultracataclasite is abrupt, consistent with the interpretation that the cataclasite underwent abrasive wear at the surface. It follows that if accumulation of ultracataclasite occurs at a site after abrasive wear of the cataclastic wall rock, then the contact that formed between the ultracataclasite and wall rock will remain texturally distinct and sharp because it is a fossil slip surface. The contacts of ultracataclasite and wall rock formed by such a process should have small roughness (e.g., 10^{-3}). In the exposures mapped, it is clear that the contacts bounding the ultracataclasites have a roughness ten to hundreds times greater than the roughness of the PFS. However, the present roughness of the contacts is not a primary characteristic, rather it reflects minor deformation of the wall rock after the contacts were formed via minor subsidiary faulting, folding, and cataclastic flow.

For very large displacement faults having long-lived slip surfaces and ultracataclasite layers produced through abrasive wear and progressive accumulation, contacts within and bounding the ultracataclasite units should be sharp and continuous unless they were disrupted by subsequent deformation. Not only do the Punchbowl and North Branch San Gabriel faults display structures consistent with such a model, the continuity of ultracataclasite contacts further attests to the occurrence of relatively minor wall-rock deformation after the contacts formed.

Implications for Mechanical and Fluid-Flow Properties

A fault-core damage-zone structure is typical of many large-displacement, strike-slip faults in crystalline and well lithified sedimentary rock at shallow to intermediate depths where deformation is dominantly brittle (e.g., Ohtani et al. 2000; Chester and Logan 1986; Wallace and Morris 1986; Flinn 1977) and has been used to characterize the permeability and mechanical-property structure of fault zones in the continental crust (e.g., Gudmundsson et al. 2001; Lockner et al. 2000; Moore et al. 2000; Seront et al. 1998; Evans et al. 1997; Caine et al. 1996; Bruhn et al. 1994). The fault-core and damage-zone characterization also is compatible with descriptions of some dip-slip faults (e.g., Cowan 1999; Manatschal 1999; Lister and Davis 1989; Kennedy and Logan 1998; Wojtal and Mitra 1986) and strike-slip faults from deeper crustal levels where intracrystalline plastic mechanisms contribute significantly to deformation (e.g., Stewart et al. 1999; White and White 1983). Localization of slip to narrow, meters-thick zones has been documented in a number of these fault zones, yet few studies have suggested that kilometers of displacement are accommodated by sliding on discrete, continuous, long-lived slip surfaces within fault cores (e.g., Chester and Chester 1998).

Structures produced by extreme localization of slip in detachment faults of metamorphic core complexes (e.g., Lister and Davis 1989) and other normal faults (e.g., Manatschal 1999; Cowan 1999) are similar to the Punchbowl and North Branch San Gabriel faults in that they all have thin ultracataclasite layers and slip surfaces that juxtapose distinctly different rock types and, in at least some cases, are bounded by a damage zone cut by a subsidiary fault fabric that records shortening at a high angle to the fault (Axen and Selverstone 1994; Chester et al. 1993;

Chester and Logan 1987; Reynolds and Lister 1987). Detachment faults are late-stage features in the upper several kilometers of the crust, representing the last in a succession of normal faults and ductile-shear zones that cut through the extending crust (Lister and Davis 1989). The depth range over which extreme localization occurs in dip-slip and strike-slip faults is uncertain.

A critical question regarding the significance of fault structure to earthquake mechanics is whether slip on the PFS in the Punchbowl and North Branch San Gabriel faults was associated with nucleation and propagation of earthquake ruptures. By analogy with nearby active faults of the San Andreas system in the Transverse Ranges of southern California, we have hypothesized that these faults were seismic at least at some time during their history (Chester et al. 1993). If true, the composite structure of each fault zone records the passage of numerous earthquake ruptures. The overall slip distribution of the two faults is consistent with this hypothesis, but some structural details of the fault core may appear inconsistent with dynamic rupture. Dynamic, propagating earthquake ruptures are expected to generate large-magnitude transient stress at rupture tips, which may lead to rupture branching and structural disorder (e.g., Reches 1999). In addition, localization of seismic slip to a PFS should generate significant heat during an earthquake at several kilometers depth if faults follow laboratory friction laws (e.g., Kanamori and Heaton 2000). The laterally continuous and layered structure of the ultracataclasite and PFS is not consistent with a disordered structure and significant branching events. Furthermore, the Punchbowl and San Gabriel faults do not display pseudotachylytes or other evidence of locally elevated temperature.

Pseudotachylytes are the only feature in ancient, exhumed faults considered diagnostic of the mode of slip (e.g., Sibson 1989). Pseudotachylytes in fault zones are the product of frictional melting, which generally requires seismic slip rates at moderate pressure and localization of slip to surfaces or narrow zones. Although, the extreme localization of slip and inferred depth of faulting in the Punchbowl and North Branch San Gabriel faults is consistent with pseudotachylyte development during seismic slip, the lack of pseudotachylyte cannot be taken as evidence for aseismic creep.

Pseudotachylyte and other fault rocks were recovered from the main-shear zone of the Nojima fault in the Hirabayashi (Awaji Island) borehole one year after the 1995 Hyogo-ken Nanbu (Kobe) earthquake (e.g., Boullier et al. 2001; Tanaka et al. 2001; Ohtani et al. 2000). The Nojima fault is a right-lateral reverse-oblique slip fault with unknown total displacement. The maximum magnitude of slip during the 1995 Kobe earthquake was 2.15 m with 1.8 m right slip and 1.3 m reverse slip. The borehole crossed the main-shear zone of the fault at a depth of approximately 624 m, and samples from the fault zone were recovered over the depth range 152.2 to 746.7 m. Several lines of evidence suggest the 1995 Kobe earthquake rupture occurred in the main-shear zone (Ohtani et al. 2000).

On the basis of mineralogy, fluid inclusions, and microstructures, most of the fault rocks recovered are interpreted to have formed at depths greater than the cored interval (Boullier et al. 2001; Ohtani et al. 2000). Deformation and alteration document a damage-zone and fault-core structure similar to the Punchbowl and

North Branch San Gabriel fault zones (Tanaka et al. 2001; Ohtani et al. 2000). Several stages of faulting were recorded by a variety of fault-rock types, including pseudotachylyte, ultracataclasite, cataclasite, breccia, and gouge. The pseudota- chylyte appears the most ancient, and it apparently was formed at significantly greater depth (5–10 km) than the cored interval (Boullier et al. 2001; Tanaka et al. 2000). The pseudotachylyte occurs as millimeters-thick dark gray or brown bands separated by thin layers of brecciated pseudotachylyte. Locally, bands are folded, boudinaged, and embedded in homogeneous glassy pseudotachylyte. The banded pseudotachylyte is cut and offset by later mineral-filled fractures. Although there is some record of coseismic and postseismic deformation in the form of fresh fractures and mineralized veins, to a large extent the fault rocks were not greatly disrupted during the 1995 Kobe earthquake. Samples from the Hirabayashi bore- hole show that very delicate, ancient, deformation features, such as the banded pseudotachylyte, may be preserved in a seismic fault even very close to an earthquake-rupture surface. Accordingly, preservation of ordered, delicate struc- tures in the Punchbowl and North Branch San Gabriel fault cores cannot be used as evidence of aseismic creep.

The extreme localization of slip to a PFS is consistent with the geometry and slip distribution expected for faults that display slip- or rate-weakening mechanical properties necessary for earthquake nucleation (Chester and Chester 1998). Main- tenance of a PFS and confinement of slip to this surface over the life of a fault, as inferred for the Punchbowl and North Branch San Gabriel faults, imply that the surface was extremely weak. Whether the weakness was absolute or only relative to surrounding rock is an important question to resolve. There are several hypothesized mechanisms of earthquake rupture that require localized slip and significant dynamic weakening, such as elastohydrodynamic weakening (Brodsky and Kanamori 2001). Such mechanisms have been invoked to explain character- istics of earthquakes and the low apparent strength of the San Andreas fault (e.g., Kanamori and Brodsky 2001; Kanamori and Heaton 2000).

Geophysical characteristics of a fault zone may be significantly different from those of the bounding host rock (e.g., Ben-Zion and Sammis 2003) because of the presence of fractures, subsidiary faults, mineral alteration, and cementation. Elec- tromagnetic and seismic imaging studies suggest that faults are tabular zones of reduced resistivity and seismic velocity (e.g., Eberhart-Phillips et al. 1995). Wave- form modeling of seismic fault-zone-guided waves suggests that these tabular zones of reduced velocity are similar in dimension to the damage zones of the Punchbowl and North Branch San Gabriel faults (Li et al. 1994; Ben-Zion and Malin 1991). Observations that the fracture and fault density increase toward fault cores suggest that elastic properties will vary similarly; for example, Young's modulus should decrease gradually from the boundaries of a fault zone toward the fault core (Chester and Logan 1986).

On the basis of direct measurement of mechanical and fluid-flow properties of fault-rock samples from the damage zone of the Punchbowl fault, Chester and Logan (1986) suggest that fault zones in crystalline rock should display a gradual increase in permeability and relative ductility, and decrease in strength across the damage zone to the core. The boundary between the fault core and damage zone should be characterized by a sharp reduction in strength and permeability, and increase in ductility. Direct measurements on other fault rocks support this general characterization for crystalline host rocks (e.g., Forster and Evans 1991; Evans et al. 1997; Seront et al. 1998; Lockner et al. 2000; Moore et al. 2000). Using a damage-zone fault-core structure and classifying it as a barrier-conduit system, Caine et al. (1996) predict fluid-flow properties across and along such a fault zone. In this permeability model, the relatively permeable damage zone acts as a conduit for fluid flow along the fault, and the low-permeability fault core serves as a barrier for cross-fault flow. As noted by Caine et al. (1996), a number of intrinsic and extrinsic parameters influence the fault-permeability structure, including porosity and degree of lithification. Deformation of porous rocks and poorly lithified sediments may instead lead to reduction in permeability of the damage zone relative to the undeformed host (e.g., Antonellini and Aydin 1995; Rawling et al. 2001; Shipton et al. 2002; Hesthammer and Fossen 2000). Faults in poorly lithified sediments contain mixed zones, which are distinct in character and genesis from the components of the damage-zone and fault-core model described for crystalline and well lithified sedimentary rock; and they illustrate the importance of host lithology on fault structure and faulting mechanisms (e.g., Goodwin et al. 1999; Rawling et al. 2001).

A simple, one-dimensional model of fault structure only provides a basic understanding of the Punchbowl and North Branch San Gabriel fault systems. Critical to understand are spatial (three-dimensional) and temporal variations in fault-zone properties, including changes related to the seismic cycle. Localized slip in a fault core and more distributed fracturing as a result of rupture tip propagation may enhance both across- and along-fault fluid-flow during and immediately following seismic rupture (e.g., Sibson 1992). Creep compaction, healing, and sealing in a fault zone during interseismic periods may lead to significant reductions in permeability, strength recovery, and changes in elastic properties. Additional work directed at understanding these changes will significantly enhance our understanding of the role of faults in rupturing the lithosphere.

Conclusions

The North Branch San Gabriel and Punchbowl faults consist of one or more tabular zones of highly deformed rock formed by concentrated shear located within a much thicker zone of fractured and faulted rock referred to as fault core(s) and damage zone, respectively. The intensity of damage increases toward the fault core, the thickness of the damage zone varies laterally, and the transition outward from damage-zone rock to undeformed host rock is gradational. Localization of slip within the fault core is recorded by the presence of discrete layers of gouge or ultracataclasite and mesoscopic slip surfaces. Almost all shear displacement across the faults occurs in the ultracataclasite and thus, at the macroscopic scale, the ultracataclasite layer in the fault core represents the fault surface.

The North Branch San Gabriel fault zone contains a single fault core located near the center of the damage zone. The Punchbowl fault zone contains paired fault cores separated by a sliver of damaged rock that is tens to several hundreds of meters thick. For the fault zone containing paired fault cores, the distribution of damage outside the sliver is similar to that observed in the fault zone containing a single fault core. Within the sliver, however, the damage intensity is greater than that expected by simple linear superposition of the damage expected for each of the paired fault cores individually. We infer that the increased intensity and extent of damage in the sliver between paired faults reflect enhanced stress concentration from mechanical interaction of the faults.

The large-displacement, strike-slip faults described herein display an ordered structure that is relatively continuous along strike, albeit with some variability in relative intensity and dimensions. Elements that define the fault core, such as fine grain size and mineral alteration, reflect high-shear strain, extreme comminution, and enhanced fluid-rock reactions. Elements that define the damage zone, such as fractures and subsidiary faults, reflect stress cycling throughout fault-displacement history. Slip was localized and basic elements of fault structure were established early in the displacement history. The ordered damage zone and fault-core structure noted herein, which are not characteristic of small displacement faults, imply that damage continues to occur throughout fault history in response to stress cycling.

Particle-size distributions of host-rock and fault-rock samples from three different traverses are similar when grouped by structural position. Protocataclasites and cataclasites from the damage zone have fractal particle-size distributions and geometrical arrangement of particles consistent with constrained comminution (Sammis et al. 1987), and they may be classified as crush breccias (Sibson 1986). The size distribution and arrangement of particles in the fine-grained foliated cataclasites and ultracataclasites of the fault core are distinct from the other fault rocks but could be consistent with constrained comminution if fracturing also depended on absolute particle size. Structural relations, including the sharp contact between the ultracataclasite layer and bounding cataclasite, and association of ultracataclasites with mesoscopic slip surfaces, lead us to conclude that the rocks in the fault cores are better classified as attrition breccias (Sibson 1986) that they formed by a process other than constrained comminution.

The Prominent Fracture Surface in the Punchbowl and North Branch San Gabriel fault cores are much more planar than the contacts of the ultracataclasite with the surrounding cataclastic rock, and always occur along a boundary or within the layer. Progressive accumulation of ultracataclasite from abrasive wear along a slip surface throughout faulting history can explain the uniform and extremely fine grain size of the ultracataclasite, layered structure of ultracataclasite units and truncation of layers at the slip surface, and sharp contact between the ultracataclasite and surrounding cataclastic rocks. Longevity and relative stability of the slip surface are evidenced by the ordered accumulation of progressively younger ultracataclasite inward toward the slip surface. Although not a common feature, the incorporation of slivers of wall rock into the ultracataclasite layer by branching of the slip surface occurred within both fault zones.

An important question remains as to whether the extreme localization of slip that occurred in the Punchbowl and North Branch San Gabriel fault at 2 to 5 km depth is a feature common to most large-displacement faults. Extreme localization of slip to a narrow zone, with or without localization to continuous slip surfaces, has been documented for other strike-slip faults and low-angle (detachment) normal faults. The maintenance of the slip surface and confinement of slip to this surface over the life of the Punchbowl and North Branch San Gabriel faults imply that the slip surface was extremely weak. Whether the weakness is absolute or only relative to the surrounding rock, and whether the observed localization is associated with seismic slip, are important questions to resolve.

Strength, ductility, elasticity, permeability, and other properties will be strongly affected by the presence of fractures, subsidiary faults, mineral alteration, and cementation. The fault-core damage-zone description of a large-displacement fault in well-lithified sedimentary and crystalline rock provides a useful framework to describe the mechanical properties, fluid-flow properties, and geophysical characteristics of fault zones. There remains much uncertainty regarding temporal variation in properties over the life of faults as well as over the seismic cycle. Additional study of faults in all tectonic settings, depth ranges, and host-rock lithology is needed to further our characterization and understanding of the structure of, and fundamental physical processes operative in, crustal fault zones.

Acknowledgments

Thoughtful reviews were provided by Laurel Goodwin, Tim Dixon, Chris Scholz, and the editors of this volume. This research was partially supported by awards 98HQGR00003 and 99HQGR0047 from the U.S. Geologic Survey, Department of Interior.

References

An, L. J. and C. G. Sammis. 1994. Particle size distribution of cataclastic fault materials from southern California: A 3-D study. *Pure Appl. Geophys.* 143:203–227.

Anders, M. H. and D. V. Wiltschko. 1994. Microfracturing, paleostress and the growth of faults. *J. Struct. Geol.* 16:795–815.

Anderson, J. L., R. H. Osbourne, and D. F. Palmer. 1983. Cataclastic rocks of the San Gabriel fault: An expression of deformation at deeper levels in the San Andreas fault zone. *Tectonophysics* 98:209–251.

Antonellini, M. and A. Aydin. 1995. Effect of faulting on fluid flow in porous sandstones: Geometry and spatial distribution. *Am. Assoc. Pet. Geol. Bull.* 79:642–671.

Axen, G. J. and J. Selverstone. 1994. Stress state and fluid-pressure level along the Whipple Detachment fault, California. *Geology* 22:835–838.

Ben-Zion, Y. and P. Malin. 1991. San Andreas fault zone head waves near Parkfield, California. *Science* 251:1592–1594.

Ben-Zion, Y. and C. G. Sammis. 2003. Characterization of fault zones. *Pure Appl. Geophys.* 160:677–715.

Biegel, R. L., C. G. Sammis, and J. H. Dieterich. 1989. The frictional properties of a simulated gouge having a fractal particle size distribution. *J. Struct. Geol.* 11:827–846.

Blenkinsop, T. G. 1991. Cataclasis and processes of particle size reduction. *Pure Appl. Geophys.* 136:59–86.

Boullier A. M., T. Ohtani, K. Fujimoto, H. Ito, and M. Dubois. 2001. Fluid inclusions in pseudo-tachylytes from the Nojima fault, Japan. *J. Geophys. Res.* 106:21965–21977.

Brodsky. E. E. and H. Kanamori. 2001. Elastohydrodynamic lubrication of faults. *J. Geophys. Res.* 106: 16357–16374.

Bruhn, R. L., W. T. Parry, W. A. Yonkee, and T. Thompson. 1994. Fracturing and hydrothermal alteration in normal fault zones. *Pure Appl. Geophys.* 142:610–644.

Brune, J. N., T. L. Henyey, and R. F. Roy. 1969. Heat flow, stress and rate of slip along the San Andreas fault, California. *J. Geophys. Res.* 74:3821–3827.

Caine, J. S., J. P. Evans, and C. B. Forster. 1996. Fault zone architecture and permeability structure. *Geology* 24:1025–1028.

Chester, F. M. and J. S. Chester. 2000. Stress and deformation along wavy frictional faults. *J. Geophys. Res.* 105:23421–23430.

Chester, F. M. and J. S. Chester. 1998. Ultracataclasite structure and friction processes of the San Andreas fault. *Tectonophysics* 295:199–221.

Chester, F. M. and J. M. Logan. 1986. Implications for mechanical properties of brittle faults from observations of the Punchbowl fault zone, California. *Pure Appl. Geophys.* 124:79–106.

Chester, F. M. and J. M. Logan. 1987. Composite planar fabric of gouge from the Punchbowl fault, California. *J. Struct. Geol.* 9:621–634.

Chester, F. M., J. P. Evans, and R. L. Biegel. 1993. Internal structure and weakening mechanisms of the San Andreas fault. *J. Geophys. Res.* 98:771–786.

Christie, J. M. 1960. Mylonitic rocks of the Moine thrust-zone in the Assynt region, north-west Scotland. *Trans. Edinb. Geol. Soc.* 18:79–93.

Cowan, D. S. 1999. Do faults preserve a record of seismic slip? A field geologist's opinion. *J. Struct. Geol.* 21:995–1001.

Cowie, P. A. and C. H. Scholz. 1992. Physical explanation for the displacement-length relationships of faults using a post-yield fracture mechanics model. *J. Struct. Geol.* 14:1133–1148.

Cowie, P. A., R. J. Knipe, and I. G. Main. 1996. Scaling laws for fault and fracture populations: Analyses and applications. *J. Struct. Geol.* 18:R5–R11.

Davison, I. 1994. Linked fault systems; extensional, strike-slip and contractional. In P. L. Hancock, ed., *Continental Deformation*, pp. 121–142. Tarrytown, NY: Pergamon Press.

Dibblee, T. W., Jr. 1968. Displacements on the San Andreas fault system in the San Gabriel, San Bernardino, and San Jacinto Mountains, southern California. In *Proceedings of the Conference on Geological Problems of the San Andreas Fault System,* pp. 260–276. Stanford Univ. Publ. Geol. Sci. 11.

Eberhart-Phillips, D., W. D. Stanley, B. D. Rodriguez, and W. J. Lutter. 1995. Surface seismic and electrical methods to detect fluids related to faulting. *J. Geophys. Res.* 100:12919–12936.

Ehlig, P. L. 1981. Origin and tectonic history of the basement terrane of the San Gabriel Mountains, central Transverse Ranges. In W. G. Ernst, ed., *The Geotectonic Development of California: Rubey*, pp. 253–283. New York: Prentice Hall.

Evans, J. P. and F. M. Chester. 1995. Fluid-rock interaction in faults of the San Andreas system: Inferences from San Gabriel fault rock geochemistry and microstructures. *J. Geophys. Res.* 100:13007–13020.

Evans, J. P., C. B. Forster, and J.V. Goddard. 1997. Permeability of fault-related rocks, and implications for hydraulic structure of fault zones. *J. Struct. Geol.* 19:1393–1404.

Evans, J. P., Z. K. Shipton, M. A. Pachell, S. J. Lim, and K. Robeson. 2002. The structure and composition of exhumed faults, and their implications for seismic processes. In G. B. Bokelman and R. L. Kovach, eds., *Proceedings of the Third Conference on Tectonic Problems of the San Andreas Fault System,* pp. 67–81. Stanford, CA: Stanford University.

Flinn, D. 1977. Transcurrent faults and associated cataclasis in Shetland. *J. Geol. Soc. Lond.* 133:231–248.

Forster, C. B. and J. P. Evans. 1991. Hydrology of thrust faults and crystalline thrust sheets: Results of combined field and modeling studies. *Geophys. Res. Lett.* 18:979–982.

Goodwin, L. B., P. S. Mozley, J. C. Moore, and W. C. Haneberg. 1999. Introduction. In W. C. Haneberg, P. S. Mozley, J. C. Moore, and L. B. Goodwin, eds., *Faults and Subsurface Flow in the Shallow Crust*, pp. 1–5. Geophysics Monograph Ser. 113. Washington, DC: American Geophysical Union.

Gudmundsson, A., S. S. Berg, K. B. Lyslo, and E. Skurtveit. 2001. Fracture networks and fluid transport in active fault zones. *J. Struct. Geol.* 23:343–353.

Hesthammer, J. and H. Fossen. 2000. Uncertainties associated with fault seal analysis: Petroleum. *Geoscience* 6:37–45.

Hickman, S. H. 1991. Stress in the lithosphere and the strength of active faults. *Rev. Geophys. Suppl.* 29:759–775.

Johnson, A. M. and R. C. Fletcher. 1994. *Folding of Viscous Layers*. New York: Columbia University Press.

Kanamori, H. and E. E. Brodsky. 2001. The physics of earthquakes. *Phys. Today* 54(6):34–40.

Kanamori, H. and T. Heaton. 2000. Microscopic and macroscopic physics of earthquakes. In J. Rundle, D. Turcotte, and W. Klein, eds., *Geocomplexity and the Physics of Earthquakes*, pp. 147–183. Geophysics Monograph Ser. 120. Washington, DC; American Geophysical Union.

Kennedy L. A. and J. M. Logan. 1998. Microstructures of cataclasites in a limestone-on-shale thrust fault: implications for low-temperature recrystallization of calcite. *Tectonophysics* 295:167–186.

Kohlstedt, D. L., B. Evans, and S. J. Mackwell. 1995. Strength of the lithosphere: Constraints imposed by laboratory experiments. *J. Geophys. Res.* 100:17587–17602.

Li, Y. G., K. Aki, D. Adams, A. Hasemi, and W.H.K. Lee. 1994. Seismic guided-waves trapped in the fault zone of the Landers, California, earthquake of 1992. *J. Geophys. Res.* 99:11705–11722.

Lister, G. S. and G. A. Davis. 1989. The origin of metamorphic core complexes and detachment faults formed during tertiary continental extension in the northern Colorado River region, USA. *J. Struct. Geol.* 11:65–94.

Little, T. A. 1995. Brittle deformation adjacent to the Awatere strike-slip fault in New Zealand: Faulting patterns, scaling relationships, and displacement partitioning. *Geol. Soc. Am. Bull.* 107:1255–1271.

Lockner, D., H. Naka, H. Tanaka, H. Ito, and R. Ikeda. 2000. Permeability and strength of core samples from the Nojima fault of the 1995 Kobe Earthquake. In H. Ito, K. Fujimoto, H. Tanaka, and D. Luckner, eds., *Proceedings of the International Workshop on the Nojima Fault Core and Borehole Data Analysis*, pp. 147–152. U.S. Geological Survey Open-file Report 00-129.

Manatschal, G. 1999. Fluid- and reaction-assisted low-angle normal faulting: Evidence from rift-related brittle fault rocks in the Alps (Err Nappe, eastern Switzerland). *J. Struct. Geol.* 21:777–793.

Marone, C. and C. H. Scholz. 1989. Particle-size distribution and microstructures within simulated fault gouge. *J. Struct. Geol.* 11:799–814.

Martel, S. J. 1990. Formation of compound strike-slip fault zones, Mount Abbot Quadrangle, California. *J. Struct. Geol.* 12:869–882.

Moody, J. B. and E. M. Hundley-Goff. 1980. Microscopic characteristics of orthoquartzite from sliding friction experiments: II. Gouge. *Tectonophysics* 62:301–319.

Moore, D. E., D. A. Lockner, H. Ito, and R. Ikeda. 2000. Correlation of deformation textures with laboratory measurements of permeability and strength of Nojima fault zone core samples. In H. Ito, K. Fujimoto, H. Tanaka, and D. Luckner, eds., *Proceedings of the International Workshop on the Nojima Fault Core and Borehole Data Analysis*, pp. 159–165. U.S. Geological Survey Open-file Report 00-129.

Morton, D. M. and J. C. Matti. 1987. The Cucamonga fault zone, geologic setting and Quaternary history. U.S. Geological Survey Professional Paper 1339.

Oakeshott, G. B. 1971. Geology of the epicentral area. *Calif. Div. Mines Geol.* 196:19–30.

Ohtani, T., K. Fujimoto, H. Ito, H. Tanaka, N. Tomida, and T. Higuchi. 2000. Fault rocks and past to recent fluid characteristics from the borehole survey of the Nojima fault ruptured in the 1995 Kobe earthquake, southwest Japan. *J. Geophys. Res.* 105:16161–16171.

Powell, R. E., R. J. Weldon II, and J. C. Matti, eds. 1993. The San Andreas Fault System: Displacement, palinspastic reconstruction, and geologic evolution. *Geol. Soc. Am. Mem. 178.*

Power, W. L. and T. E. Tullis. 1991. Euclidean and fractal models for the description of rock surface roughness. *J. Geophys. Res.* 96:415–424.

Rawling, G. C., L. B. Goodwin, and J. L. Wilson. 2001. Internal architecture, permeability structure, and hydrologic significance of contrasting fault-zone types. *Geology* 29:43–46.

Reches, Z. 1999. Mechanisms of slip nucleation during earthquakes. *Earth Planet. Sci. Lett.* 170:475–486.

Reches, Z. and D. A. Lockner. 1994. Nucleation and growth of faults in brittle rocks. *J. Geophys. Res.* 99:18159–18173.

Reed, J. J. 1964. Mylonites, cataclasites, and associated rocks along the Alpine fault, South Island, New Zealand. *New Zealand J. Geol. Geophys.* 7:645–684.

Reynolds, S. J. and G. S. Lister. 1987. Structural aspects of fluid-rock interactions in detachment zones. *Geology* 15:362–366.

Robertson, E. C. 1983. Relationship of fault displacement to gouge and breccia thickness. *Mining Eng.* 35:1426–1432.

Rudnicki, J. W. 1980. Fracture mechanics applied to the earth's crust. *Annu. Rev. Earth Planet. Sci.* 8:489–525.

Rutter, E. H., R. H. Maddock, S. H. Hall, and S. H. White. 1986. Comparative microstructures of natural and experimentally produced clay-bearing fault gouges. *Pure Appl. Geophys.* 124: 3–30.

Sammis, C. G., G.C.P. King, and R. L. Biegel. 1987. The kinematics of gouge deformation. *Pure Appl. Geophys.* 125:777–812.

Savage, J. C. 1994. Evidence for near-frictionless faulting in the (M-6.9) Loma Prieta, California, Earthquake and its aftershocks: Comment. *Geology* 22(3):278–279.

Schmid, S. M. and M. R. Handy. 1991. Towards a genetic classification of fault rocks: Geological usage and tectonophysical implications. In D. W. Muller, J. McKenzie, and H. Weissert, eds., *Controversies in Modern Geology*, pp. 339–361. New York: Academic Press.

Scholz, C.H. 1987. Wear and gouge formation in brittle faulting. *Geology* 15:493–495.

Scholz, C. H. 1990. *The Mechanics of Earthquakes and Faulting*. New York: Cambridge University Press.

Scholz, C. H. and C. A. Aviles. 1986. The fractal geometry of faults and faulting. In S. Das, J. Boatwright, and C. Scholz, eds., *Earthquake Source Mechanics*, pp. 147–156. Geophysics Monograph. Washington, DC: American Geophysical Union.

Scholz, C. H., N. H. Dawers, J. Z. Yu, M. H. Anders, and P. A. Cowie. 1993. Fault growth and fault scaling laws: Preliminary results. *J. Geophys. Res.* 98:21951–21961.

Scholz, C. H. and T. C. Hanks. 2003. Fiction, friction, and the San Andreas fault. In G. D. Karner, B. Taylor, N. W. Driscoll, and D. L. Kohlstedt, eds., *Rheology and Deformation in the Lithosphere at Continental Margins*. New York: Columbia University Press.

Schulz, S. E. and J. P. Evans. 1998. Spatial variability in microscopic deformation and composition of the Punchbowl fault, southern California: Implications for mechanisms, fluid-rock interaction, and fault morphology. *Tectonophysics* 295:223–244.

Schulz, S. E. and J. P. Evans. 2000. Mesoscopic structure of the Punchbowl Fault, southern California and the geologic and geophysical structure of active strike-slip faults. *J. Struct. Geol.* 22:913–930.

Segall, P. and D. D. Pollard. 1980. Mechanics of discontinuous faults. *J. Geophys. Res.* 85:4337–4350.

Segall, P. and D. D. Pollard. 1983. Nucleation and growth of strike slip faults in granite. *J. Geophys. Res.* 88:555–568.

Seront, B., T. F. Wong, J. S. Caine, C. B. Forster, R. L. Bruhn, and J. T. Fredrich. 1998. Laboratory characterization of hydromechanical properties of a seismogenic normal fault system. *J. Struct. Geol.* 20:865–881.

Shipton, Z. K., J. P. Evans, K. R. Robeson, C. B. Forster, and S. Snelgrove. 2002. Structural heterogeneity and permeability in faulted eolian sandstone: Implications for subsurface modeling of faults. *Am. Assoc. Pet. Geol. Bull.* 86: 863–883.

Sibson, R. H. 1977. Fault rocks and fault mechanisms: *J. Geol. Soc. Lond.* 133:191–213.

Sibson, R. H. 1986. Brecciation processes in fault zone: Inferences from earthquake rupturing. *Pure Appl. Geophys.* 124:159–175.

Sibson, R. H. 1989. Earthquake faulting as a structural process. *J. Struct. Geol.* 11:1–14.

Sibson, R. H. 1992. Implications of fault-valve behaviour for rupture nucleation and recurrence. *Tectonophysics* 211:283–293.

Snoke, A. W. and J. Tullis. 1998. An overview of fault rocks. In A. W. Snoke, J. Tullis, and V. R. Todd, eds., *Fault-Related Rocks: A Photographic Atlas*, pp. 3–13. Princeton, NJ: Princeton University Press.

Stewart, M., R. A. Strachan, and R. E. Holdsworth. 1999. Structure and early kinematic history of the Great Glen Fault Zone, Scotland. *Tectonics* 18:326–342.

Swanson, M. T. 1989. Sidewall ripouts in strike-slip faults. *J. Struct. Geol.* 11:933–948.

Swanson, M. T. 1992. Fault structure, wear mechanisms and rupture processes in pseudotachylyte generation. *Tectonophysics* 204:223–242.

Tanaka, H., K. Fujimoto, T. Ohtani, and H. Ito. 2001. Structural and chemical characterization of shear zones in the freshly activated Nojima fault, Awaji Island, southwest Japan. *J. Geophys. Res.* 106:8789–8810.

Vermilye, J. M. and C. H. Scholz. 1998. The process zone: A microstructural view of fault growth. *J. Geophys. Res.* 103:12223–12237.

Wallace, R. E. and H. T. Morris. 1986. Characteristics of faults and shear zones in deep mines. *Pure Appl. Geophys.* 124:107–125.

Waters, A. C. and C. D. Campbell. 1935. Mylonites from the San Andreas fault zone. *Am. J. Sci.* 29:473–503.

White, J. C. and S. H. White. 1983. Semi-brittle deformation within the Alpine fault zone, New Zealand. *J. Struct. Geol.* 5:579–589.

Wilson, J. E., J. S. Chester, and F. M. Chester. 2003. Microfracture analysis of fault growth and wear processes, Punchbowl Fault, San Andreas system, California. *J. Struct. Geol.,* in press.

Wojtal, S. and G. Mitra. 1986. Strain hardening and strain softening in fault zones from foreland thrusts. *Geol. Soc. Am. Bull.* 97:674–687.

Zoback, M. D. and G. C. Beroza. 1993. Evidence for near-frictionless faulting in the 1989 (M-6.9) Loma Prieta, California, Earthquake and its aftershocks. *Geology* 21:181–185.

CHAPTER NINE

The Strength of the San Andreas Fault: A Discussion

Christopher H. Scholz and Thomas C. Hanks

Introduction

Fault mechanics is based on two central premises: the Coulomb failure criterion, which defines the geometrical relations of faults with the orientations of the principal stresses that formed them (Anderson 1951), and a simple friction law for rock (Byerlee 1978) that defines the stresses needed to produce continued slip on them.

There has been a long-standing controversy, lasting for more than three decades, as to whether the Byerlee friction law actually applies to faults, or at least to all faults. This controversy was first centered on the San Andreas fault (SAF) of California, for which evidence has been marshaled (e.g., Lachenbruch and Sass 1992) to suggest that this fault is very weak, with a friction coefficient $\mu \leq 0.1$ (based on heat-flow and stress-orientation data) or ≤ 0.2 (based on heat-flow data alone), either of which is far less than the $0.6 < \mu < 0.8$ required by laboratory measurements made with suitable geological materials. This controversy has subsequently encouraged proposals that other classes of faults could be similarly weak, also placing them beyond the constraints of Anderson-Byerlee fault mechanics (e.g., Wernicke 1995).

The history of this controversy is long and convoluted and deserves a brief recapitulation. The traditional view of seismologists, up to 1966, was that the stress drop in an earthquake was the total stress. This view was initiated by Tsuboi (1933, 1956), who identified the geodetically observed coseismic strain drop (about $1-2 \times 10^{-4}$) as the "ultimate strain" that the Earth could sustain. Thus the earthquake stress drop was identified as the strength of faults (Chinnery 1964). This view was challenged by Brace and Byerlee (1966), who proposed that earthquakes were the result of a stick-slip frictional instability. In this view, the earthquake stress drop corresponded to the difference between static and dynamic friction and was therefore only a small fraction of the absolute shear stress acting on the fault.

This new idea was contested almost immediately by a project to measure heat flow in the vicinity of the SAF to "See if there really is any friction on faults"

(J. N. Brune, personal communication to C.H.S., 1968). The famous result was that no heat-flow anomaly was observed in the vicinity of the SAF (Brune et al. 1969). Those authors concluded that the shear stress averaged over the seismogenic thickness could therefore be no larger than that necessary to account for earthquake stress drops, about 20 MPa. This argument was subsequently taken up by Lachenbruch and Sass, who, in a series of papers, added greatly to the number of heat-flow measurements and to the depth of their interpretation (Lachenbruch and Sass 1973, 1980, 1992).

Attempts to verify this conclusion with stress measurements have been mixed. Borehole stress measurements to a depth of 0.85 km in the Mojave block showed that shear stresses increase linearly with depth at a rate that would predict thickness-averaged stresses several times larger than expected from the heat-flow measurements (McGarr et al. 1982). Measurements made to 3.5 km at Cajon Pass showed that the crust adjacent to the SAF was strong, obeying Byerlee friction (Zoback and Healy 1992). The same authors claimed that these results also meant that the SAF was weak, but this argument was refuted by Scholz and Saucier (1993), who showed that the Cajon Pass stress measurements were insensitive to the strength of the SAF, and that the measurements were likely reflecting the strength of the Cleghorn fault, a minor left-lateral fault immediately adjacent to the borehole.

Mount and Suppe (1987) and Zoback et al. (1987) described observations from central California that seemed to indicate that the maximum principal stress direction was nearly orthogonal to the SAF, which would indicate that the fault was extraordinarily weak. This often cited result has been the strongest purported stress evidence for a weak SAF.

In the meantime, a school of geologists has been interpreting structures in the western United States as "detachments": low-angle normal faults that remain active with dips that are nearly horizontal (e.g., Wernicke and Burchfiel 1982; Wernicke 1995). Slip on normal faults with such low dips is prohibited by Anderson-Byerlee mechanics. On the other hand, they could be active in the ductile regime. There is no direct evidence, such as from earthquakes, that they do slip at such low angles within the frictional regime.

With this background in mind, we address the following questions in the remainder of this chapter. Are continental faults typically strong or weak? If the SAF is weak, is it then a common or an exceptional case, and if exceptional, what are the conditions that allow it to behave in this unusual manner? If, on the other hand, the SAF is strong, what are the basic flaws in the arguments that it is weak?

Some Basic Properties of Friction

The central issue is whether faults obey simple friction laws, and, if so, what is the friction coefficient associated with fault slip? The basic laws governing friction, known as Amonton's laws, have been known for more than 300 years. The first of these laws is that the friction depends on the normal load, written in the modern form in terms of stresses as

$$\tau = \mu(\sigma_n - p) \tag{9.1}$$

where τ is the shear stress necessary to drive a fault subjected to a normal stress σ_n and pore pressure p; μ is the friction coefficient. To avoid confusion, we refer to τ as the strength of the fault and μ as its friction. The second law of Amonton is that friction is independent of the area of the surfaces in contact. This seemingly counterintuitive law arises because the friction is determined by the real area of contact formed by the yielding of contact points and not by the apparent area. The real area of contact is solely determined by the ratio of normal load to yield strength and not by the apparent area of contact (see Scholz 1990:46–48). Thus the dimensionless parameter μ is scale independent. The μ measured at the laboratory scale should therefore be the same at the crustal scale.

A general rule about friction is that it is surprisingly independent of the material or temperature. This is because friction is the ratio of two strengths of the same material: (1) the yield strength that determines the real contact area and (2) the shear strength necessary to shear off the contact junctions. Thus, steel has the same friction as lead because, at a given normal load, the lead will form a proportionately greater contact area that will just offset its greater weakness in shear. Similarly, both strengths will be affected by temperature in the same way, canceling out the effect of temperature on friction. Byerlee (1978) was the first to demonstrate this material independence of friction for rocks. He found that rock, independent of lithology, had a friction coefficient in the range 0.6–1.0, a finding now referred to as Byerlee friction, or Byerlee's law. Similarly, Blanpied et al. (1998) demonstrated the temperature independence of rock friction by showing that friction of quartzite and granite remain at about 0.6 up to the onset of ductility at 350°C. The material independence of friction is, however, a rule, not a law, because there are exceptional materials with much lower friction. Those that may be of geologic importance will be discussed later.

In engineering, friction is usually something to be minimized, which is done by means of lubrication, of which there are two general types. The first is to separate the surfaces with a solid lubricant, such as soap or graphite, which prevents contact between the surfaces and at the same time is very weak in shear. The second is to use a viscous liquid like oil, which by viscous or hydrodynamic forces results in the development of a high fluid pressure between the surfaces, thus reducing the effective normal stress. For brittle materials like rock, however, the pore pressure is limited by the hydrofracturing constraint:

$$p \leq \sigma_3 \tag{9.2}$$

where σ_3 is the least principal stress. If p exceeds σ_3, hydrofracturing will occur and the fluid pressure will be drained. In a later section we will examine whether either of these lubrication methods can explain the San Andreas heat-flow constraint.

Friction in the Continental Crust

Stresses measured to a depth of 8 km at the KTB borehole in Germany are shown in figure 9.1. They show that stresses increase with depth at a rate that indicates that they are governed by friction on optimally oriented faults with a friction coefficient $0.6 < \mu < 0.7$. This same result has been universally obtained from all stress measurements made in deep (>1 km) boreholes (McGarr and Gay 1978; Zoback and Healy 1984, 1992; Brudy et al. 1997; Lund and Zoback 1999; Townend and Zoback 2000). These boreholes sampled a variety of tectonic settings, ranging from cratonic shields to plate-boundary regions and in extensional, compressive, and strike-slip settings. They include the Cajon Pass borehole, located only 4 km from the San Andreas fault. These observations taken together indicate that the entire continental crust is loaded close to its failure point, that everywhere the friction on faults is in the Byerlee range, and there exists no mechanism other than earthquakes to relax the accumulated stresses.

It is further observed in all these boreholes that the pore pressure is, at all depths, very close to hydrostatic because of the high permeability of the crust. The crustal scale permeability is three to four orders of magnitude higher than that measured for core samples, presumably because of the presence of joints and faults (Townend and Zoback 2000; note that stress data in their figure 1 and the permeability data in their figure 3 feature data from Cajon Pass, so that their conclusions apply equally to the SAF). Thus, not only do faults generally have high friction, but they are also not, in general, subjected to weakening by elevated pore pressure.

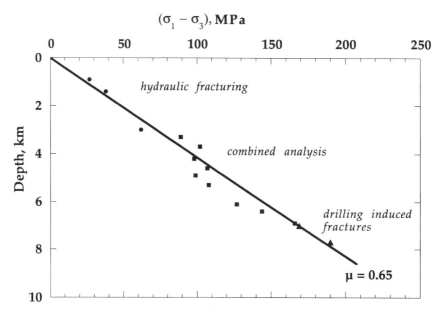

Figure 9.1 The state of stress measured in the KTB borehole, Germany (after Brudy et al., 1997).

Another friction indicator is the dip angle of active dip-slip faults. In such cases, either the maximum or minimum principal stress is vertical and the other horizontal so we can determine the angle θ between the fault and σ_1, the maximum principal stress. This angle has an optimum at $\theta_{opt} = 1/2 \cot^{-1}(\mu)$ and a maximum (lock-up) value of $\theta_{lu} = \cot^{-1}(\mu)$, which applies when equation (9.2) is an equality (see Sibson 1985).

To determine this angle, one requires a fault that experienced a large earthquake in which the fault-dip angle is known unequivocally from some information additional to the focal mechanism, such as aftershock locations, geodetic data, or surface faulting. Compilations of such data for both normal and thrust faults are shown as histograms in figure 9.2 where the optimum and lock-up angles for $\mu = 0.6$ are also indicated. The data agree very well with that friction level: there

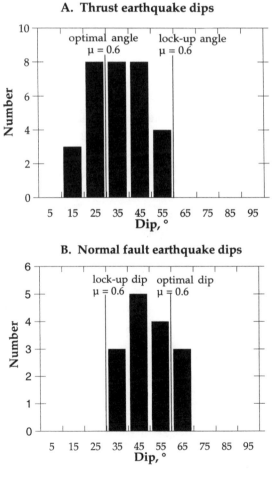

Figure 9.2 Histograms of dips of active dip-slip faults. The optimum and lock-up angles for $\mu = 0.6$ are shown. (A) Thrust faults, data from Sibson and Xie (1998). (B) Normal faults, data from Jackson (1987). A more recent compilation by Collineti and Sibson (2001) does not alter the results of the earlier one.

are no violations of the lock-up angle which would indicate a lower value of friction, just as is the case with the borehole stress measurements. (A reviewer pointed out that the data could admit faults with low μ and low p, which is true, but this is, for the ensemble, special pleading.) The fact that there are a few earthquakes that occurred with dips near the lock-up angle does indicate, however, that in some cases p approaches σ_3. Sibson (1987) has shown, from veining evidence, that repeated hydrofracturing has occurred in some high-angle reverse faults, indicating that $\sigma_3 = p$, the lithostat, in those cases. High pore pressures may also occur in subduction zones, particularly in their accretionary prisms (Le Pichon et al. 1993). According to Hubbert and Rubey (1959), neither of those cases, both thrust faults, violate friction laws.

Abers (1991) analyzed a M 6.8 normal faulting earthquake in the Woodlark Basin for which he found a likely fault plane with a dip between 10° and 25°. This event was featured in Wernicke's (1995) argument for active low-angle normal faulting, where the dip was quoted as 17°. Ultimately, the dip of the proposed causative fault was determined by seismic imaging and drilling to be 27 ± 3° (Taylor et al. 1999). This is not significantly different from the 29° lock-up angle expected for $\mu = 0.6$. Because this fault is near the lock-up angle, $p \approx \sigma_3$, as suggested from seismic reflection modeling (Floyd et al. 2001), but the friction is not required to be unusually low, nor is any special mechanism required to allow the hydrofracture constraint to be violated. The gouge from this fault is serpentinite, with small amounts of chlorite and talc (Taylor and Huchon, 2002). The latter was volumetrically minor, and thus unlikely to significantly reduce the friction coefficient. This fault is not an exception to Anderson-Byerlee mechanics.

Is the San Andreas Fault an Exception?

These same stress indicators can also be used to address the question as to whether the San Andreas fault is an exceptional case, as was done recently by Scholz (2000). Borehole-stress measurements in California, as outlined above, yielded mixed results, with the Cajon Pass measurements indicating that a minor fault in the adjacent crust obeys Anderson-Byerlee mechanics, but providing no information about the strength of the San Andreas fault itself. The Mojave measurements extended only to 0.85 km depth, but if extrapolated to crustal depths, indicated a strength several times that permitted by the heat-flow constraint.

Stress-orientation data are more plentiful in California. If the SAF is weak relative to the adjacent crust, it will be nearly a principal plane, so the maximum compressive stress direction will become nearly fault normal as the fault is approached. If, on the other hand, the fault is strong, then the maximum principle stress direction will approach the fault at an angle less than the lock-up angle (60° for $\mu = 0.6$). Everywhere in southern California, stresses inverted from earthquake focal mechanism data show the second of these two patterns (Hardebeck and Hauksson 1999), in which the σ_1 direction rotates to less than 60° as it approaches within 20 km of the SAF and its major branches. This means that the SAF cannot be weak relative to other minor faults in southern California (Scholz 2000). Because, as shown previously, minor faults in general are strong, so must be the SAF. The crustal averaged shear stress on the SAF in the "Big Bend" region was so calculated to be in the range 90–150 MPa, five to six times larger than expected from the heat-flow constraint (Scholz 2000). Townend and Zoback (2001) subsequently rebinned the Hardebeck and Hauksson data and claimed that they did not show the stress rotation, but Hardebeck and Hauksson (2001) showed that if the rebinned results were displayed as fault-normal profiles they displayed the same stress rotation as their original 1999 results. Hardebeck and Hauksson (2001) agreed with Scholz (2000) that this stress rotation did mean that the strength of the SAF was not weak relative to the surrounding crust but, on the basis of their inferred stress rotations caused by the Landers earthquake, they concluded that the surrounding crust was also weak (~10 MPa). This represents a more extreme (and more difficult to explain) position that all faults in southern California are weak, but one that is contradicted by the Cajon Pass stress measurements, which showed that the adjacent crust there is strong.

The situation in the Coast Ranges of central and northern California is more complicated. Because the present-day plate motion there is oblique to the SAF, strain partitioning occurs, with SAF accommodating the transverse component and the fault-normal component being accommodated by folding, thrusting, and uplift of the adjacent crust (Page et al. 1998). The fault-normal compression found by Mount and Suppe (1987) occurred in an active anticline with a fold axis subparallel to the SAF. They interpreted the fold as having been formed in this orientation, which implies that this compression direction represents the regional stress and the SAF must be consequently weak. Miller (1998), however, has shown from paleomagnetic data that since their formation those folds have rotated 20–30° clockwise, having been formed in an orientation appropriate to ordinary wrench tectonics. This suggests that the stresses within the folds result from the folding itself and are not representative of the regional stress. This is supported by focal-mechanism data for the same area that indicate σ_1 directions less than the lock-up angle everywhere but within the folds (Hardebeck and Hauksson 1999).

Several places also exist in northern California where there are active thrust faults which are subparallel and adjacent to the SAF (Oppenheimer et al. 1988; Zoback et al. 1999). This situation is shown in figure 9.3, which is another form of the strain partitioning discussed in the previous paragraph. The "California interpretation," as given by Oppenheimer et al. and Zoback et al., is that the stress field is uniform with σ_1 normal to the thrust-fault strike and that the strike-slip fault is nearly frictionless. Although this is a permissible model, it is not a unique one and, therefore, cannot be used to prove the frictionless fault assumption (Molnar 1992). Because the strike-slip fault can accommodate only the transverse component of the convergence, there is a kinematic requirement that the fault-normal component be accommodated by other means, such as the thrusting shown. This requirement would exist regardless of the friction on the strike-slip fault. This form of strain partitioning is also common at subduction zones where convergence is oblique to the trench (Fitch 1972; McCaffrey 1994). In that case, a strike-slip

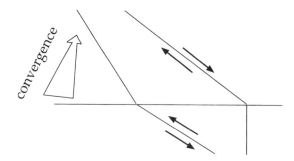

Figure 9.3 A perspective drawing of a model of strain partitioning associated with oblique plate convergence at a transcurrent boundary.

fault forms in the same configuration as in figure 9.3 and the motions are partitioned in the same way, even though in that case the thrust preexisted the strike-slip fault. One may propose that such strain partitioning occurs because it is a lower-energy configuration. Michael (1990) has posed the problem in that way and found that such strain partitioning is favored in all but the most extreme cases, without a requirement that the strike-slip fault be relatively weak with respect to the thrust fault. McCaffrey (1994) reached the same conclusion using a force-balance approach. In this strain-partitioning scenario, the σ_1 direction remains constant, in contrast to the interpretation of Provost and Houston (2001), in which σ_1 rotates from nearly 90° from the SAF at the thrust to 45° at the SAF.

The "creeping section" of the SAF, however, may be weaker than other parts of the SAF. Its stable sliding behavior, rare among crustal faults, is the strongest argument for its weakness, because it may indicate that the effective normal stress there is low enough for the fault to be in the stable sliding regime (cf., Scholz 1998), which would argue for high pore pressures there. Berry (1973) has provided evidence for high pore pressures at shallow depths in this region, and there is associated with the creeping section a unique geological situation that may allow high pore pressures to be retained at seismogenic depths (Irwin and Barnes 1975). On the other hand, as described below, there is another explanation of the stable sliding behavior of this region that does not require it to be weak.

Lubrication and Other Mechanisms for Reducing Fault Friction

If the interpretation of the heat flow data is correct in that the SAF is very weak vis-à-vis the Anderson-Byerlee state, then it must clearly be an exception. The question then is what is different about the SAF that would result in such a strikingly different mechanical behavior? One difference seems to be its great offset, compared with other faults, that results in its having a proportionately wider gouge zone (Scholz 1987). Attempts to explain its weakness have therefore concentrated on the properties of the gouge zone, for which there are two possibilities: the ubiquitous presence in the gouge zone of an intrinsically weak material, or the existence of some sort of seal that would allow pore pressure in the fault zone to greatly exceed that allowed by equation (9.2).

Solid Lubrication: Weak Fault Gouge

If equation (9.2) is in force, the "heat-flow constraint" requires the fault zone to contain a material with an intrinsic friction coefficient $\mu \leq 0.2$ (Lachenbruch and Sass 1992). Much effort has been expended to find plausible geological materials that might satisfy this criterion. Only two potential candidates have been found: montmorillonite and chrysotile. These two minerals have apparent low friction because they adsorb water layers that act as "bound" pore pressure (Morrow et al. 2000).

Montmorillonite is the weakest of the clays, with a $\mu \approx 0.2$ at low temperature and pressure (Morrow et al. 1992). However, montmorillonite breaks down to illite at 150–200°C, the latter being nonadsorptive and consequently considerably stronger. Thus, Morrow et al. (1992) found that a fault model that incorporates this transformation yields a depth-averaged friction of about 0.4. Thus, such a clay gouge zone is not compatible with the heat-flow constraint. Furthermore, this model assumes that fault gouge is composed entirely of these minerals throughout the schizosphere. Clay gouges found at the surface are commonly composed of such minerals; however, it seems that they are a product of near-surface weathering. A geochemical study of natural "clay fault gouge" on several faults in the San Andreas system, which have been exhumed from a depth of several kilometers, shows that their compositions are dominated by clay-size particles of the host-rock minerals rather than by clay minerals (Kirschner and Chester 1999). Because it is only the clay minerals that are weak, the friction of the natural gouge is likely to be closer to the Byerlee range.

Chrysotile, the rarest form of serpentinite, has a friction of about $\mu = 0.2$ at room temperature (Reinen et al. 1991), but it strengthens with temperature, reaching the strength of the more common forms, lizardite and antigorite, $0.4 \leq \mu \leq 0.5$ at 200°C (Moore et al. 1996, 1997). The friction of these minerals tends to have a neutral velocity dependence, which would promote stable sliding. The presence of serpentinite along the "creeping" section of the SAF is an alternative explanation of this anomalous behavior (Reinen et al. 1994), but if this is the case the creeping section would not be anomalously weak.

Thus the ubiquitous presence of clay or serpentinite minerals in the gouge zone could reduce friction somewhat from the typical Byerlee values, but not by enough to satisfy the heat-flow constraint. In either case, the fault would not be seismogenic because both materials are either velocity strengthening or neutral to it.

Fluid-Pressure Lubrication: Fault Seals

The alternative is that a seal is present in or about the gouge zone that can allow the pore pressure within the gouge zone to significantly exceed σ_3, thus avoiding the requirement for an unrealistically low value for μ. Although this possibility has been proposed by several authors (e.g., Byerlee 1990; Axen 1992), the difficult question as to how such a seal could be both impervious and immune to hydrofracturing has not been addressed. In the absence of any geological material possessing these properties, Rice (1992) has suggested an alternative. He proposed that the fault contains a Coulomb-plastic fault gouge of very low permeability, such that it maintains a pore-pressure gradient from a value of $p > \sigma_3$ in the interior to $\leq \sigma_3$ at the wall-rock interface so that no hydrofracturing occurs there. This is thus a dynamic seal; to maintain this gradient there must be outward flow of the pore fluid. A source is therefore required to replenish the pore fluid in the interior of the gouge zone: Rice suggests this source is the mantle. Because the gouge-zone width is approximately 100 m (Chester et al. 1993) and the distance to the mantle is approximately 10 km, this material must have an anisotropy between axial and lateral diffusivity on the order of 10^4 and also must be capped on top with a material that is impermeable in the vertical direction. Although clay gouge is a candidate for a very-low-permeability material, the requirement for this degree of permeability anisotropy in a single material does not seem realistic. Faulkner and Rutter (2001) have described a possible example, in which the fault gouge is composed of phyllitic minerals derived from the country rock, which is neither common nor applicable to the SAF. Thus, the possibility for such a hydrological deus ex machina to evolve in a natural fault seems remote.

Low-Dynamic Friction Models

Many suggestions have been made of novel ways in which friction may be reduced to a low value during dynamic slip, thereby allowing slip to occur with very little friction heating. These include dynamic pore-fluid overpressure from frictional heating (Sibson 1980), acoustic fluidization (Melosh 1996), and opening-mode displacements (Andrews and Ben-Zion 1997). Because these mechanisms do not provide a mechanism for static friction to be low, they predict earthquake stress drops in the 100-MPa range, which are not observed. Other schemes have been investigated in which rupture begins at a strong point where static friction is reached, but then propagates with a very-low-dynamic friction (e.g., Andrews and Ben-Zion 1997). It is hard to imagine how such strong points could survive in such a regime—they would be ruptured away and quickly reduced to the low-stress norm. Furthermore, one would not then expect that earthquake rupture would be so readily stopped when reaching the rupture zone of a previous earthquake, which is so typically observed (e.g., Sykes et al. 1971). McGarr (1999) reports that the seismic efficiency is always small, ≤ 0.06, a number that should also be close to the value of the ratio between stress drop and total stress.

One also might ask, if the SAF is extraordinarily weak vis-à-vis other crustal faults, why does it produce earthquakes indistinguishable from those produced by other faults? The stress drops of intraplate earthquakes are greater, by about a factor of 2 or 3, than interplate earthquakes, such as those that occur on the SAF

(Scholz et al. 1986; Kanamori and Allen 1986; Hanks and Johnston 1992; Scholz 1994), but this is a rather minor difference when compared with the order-of-magnitude weakness difference of the SAF required by the heat-flow constraint. In fact, the observed differences are quite similar to those expected from lab studies of frictional healing (Beeler et al. 2001). A very weak fault would be expected instead to be aseismic, either because of the velocity-strengthening behavior of intrinsically weak materials such as clay, or because the effective normal stress is such that the fault is below the friction stability transition. Lubrication, after all, is used not only to reduce friction but to eliminate squeaking.

Heat Generation and Transport

A strong SAF would provide significant quantities of frictionally generated heat in the upper (seismogenic) crust in the time since the fault took its modern form with the opening of the Gulf of California at 5.5 Ma, and for more than thirty years the absence of a heat-flow anomaly localized to the SAF has been the decisive argument that it is weak (Brune et al. 1969; Lachenbruch and Sass 1973). It is curious that this argument has proven to be so persuasive to an entire generation of Earth scientists. It is a negative argument; no localized anomaly of any size is observed, suggesting that either the model (conduction) is wrong or that there is no heat production (and stress). Despite the difficulty in explaining it, the latter interpretation has been almost universally accepted. Yet it has never been demonstrated that the Earth's crust can maintain a crustal-scale lateral heat-flow anomaly by conduction alone. What has been observed in such cases is that heat transport is dominated by advection of circulating water. The most well known example is the lack of a conductive heat-flow anomaly in the vicinity of midocean ridges due to circulation of water through the oceanic crust (Anderson et al. 1977). Conductive anomalies are also not observed adjacent to volcanic sources, again because of circulating water, often at temperatures only a few degrees above ambient (Manga 1998).

Figure 9.4 is a version of figure 11 of Lachenbruch and Sass (1980), hereafter LS80, showing the 81 "core" heat-flow observations in the vicinity of the SAF from the Mendocino Triple Junction (MTJ) to the Salton Sea, although all but six of them are north of the Transverse Ranges. Also shown are model conductive heat-flow anomalies, with a steady-state maximum of 1 HFU, arising from a planar frictional source of heat specified by τ_{ave} = 51.8 MPa and $2v$ = 4 cm/yr (or τ_{ave} = 103.5 MPa and $2v$ = 2 cm/yr), where τ_{ave} is the average frictional stress over a seismogenic depth of 14 km and $2v$ is the displacement rate across the SAF.

What is observed is a broad heat-flow anomaly centered on the SAF. Much of the scatter in figure 9.4 is due to the systematic diminution of the heat-flow anomaly as the MTJ is approached from the south, over the last 200 km of a long-strike distance, as shown in figure 9.5, a reproduction of figure 12 of LS80.

Because the MTJ is the point at which net slip of the SAF is zero, the dying

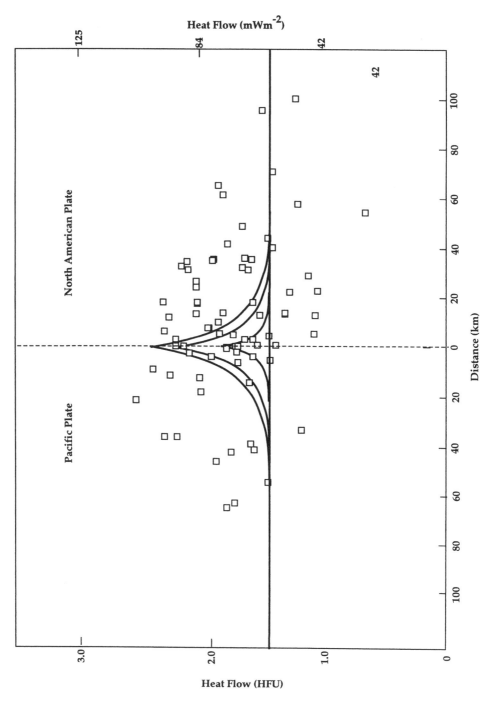

Figure 9.4 Heat-flow measurements in a profile normal to the San Andreas fault. Curves are model anomalies assuming heat flow is by conduction only, for frictional heating with $\tau_{ave} = 51.8$ MPa and $2v = 4$ cm/yr (or $\tau_{ave} = 103.5$ MPa and $2v = 2$ cm/yr), where τ_{ave} is the average frictional stress over a seismogenic depth of 14 km and $2v$ is the displacement rate across the SAF. Lower model is for 0.3 m.y., middle for 2.4 m.y., and upper is steady state. After Lachenbruch and Sass (1980).

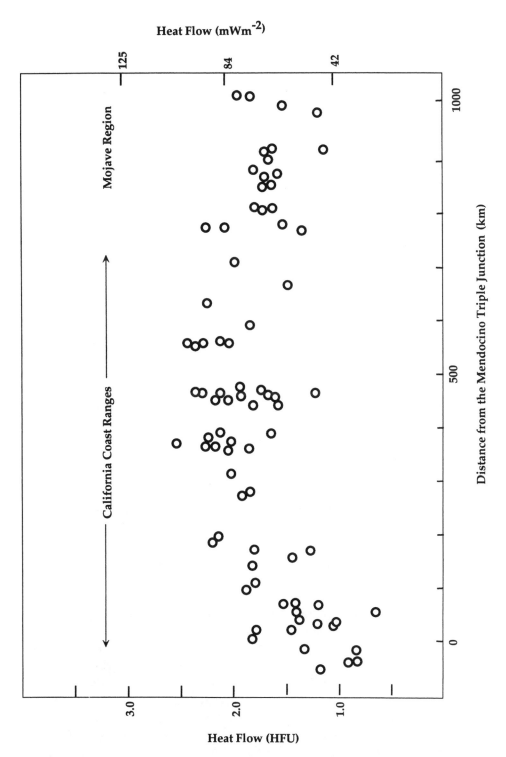

Figure 9.5 The heat-flow data of figure 9.4 plotted parallel to the SAF as a function of distance from the Mendicino triple junction. After Lachenbruch and Sass (1980).

off of the heat-flow anomaly as it is approached, as shown in figure 9.5, strongly suggests that the heat-flow anomaly is somehow intimately connected with the presence of the SAF. One way, of course, is if the heat flow is generated by shear heating.

LS80, having dismissed frictional generation of heat in any significant amounts along the SAF, proceeded to explore two other mechanisms by which the broad-scale heat-flow anomaly might be explained and, moreover, why it disappears at the MTJ. The first of these is the slab-window model, presented and discussed on pages 6209–6213 of LS80. It is intended to explain the rise in heat flow along SAF for the first several hundred kilometers south of the MTJ (figure 9.5) in the absence of significant frictional heating along the SAF. The essence of the slab-window model is a one-sided spreading center south of the MTJ, in which the "slab window" is filled with upwelling asthenospheric material. LS80 notes that "extreme conditions" are required for this model to fit the heat-flow data, namely temperatures of $\sim$1,100°C for the upwelling material emplaced at just 20 km depth.

Considerable effort has been directed to exploring the geophysical structure of the crust and uppermost mantle in the slab-window domain south of the MTJ, but the presence of the slab window is neither so clear nor so dramatic as suggested by the simple geometric models for it. Upper-mantle densities in the Coast Ranges south of the MTJ are consistent with some thermal elevation, and strong (but volumetrically insignificant) seismic reflectors in the lower crust are thought to be sills of partial melt derived from the upper mantle (Levander et al. 1998). Even so, the zone of continuous, high upper-mantle *P*-wave velocities south of the MTJ from $\approx$100 km west of the SAF to almost 100 km east of it, as revealed by the Mendocino Triple Junction Seismic Experiment, means that any massive asthenospheric upwelling is either deeper ($\geq$30–35 km) or farther east ($\geq$100 km) than anticipated in the slab-window model of LS80 (Henstock et al. 1997; ten Brink et al, 1999; Henstock and Levander 2000).

The implications of the these findings for the slab window model are best summarized by LS80: "If field evidence for extreme conditions of the sort implied by this model cannot be found [which seems to be the case at this time, 20 years later], it is likely that an appreciable contribution from shear-strain heating will have to be invoked [to explain the broad-scale heat-flow anomaly associated with the SAF]." This heat will arise from the "brittle" model of shear-strain heating in the upper crust, originally put forward by Lachenbruch and Sass (1973) and considered in greater detail on pages 6213–6217 of LS80.

The broad-scale heat-flow anomaly puts two important constraints on how the brittle model manufactures heat: the source of the heat must be shallow, within the upper crust, and shear stresses within this model must be high to match the large quantity of heat contained in the broad-scale heat-flow anomaly. LS80 imposes a third constraint, that the SAF be weak. Thus, the brittle model (see figures 27 and 28 of LS80) generates heat at the base of the seismogenic crust ($h = 14$ km) across a thin decoupling zone. This upper layer moves faster than the material beneath it, which in turn resists the motion of the upper seismogenic crust. This curious mechanical situation arises from the condition that frictional stress on the

SAF be low, while the lateral driving stress on vertical planes parallel to but away from (at $y = y^* = 42$ km in the LS80 brittle model) the SAF be high. The high shear stress on vertical planes away from the SAF arises in turn from the amount of heat to be generated by the quantity $\tau_b^* \Delta v$ in the decoupling zone (where τ_b is the basal shear resisting motion of the seismogenic layer and Δv is the velocity jump across the decoupling zone) to match the broad-scale heat-flow anomaly.

The LS80 brittle model was constructed to explain the broad-scale heat-flow anomaly associated with the SAF, but there is little in the way of underpinning evidence for this edifice. This model requires, for example, large gradients of horizontal stress on vertical planes away from the SAF, several megapascals per kilometer depending on the choice of Δv and the maximum lateral shear at $y^* = 42$ km. In situ stress determinations near and away from the SAF at Palmdale (Zoback et al. 1980) do not support the existence of such large stress gradients (McGarr et al. 1982). Such large stress gradients would also require that the SAF system be fundamentally strong, even if the SAF itself is weak, and raise the question as to why there is no localized heat-flow anomaly associated with the Hayward-Rodgers Creek fault, for example, 20 km east of the SAF in northern California. Finally, these large stress gradients should give rise to systematic changes in earthquake focal-mechanism types and orientations within $y^* = 42$ km of the SAF that are not observed. Likewise, the proposed decoupling zone at $h = 14$ km capable of generating as much heat as it does to match the broad-scale heat-flow anomaly should make for a profound geophysical discontinuity at this midcrustal depth, but we know of no systematic evidence to support its existence over the broad tracts of coastal California it should underlie. The evidence, rather, favors the SAF continuing through the lower crust as a steeply dipping ductile-shear zone rather than being decoupled horizontally (Gilbert et al. 1994; Henstock et al. 1997; Parsons 1998).

The brittle model has not stood the test of time, and the slab-window effect appears insufficient to generate the entire heat-flow anomaly associated with the SAF. One can see from figure 9.4 that the broad heat-flow anomaly is several times larger than what would be expected from frictional heating from a strong fault, so such shear heating could hide within the observed anomaly. All that is required is a mechanism that would obscure the narrow peaked anomaly expected from the purely conduction model. One factor is that the SAF system is not a single fault along most of its length. In northern California it is a 100-km-wide shear zone consisting of a half-dozen or more subparallel faults all generating heat. What's more, the slip rates of these faults vary over geologic time—the San Andreas fault on the San Francisco peninsula, for example, has been active for only the past 2 m.y. (Wakabayashi 1999). A more general explanation, however, is that heat transport is not entirely by conduction but also by advection of water through fractures in the upper crust.

The arguments for and against water in the upper crust affecting heat-flow measurements in the vicinity of the SAF are not new (Hanks 1977; O'Neil and Hanks 1980; Lachenbruch and Sass 1980). Indeed the "stress paradox" of the SAF was originally put in terms of no matter which way the paradox was resolved—

high stress or low stress—the action of water would somehow be involved (Hanks and Raleigh 1980). That water can and does move freely through the upper crust in tectonically active areas is supported by a vast array of geochemical and geophysical observations. If this is not the case for the upper crust traversed by the SAF, it must be quite a special exception.

Direct evidence for massive, meteoric water circulation to depths of ≈ 10 km in tectonically active, continental areas exists in the relative depletion of D and ^{18}O in the rocks through which the water passed (Taylor 1977). The condition for this to occur are natural enough: a heat source, generally plutonic, to drive the circulation system and a medium permeable enough to permit the passage of water, often along joints, fractures, and faults (e.g., O'Neil et al. 1977). In any number of cases, the water/rock ratios required to yield the observed depletions of D and ^{18}O are unity or greater. Although the timescales for and rates of this water circulation are not known from the stable isotope data, its volume alone suggests that it must be an important heat-transport mechanism. Moreover, that the circulating fluid is meteoric water constrains fluid pressures to be, on average through the system volume, not much in excess of hydrostatic; for greater fluid pressures with depth, circulation of meteoric water would not be possible.

In his review of laboratory, in situ, and inferential estimates of permeability for crystalline and argillaceous rocks, Brace (1980) finds that the permeability observations are such that "pore pressure much greater than hydrostatic seems ruled out in terrains of outcropping crystalline rocks . . . anomalously high pore pressures seem to require everywhere a thick blanket of clay-rich rocks." Brace (1980) also finds that the permeability of crystalline rocks (mainly through joints and fractures) is large enough to allow fluid-flow rates to greatly outstrip heat transport via thermal conduction, at least to depths of 5 km or so. With a significantly expanded, more modern database, Manning and Ingebritsen (1999) "recognize 10^{-16} m^2 as the approximate threshold value for significant advective heat transport in a fairly wide range of upper crustal contexts." Their summary relation for permeability (k, in square meters) with depth (z, in kilometers), $\log k = -3.2 \log z - 14$, allows for permeabilities greater than or equal to the threshold value for depths less than 5 km.

Barton et al. (1995), Ito and Zoback (2000), and Townend and Zoback (2000) have concluded from a number of deep-drilling and induced-seismicity experiments that the brittle crust is critically stressed with respect to Byerlee friction, that the permeability of critically stressed fractures and faults is much higher than for fractures and faults not optimally oriented for failure in the present stress field, and that pore pressures are close to hydrostatic, even to 9 km depth in the KTB hole. In addition, Townend and Zoback (2000) report in situ bulk permeabilities in the range of 10^{-17} to 10^{-16} m^2. These high permeabilities do not allow pressure perturbations away from hydrostatic to last for long, which leads Townend and Zoback (2000) to a "counterintuitive result—that faulting makes the crust inherently stronger by preventing high pore pressures from developing at depth."

In view of the geochemical and geophysical evidence briefly summarized in the preceding three paragraphs, one may anticipate that the surface expression of water circulation at depth would be controlled by active faults and their associated fracture systems. This seems to be the case. Figure 9.6 shows, with a few peripheral exceptions, a strong spatial correlation throughout the western United States between modern hot springs (water temperature exceeding local mean annual air temperature by 8°C or more) and surface traces of faults active in the Quaternary. In the absence of local, very shallow heat sources, this water probably comes from depths of a kilometer or more. Because figure 9.6 is spatially dominated by the Basin and Range province, an extending region with a thinned crust and local volcanism, its relevance to the matters of heat generation along and transport away from the SAF is plainly limited. Our point is simply that any tendency of water to reach the Earth's surface is greatly abetted by the presence of faults and fractures that also reach the Earth's surface.

One simple possibility is that, because most of the heat-flow measurements were made in shallow (~100 m) boreholes, topographically generated ground-water flow may disturb the heat-flow anomaly sufficiently for it to be obscured

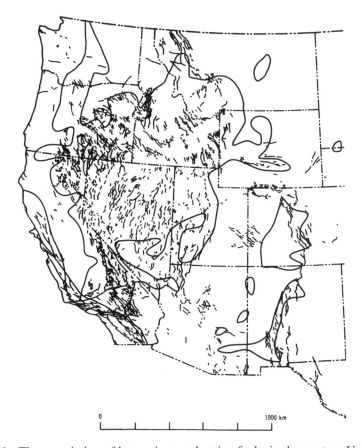

Figure 9.6 The association of hot springs and active faults in the western United States. The contours indicate the boundaries of hot spring provinces. Modified from Eaton (1980).

(Williams and Narisimhan 1989). This possibility was modeled by Saffer et al. (2003). They found that this effect would obscure the anomaly in an area of moderate relief like Parkfield, but that in the Mojave region, with very flat topography northeast of the fault, a model of a strong vertical SAF is inconsistent with the low heat-flow values northeast of the fault. Because their regional heat-flow baseline is arbitrary, however, one could shift it up or down to "fit" any model. Moreover, that area is in a compressive bend of the fault, with the San Gabriel Mountains uplifted to the southwest of the SAF, so it is likely that the fault there dips 70–80° to the southwest, just as it does in the Loma Prieta region where the Santa Cruz Mountains are being uplifted, and in the San Gorgonio Pass region where the SAF dips to the north below the uplifting San Bernardino Mountains (Seeber and Armbruster 1995). That dip alone may explain the low heat flow to the northeast of the fault.

All of these heat-production models are unrealistic because they assume a steady-state heat source, whereas, except for the creeping sections, all heat production would occur as transients at the time of large earthquakes. Because the temperatures and temperature gradients are so much higher in transient than steady-state heating, they are much more likely to drive fluid flow. Large hydrological disturbances are known to follow earthquakes (Muir-Wood and King 1993). These have been modeled as poroelastic effects, but they could easily include a thermal driving effect. To get a rough estimate of the relative sizes of effects, consider the M 7.3 Hegben Lake earthquake, in which 0.5 km^3 of water was observed to be expelled at the surface in the first two months after the earthquake (Muir-Wood and King 1993). Suppose that this represents about 10 percent of the total groundwater disturbed at depth. Then assuming that the earthquake occurred on a favorably oriented normal fault with $\mu = 0.6$, we can calculate the entire work of faulting at about 7×10^{16} J. This heat would be sufficient to heat this volume of water by only about 3°C, which simply says that there is plenty of water flowing about in the immediate postseismic period to redistribute the heat away from the fault, which would be its normal tendency.

Finally, the effects of water circulation on heat flow at or near the Earth's surface have been documented many times. Perhaps the most famous example of water circulation in tectonically active areas affecting the near-surface thermal regime occurs at oceanic ridges, as mentioned earlier. Active volcanic systems in continental areas are well known for their interactions with water, and any geothermal play, almost by definition, involves significant advection of the Earth's heat by water. In a review of heat flow in the United States, Lachenbruch and Sass (1977) note: "Heat convected by moving groundwater requires careful attention in heat flow studies; it can perturb or completely dominate the regional heat flux associated with the crustal regime at depth." Lachenbruch and Sass (1977) also describe the very modest groundwater-flow velocities they think are responsible for the "Eureka Low," an areally extensive, anomalously low heat-flow region within the otherwise high heat-flow Basin and Range province. Similarly, in a detailed study of the thermal regime of Yucca Mountain and environs, situated on the southwestern margin of the Eureka Low, Sass et al. (1995) found that "the thermal regime was indeed distorted by the effects of water movement and provides data complementary to conventional hydrologic studies."

LS80 dismissed the case for fluid flow in the Earth's upper crust as having any significant impact on the heat-flow data shown in figure 9.4, although they did so without the benefit of the many more recent studies cited in the previous five paragraphs. In essence, their argument is based on the low rate of discharge from known hot springs along and near the SAF and the similarity of the heat-flow data in the terrains east and west of the fault that they thought should have different hydrologic regimes. However, flow of waters only a few degrees above ambient can transport large thermal fluxes without requiring massive hot springs activity (Manga 1998).

Summary and Conclusions

We have discussed three separate but related themes in this chapter. First, we have reviewed the failure of all attempts to explain how the SAF could be as weak as it is purported to be by the heat-flow argument. Second, we have attempted to show that stress measurements in California do not support the low-strength hypothesis: they are either equivocal or contradict it. Finally, we point out that the heat-flow issue is not concerned with whether there is a SAF heat-flow anomaly or its size, but is solely about its shape. Its shape depends entirely on an assumed conduction model—a model that has never been demonstrated for any crustal scale heat source. The "slab window effect" of LS80 has been found inadequate to explain all of the broad heat-flow anomaly associated with the SAF, and their "brittle" model is no longer considered viable. The source of this excess heat flow remains a mystery, but it is certainly large enough to contain the frictional heat generated by a strong SAF. It does not seem to us that the heat-flow data prove anything, one way or another, about the magnitude of shear stresses operating on the SAF, and it is our view that the widespread belief that the heat-flow data prove that the SAF is weak is misplaced.

References

Abers, G. A. 1991. Possible seismogenic shallow-dipping normal faults in the woodlark-dentrecasteaux extensional province, Papua-New-Guinea. *Geology* 19(12):1205–1208.

Anderson, E. M. 1951. *The Dynamics of Faulting*. Edinburgh, UK: Oliver and Boyd.

Anderson, R. N., M. G. Langseth, and J. G. Sclater. 1977. The mechanisms of heat transfer through the floor of the Indian Ocean. *J. Geophys. Res.* 82:3391–3409.

Andrews, D. J. and Y. BenZion. 1997. Wrinkle-like slip pulse on a fault between different materials. *J. Geophys. Res. Solid Earth* 102(B1):553–571.

Axen, G. J. 1992. Pore pressure, stress increase, and fault weakening in low-angle normal faulting. *J. Geophys. Res. Solid Earth* 97(B6):8979–8991.

Barton, C. A., M. D. Zoback, and D. Moos. 1995. Fluid-flow along potentially active faults in crystalline rock. *Geology* 23(8):683–686.

Beeler, N. M., S. H. Hickman, and T. F. Wong. 2001. Earthquake stress drop and laboratory-inferred interseismic strength recovery. *J. Geophys. Res.* 106:30701–30713.

Berry, F.A.F. 1973. High fluid potentials in California Coast Ranges and their tectonic significance. *Am. Assoc. Pet. Geol.* 57:1219–1249.

Blanpied, M. L., C. J. Marone, D. A. Lockner, J. D. Byerlee, and D. P. King. 1998. Quantitative measure of the variation in fault rheology due to fluid-rock interactions. *J. Geophys. Res.* 103(B5):9691–9712.,

Brace, W. F. 1980. Permeability of crystalline and argillaceous rocks. *Int. J. Rock Mech. Mining Sci.* 17(5):241–251.

Brace, W. F. and J. D. Byerlee. 1966. Stick slip as a mechanism for earthquakes. *Science* 153:990–992.

Brudy, M., M. D. Zoback, K. Fuchs, F. Rummel, and J. Baumgartner. 1997. Estimation of the complete stress tensor to 8 km depth in the KTB scientific drill holes: Implications for crustal strength. *J. Geophys. Res. Solid Earth* 102 (B8):18453–18475.

Brune, J., T. Henyey, and R. Roy. 1969. Heat flow, stress, and rate of slip along the San Andreas fault, California. *J. Geophys. Res.* 74:3821–3827.

Byerlee, J. D. 1978. Friction of rock. *Pure Appl. Geophys.* 116:615–626.

Byerlee, J. D. 1990. Friction, overpressure, and fault normal compression. *Geophys. Res. Lett.* 17:2109–2112.

Chester, F. M., J. P. Evans, and R. L. Biegel. 1993. Internal structure and weakening mechanisms of the San-Andreas Fault. *J. Geophys. Res. Solid Earth* 98(B1):771–786.

Chinnery, M. A. 1964. The strength of the earth's crust under horizontal shear stress. *J. Geophys. Res.* 69:2085–2089.

Colletini, C. and R. H. Sibson. 2001. Normal faults, normal friction? *Geology* 29:927–930.

Eaton, G. P. 1980. Geophysical and geological characteristics of the crust of the Basin and Range province. In *Continental Tectonics*, pp. 96–113. Washington, DC: National Academy of Sciences.

Faulkner, D. R. and E. H. Rutter. 2001. Can the maintenance of overpressured fluids in large strike-slip fault zones explain their apparent weakness? *Geology* 29:503–506.

Fitch, T. J. 1972. Plate convergence, transcurrent faults, and internal deformation adjacent to southeast Asia and the western Pacific. *J. Geophys. Res.* 77:4432–4460.

Gilbert, L. E., C. H. Scholz, and J. Beavan. 1994. Strain localization along the San Andreas fault: Consequences for loading mechanisms. *J. Geophys. Res..* 99:23975–23984.

Gloyd, J. S., J. C. Mutter, A. M. Goodliffe, et al. 2001. Evidence for fault weakness and fluid flow within an active low-angle normal fault. *Nature* 411:779–783.

Hanks, T. C. 1977. Earthquake stress drops, ambient tectonic stresses and stresses that drive plate motions. *Pure Appl. Geophys.* 115(1–2):441–458.

Hanks, T. C. and A. C. Johnston. 1992. Common features of the excitation and propagation of strong ground motion for North American earthquakes. *Bull. Seism. Soc. Am.* 82:1–23.

Hanks, T. C. and C. B. Raleigh. 1980. The conference on magnitude of deviatoric stresses in the earth's crust and uppermost mantle. *J. Geophys. Res.* 85(B11):6083–6085.

Hardebeck, J. L. and E. Hauksson. 1999. Role of fluids in faulting inferred from stress field signatures. *Science* 285(5425):236–239.

Hardebeck, J. L. and E. Hauksson. 2001. Crustal stress field in southern California and its implications for fault mechanics. *J. Geophys. Res.* 106:21859–21882.

Henstock, T. J. and A. Levander. 2000. Lithospheric evolution in the wake of the Mendocino triple junction: Structure of the San Andreas Fault system at 2 Ma. *Geophys. J. Int.* 140(1):233–247.

Henstock, T. J., A. Levander, and J. A. Hole. 1997. Deformation in the lower crust of the San Andreas fault system in northern California. *Science* 278(5338):650–653.

Hubbert, M. K. and W. W. Rubey. 1959. Role of fluid pressure in the mechanics of overthrust faulting. *Bull. Geol. Soc. Am.* 70:115–166.

Irwin, W. P. and I. Barnes. 1975. Effects of geological structure and metamorphic fluids on seismic behavior of the San Andreas fault system in central and northern California. *Geology* 3:713–716.

Ito, T. and M. D. Zoback. 2000. Fracture permeability and in situ stress to 7 km depth in the KTB Scientific Drillhole. *Geophys. Res. Lett.* 27(7):1045–1048.

Jackson, J. A. 1987. Active normal faulting and crustal extension. In J.D.M. Coward and P. Hancock, eds., *Continental Extensional Tectonics*, pp. 3–18. London: Blackwell.

Kanamori, H. and C. Allen. 1986. Earthquake repeat time and average stress drop. In J.B.S. Das and C. Scholz, eds., *Earthquake Source Mechanics*, pp. 227–236. AGU Geophysics Monograph 37. Washington, DC: American Geophysical Union.

Kirschner, D. L. and F. M. Chester. 1999. Are fluids important in seismogenic faulting? Geochemical evidence for limited fluid-rock interaction in two strike-slip faults of the San Andreas system. *Eos* 80(suppl.):F727.

Lachenbruch, A. H. and J. H. Sass. 1973. Thermo-mechanical aspects of the San Andreas. In R.K.A.A. Nur, ed., *Proceedings of the Conference on the Tectonic Problems of the San Andreas Fault System*, pp. 192–205. Palo Alto, CA: Stanford University Press.

Lachenbruch, A. H. and J. H. Sass. 1977. Heat flow in the United States and the thermal regime of the crust. In *The Earth's Crust*, pp. 626–675. Washington DC: American Geophysical Union.

Lachenbruch, A. H. and J. H. Sass. 1980. Heat-flow and energetics of the San-Andreas fault zone. *J. Geophys. Res.* 85(B11):6185–6222.

Lachenbruch, A. H. and J. H. Sass. 1992. Heat-flow from Cajon Pass, fault strength, and tectonic implications. *J. Geophys. Res. Solid Earth* 97(B4):4995–5015.

Le Pichon, X., P. Henry, and S. Lallemant. 1993. Accretion and erosion in subduction zones: The role of fluids. *Annu. Rev. Earth Planet. Sci.* 21:307–331.

Levander, A., T. J. Henstock, A. S. Meltzer, B. C. Beaudoin, A. M. Trehu, and S. L. Klemperer. 1998. Fluids in the lower crust following Mendocino triple junction migration: Active basaltic intrusion? *Geology* 26(2):171–174.

Lund, B. and M. D. Zoback. 1999. Orientation and magnitude of in situ stress to 6.5 km depth in the Baltic Shield. *Int. J. Rock Mech. Mining Sci.* 36(2):169–190.

Manga, M. 1998. Advective heat transport by low-temperature discharge in the Oregon Cascades. *Geology* 26:799–802.

Manning, C. E. and S. E. Ingebritsen. 1999. Permeability of the continental crust: Implications of geothermal data and metamorphic systems. *Rev. Geophys.* 37(1):127–150.

McCaffrey, R. 1994. Global variability in subduction thrust zone fore-arc systems. *Pageoph* 142:173–224.

McGarr, A., M. D. Zoback, and T. C. Hanks. 1982. Implications of an elastic analysis of in situ stress measurements near the San Andreas fault. *J. Geophys. Res.* 87:7797–7806.

McGarr, A. 1999. On relating apparent stress to the stress causing earthquake fault slip. *J. Geophys. Res. Solid Earth* 104(B2):3003–3011.

McGarr, A. and N. C. Gay. 1978. State of stress in the earth's crust. *Annu. Rev. Earth Planet. Sci.* 6:405–436.

Melosh, H. J. 1996. Dynamical weakening of faults by acoustic fluidization. *Nature* 379(6566):601–606.

Michael, A. J. 1990. Energy constraints on kinematic models of oblique faulting: Loma Prieta vs. Parkfield-Coalinga. *Geophys. Res. Lett.* 17:1453–1456.

Miller, D. D. 1998. Distributed shear, rotation, and partitioned strain along the San Andreas fault, central California. *Geology* 26(10):867–870.

Molnar, P. 1992. Brace-Goetze strength profiles, the partitioning of strike-slip and thrust faulting at zones of oblique convergence, and the stress-heat flow paradox of the San Andreas fault. In B. Evans and T.-F. Wong, eds., *Fault Mechanics and Transport Properties of Rocks*, pp. 435–460. London: Academic Press.

Moore, D. E., D. A. Lockner, S. L. Ma, R. Summers, and J. D. Byerlee. 1997. Strengths of serpentinite gouges at elevated temperatures. *J. Geophys. Res. Solid Earth* 102(B7):14787–14801.

Moore, D. E., D. A. Lockner, R. Summers, M. Shengli, and J. D. Byerlee. 1996. Strength of chrysotile-serpentinite gouge under hydrothermal conditions: Can it explain a weak San Andreas fault? *Geology* 24(11):1041–1044.

Morrow, C., B. Radley, and J. D. Byerlee. 1992. Frictional strength and the effective pressure law for montmorillonite and illite clays. In B. Evans and T.-F. Wong, eds., *Fault Mechanics and Transport Properties of Rocks*, pp. 69–88. London: Academic Press.

Morrow, C. A., D. E. Moore, and D. A. Lockner. 2000. The effect of mineral bond strength and adsorbed water on fault gouge frictional strength. *Geophys. Res. Lett.* 27(6):815–818.

Mount, V. S. and J. Suppe. 1987. State of stress near the San-Andreas Fault: Implications for wrench tectonics. *Geology* 15(12):1143–1146.

Muir-Wood, R. and G.C.P. King. 1993. Hydrological signatures of earthquake strain. *J. Geophys. Res.* 98:22035–22068.

O'Neil, J. R., S. E. Shaw, and R. H. Flood. 1977. Oxygen and hydrogen isotope compositions as indicators of granite genesis in the New England batholith, Australia. *Contrib. Mineral. Petrol.* 62:313–328.

O'Neil, J. R. and T. C. Hanks. 1980. Geochemical evidence for water-rock interaction along the San-Andreas and Garlock Faults of California. *J. Geophys. Res.* 85 (B11):6286–6292.

O'Neil, J. R. 1985. Water-rock interactions in fault gouge. *Pageoph* 122:440–446.

Oppenheimer, D. H., P. A. Reasenberg, and R. W. Simpson. 1988. Fault plane solutions for the 1984 Morgan Hill, California, earthquake sequence: Evidence for the state of stress on the Calaveras fault. *J. Geophys. Res.* 93:9007–9026.

Page, B. M., G. A. Thompson, and R. G. Coleman. 1998. Late Cenozoic tectonics of the central and southern coast ranges of California. *Geol. Soc. Am. Bull.* 110(7):846–876.

Parsons, T. 1998. Seismic-reflection evidence that the Hayward fault extends into the lower crust of the San Francisco Bay area. California. *Bull. Seism. Soc. Am.* 88:1212–1223.

Provost, A. S. and H. Houston. 2001. Orientation of the stress field surrounding the creeping section of the San Andreas fault: Evidence for a narrow mechanically weak fault zone. *J. Geophys. Res.* 106:11373–11386.

Reinen, L. A., J. D. Weeks, and T. E. Tullis. 1991. The frictional behavior of serpentinite: Implications for aseismic creep on shallow crustal faults. *Geophys. Res. Lett.* 18(10):1921–1924.

Reinen, L. A., J. D. Weeks, and T. E. Tullis. 1994. The frictional behavior of lizardite and antigorite serpentinites: Experiments, constitutive models, and implications for natural faults. *Pure Appl. Geophys.* 143(1–3):317–358.

Rice, J. R. 1992. Fault stress states, pore pressure distributions, and the weakness of the San Andreas fault. In B. Evans and T.-F. Wong, eds., *Fault Mechanics and Transport Properties of Rocks*, pp. 475–504. London: Academic Press.

Richard, P. and P. Cobbold. 1989. Structures en fleur positives et décrochements crustaux: Modélisation et enterprétation mechanique. *C. R. Acad. Sci. Paris Ser. II* 308:553–560.

Saffer, D. M., B. A. Bekins, and S. Hickman. 2003. Topographically driven groundwater flow and the San Andreas heat flow paradox revisited. *J. Geophys. Res.* 108(B5):12-1–12-14.

Sass, J. H., W. W. Dudley, Jr., and A. H. Lachenbruch. 1995. Regional setting. In *Major Results of Geophysical Investigations at Yucca Mountain and Vicinity, Southern Nevada,* chapter 8. U.S. Geological Survey.

Scholz, C. H. 1987. Wear and gouge formation in brittle faulting. *Geology* 15:493–495.

Scholz, C. H. 1990. *Mechanics of Earthquakes and Faulting*. Cambridge, UK: Cambridge University Press.

Scholz, C. H. 1994. Reply to comments on "A reappraisal of large earthquake scaling" by C. Scholz. *Bull. Seism. Soc. Am.* 84:1677–1678.

Scholz. C. H. 1998. Earthquakes and friction laws. *Nature* 391:37–42.

Scholz, C. H. 2000. Evidence for a strong San Andreas fault. *Geology,* 28(2):163–166.

Scholz, C. H., C. Aviles, and S. Wesnousky. 1986. Scaling differences between large intraplate and interplate earthquakes. *Bull. Seism. Soc. Am.* 76:65–70.

Scholz, C. H. and F. J. Saucier. 1993. What do the Cajon Pass stress measurements say about stress on the San-Andreas Fault? In-situ stress measurements to 3.5 km depth in the Cajon Pass scientific-research borehole: Implications for the mechanics of crustal faulting. Comment. *J. Geophys. Res. Solid Earth* 98(B10):17867–17869.,

Seeber. L. and J. G. Armbruster. 1995. The San Andreas fault system through the Transverse Ranges as illuminated by earthquakes. *J. Geophys. Res.* 100:8285–8310.

Sibson, R. H. 1980. Power dissipation and stress levels on faults in the upper crust. *J. Geophys. Res.* 85:6239–6247.

Sibson, R. H. 1985. A note of fault reactivation, *J. Struct. Geol.* 7:751–754.

Sibson, R. H. 1987. Earthquake rupturing as a mineralizing agent in hydrothermal systems. *Geology* 15:701–704.

Sibson, R. H. and G. Y. Xie. 1998. Dip range for intracontinental reverse fault ruptures: Truth not stranger than friction? *Bull. Seism. Soc. Am.* 88(4):1014–1022.

Sykes, L. R. 1971. Aftershock zones of great earthquakes, seismicity gaps, and earthquake prediction for Alaska and the Aleutians. *J. Geophys. Res.* 76:8021–8041.

Taylor, B., A. M. Goodcliffe, and F. Martinez. 1999. How continents break up: Insights from Papua New Guinea. *J. Geophys. Res. I* 104:7497–7512.

Taylor, B. and P. Huchon. 2002. Active continental extension in the western Woodlark Basin: a synthesis of Leg 180 results. In P. Huchon, B. Taylor, and A. Klaus, eds., *Proc. ODP, Sci. Results, 180* (Online). Available from World Wide Web: ⟨http://www-odp.tamu.edu/publications/180_SR/synth/synth.htm⟩

Taylor, H.P.J. 1977. Water/rock interactions and the origin of H_2O in granitic batholiths. *J. Geol. Soc. Lond.* 133:509–558.

ten Brink, U. S., N. Shimizu, and P. C. Molzer. 1999. Plate deformation at depth under northern California: Slab gap or stretched? *Tectonics* 18:1084–1098.

Townend, J. and M. D. Zoback. 2000. How faulting keeps the crust strong. *Geology* 28(5):399–402.

Townend, J. and M. D. Zoback. 2001. Implications of earthquake focal mechanisms for the frictional strength of the San Andreas fault system. In R. E. Holdsworth, R. A. Strachan, J. F. Magloughlin, and R. J. Knipe, eds., *The Nature and Tectonic Significance of Fault Zone Weakening,* pp. 13–21. Geological Society of London, Special Publication 186.

Tsuboi, C. 1933. Investigation of deformation of the crust found by precise geodetic means. *Jpn. J. Astron. Geophys.* 10:93–248.

Tsuboi, C. 1956. Earthquake energy, earthquake volume, aftershock area, and strength of the earth's crust. *J. Phys. Earthquakes* 4: 63–66.

Wakabayashi, J. 1999. Distribution of displacement on and evolution of a young transform fault system: The northern San Andreas fault system, California. *Tectonics* 18:1245–1274.

Wernicke, B. 1995. Low-angle normal faults and seismicity: A Review. *J. Geophys. Res. Solid Earth* 100(B10):20159–20174.

Wernicke, B. and B. C. Burchfiel. 1982. Modes of extensional tectonics. *J. Struct. Geol.* 4(2):105–115.

Williams, C. F. and T. N. Narasimhan. 1989. Hydrogeologic constraints on heat-flow along the San Andreas fault: A testing of hypotheses. *Earth Planet. Sci. Lett.* 92:131–143.

Zoback, M. D. and J. H. Healy. 1984. Friction, faulting, and in situ stress. *Ann. Geophys.* 2(6):689–698.

Zoback, M. D. and J. H. Healy. 1992. In situ stress measurements to 3.5 km depth in the Cajon Pass scientific-research borehole: Implications for the mechanics of crustal faulting. *J. Geophys. Res. Solid Earth* 97(B4):5039–5057.

Zoback, M. D., H. Tsukahara, and S. Hickman. 1980. Stress measurements at depth in the vicinity of the San-Andreas Fault: Implications for the magnitude of shear-stress at depth. *J. Geophys. Res.* 85(B11):6157–6173.

Zoback, M. D., M. L. Zoback, V. S. Mount, J. Suppe, J. P. Eaton, J. H. Healy, D. Oppenheimer, P. Reasenberg, L. Jones, C. B. Raleigh, I. G. Wong, O. Scotti, and C. Wentworth. 1987. New evidence on the state of stress of the San-Andreas Fault System. *Science* 238(4830):1105–1111.

Zoback, M. L., Z. C. Jachens, and J. A. Olson. 1999. Abrupt along-strike change in tectonic style: San Andreas fault zone, San Francisco Peninsula. *J. Geophys. Res. Solid Earth* 104(B5):10719–10742.

CHAPTER TEN

Deformation Behavior of
Partially Molten Mantle Rocks

Yaqin Xu, M. E. Zimmerman, and D. L. Kohlstedt

Introduction

A small amount of melt influences the elastic (reversible and time-independent), anelastic (reversible but time-dependent), and plastic or viscous (irreversible and time-dependent) deformation behavior of partially molten rocks. These material properties govern seismic velocity and seismic attenuation as well as the small- and large-scale flow behavior of mantle rocks. At the same time, deformation directly affects the distribution of melt in a partially molten rock. Consequently, the material properties of the rock are affected by the process of deformation.

In this chapter, we address three aspects of deformation of partially molten rocks. First, new experimental results are presented on anelastic and viscous (plastic) deformation at low stresses, low strains, and low (seismic) frequencies. A pronounced attenuation peak, characteristic of anelastic deformation, is described for the first time in these materials. The results are discussed in the context of recently published results on seismic wave attenuation in fine-grained aggregates of olivine and olivine plus melt. Second, published results on steady-state (large-strain) deformation are reviewed. Flow laws describing the creep behavior of partially molten aggregates of mantle composition, as constrained by laboratory results and theoretical models, are summarized. Emphasis is given to the effects of melt and water, as well as differential stress, grain size, and temperature, on creep behavior. Third, the influence of large-strain deformation on the distribution of melt in partially molten rocks is described. Although it has been recognized for several years that grain-scale melt pockets become strongly aligned during deformation, it has been demonstrated only very recently that deformation-driven melt segregation can also take place over much larger length scales. This process is particularly evident in samples deformed to large strain in simple-shear experiments. The resulting melt-rich bands dramatically affect the viscosity, permeability, and seismic properties of a partially molten rock.

Low-Frequency Anelastic
and Viscoelastic Deformation

Overview

The upper-mantle seismic low-velocity zone is most pronounced beneath young oceanic crust and tectonically active regions, where the decrease in shear velocity is ~10 percent and the attenuation or internal friction is ~0.01 to 0.1 (Forsyth 1975; Hwang and Mitchell 1987; Chan et al. 1989; William et al. 1992; Mitchell 1995). Mechanisms proposed to account for low-velocity, high-attenuation zones include both relaxation processes related to partial melting of upper-mantle rocks, such as the relaxation of a viscous fluid in a single crack in shear (Walsh 1969) and the relaxation of a fluid pressure by fluid flow between cracks with different orientations (Mavko and Nur 1975; Mavko 1980), as well as solid-state processes, such as grain-boundary sliding (Jackson 1969; Jackson and Anderson 1970; Karato and Spetzler 1990) and dislocation motion (Minster and Anderson 1980, 1981). Several reviews provide a thorough background concerning attenuation processes (Goetze 1977; Karato and Spetzler 1990; Jackson 1993).

Both the experimental constraints (e.g., Tan et al. 1997; Gribb and Cooper 1998) and theoretical understanding (Gribb and Cooper 1998) of low-frequency (seismic) attenuation in fine-grained aggregates of olivine have progressed significantly during the past few years. Results on these synthetic rocks as well as those on coarse-grained peridotites have recently been critiqued by Bagdassarov (2000) and Jackson (2000). In addition, Gribb and Cooper (2000) have investigated the attenuation behavior of fine-grained aggregates of olivine plus melt of approximately leucite-normative basanite composition. Here, we discuss the current understanding of low-frequency attenuation in these samples in the context of results from our laboratory on samples of fine-grained olivine plus basalt (Xu 1997; Xu and Kohlstedt 1994, 1997). In carefully designed experiments in our laboratory and elsewhere (Tan et al. 1997; Gribb and Cooper 1998; Bunton 2001), emphasis is placed on the use of fine-grained samples to avoid microcracking caused by grain-scale anisotropy in thermal expansion and elastic moduli. In addition, samples were predried at temperatures >1300 K at oxygen partial pressures within the olivine stability field to prevent breakdown of hydrous phases and concomitant onset of melting during subsequent measurements of attenuation.

Experimental Approach

In our laboratory, aggregates of olivine plus 2, 6, or 13 volume percent basalt were fabricated by either hydrostatically hot pressing or sintering fine-grained powders of San Carlos olivine with a starting particle size of <10 μm mixed with midocean ridge basalt (MORB) with a particle size of <8 μm. The resulting samples had an average grain size of ~20 μm and <1 percent porosity. The chemical compositions of San Carlos olivine and the MORB have been reported previously (Cooper and Kohlstedt 1984). (i) Samples with 2 volume percent basalt were cut from a billet approximately the size of a hockey puck (~70 mm in diameter and ~30 mm thick) that was commercially hot pressed in an evacuated and welded stainless steel canister. Some crystallization of the amorphous phase was observed in the hot-pressed samples. Thermal cracks developed on heating these samples to 1,475 K at room pressure. The grain-size distribution was bimodal with maxima at ~20 and ~40 μm. (ii) Samples with 6 and 13 volume percent melt were prepared by sintering powders at 1,645 K in a gas mixture composed of 22.5 volume percent CO and 77.5 volume percent CO_2 for 135 h. The powders were cold pressed at ~70 MPa, and the sintered aggregates were cooled at 500 K/h. No crystallization of the melt was observed, and no thermal cracks were detected after reheating samples to 1,625 K. The average grain size was ~20 μm. (iii) Samples with 0.4 volume percent melt were prepared from basalt-free olivine powders hot pressed at 300 MPa and 1,575 K for 3 h following the procedure outlined in Hirth and Kohlstedt (1995a). This melt formed from impurities in the olivine powders, probably introduced during pulverization of olivine single crystals. Oxygen fugacity was buffered near Ni-NiO by oxidation of the Ni capsule, which produced a thin, green NiO layer between the sample and the capsule. No crystallization was observed in the melt pockets. No cracks were detected after reheating samples to 1,575 K at room pressure. The average grain size was ~20 μm.

A free-oscillation, inverted torsion pendulum was used to determine the internal friction, Q^{-1}, as well as the real and imaginary parts of the complex-shear modulus, G_1 and G_2, at intervals of 5 K in the temperature range 295 to 1625 K at frequencies between 1 and 10 Hz in one atmosphere of a gas mixture composed of 22.5 volume percent CO and 77.5 volume percent CO_2. Specimen dimensions for the free-oscillation experiments were between 35 × 5 × 1 mm and 10 × 2.5 × 0.6 mm. Torsional motion of the sample was monitored by following a solid-state laser beam reflected from a mirror on a extension arm attached to the sample. The resolution of the charge-coupled device (CCD) detector used to track the position of the laser beam limited the strain amplitude to the range 10^{-6}–10^{-4}. Details of the torsion pendulum apparatus are described in Weiner et al. (1987), Versteeg (1992), Versteeg and Kohlstedt (1994), and Xu (1997). At each temperature, the decay of the oscillation amplitude A with time t was recorded. The damping constant δ and the frequency $f = \omega/2\pi$ (ω = angular frequency) were then determined by fitting the experimental data to the relation (Norwick and Berry 1972:19–22)

$$A = A_o \exp(-\delta f t) \cos(2\pi f t + \zeta), \qquad (10.1)$$

where A_o is the amplitude at the start of data collection and ς is a constant introduced because data acquisition started before the torsion pendulum was set into motion. A nonlinear least-squares fit of the two data sets in figure 10.1 to equation (10.1) yielded the values of f and δ reported in this figure.

Three processes potentially contribute to damping in our experiments: (i) apparatus effects such as frictional sliding at the interfaces between the sample and

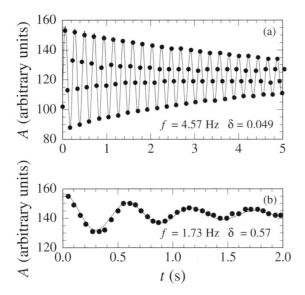

Figure 10.1 Amplitude versus time for sample of olivine plus 6 volume percent basalt. (a) Attenuation measured at 775 K. (b) Attenuation measured at 1,275 K. Values for f and δ were obtained by fitting the experimental data (solid circles) to equation (10.1). The best-fit curve is plotted as a solid line.

the grips, (ii) anelastic deformation of the sample, and (iii) plastic (viscous) deformation of the sample. To calibrate our apparatus, calibration runs were conducted on alumina single crystals. The value for internal friction of $Q^{-1} \approx 10^{-3}$ measured with the alumina single crystals is similar to that for our olivine-basalt aggregates below $\sim$1,075 K and is small compared with that for our samples above $\sim$1,075 K. Thus, the internal friction peaks, the rise in high-temperature background, and the associated decrease in shear modulus observed for olivine-basalt aggregates cannot be attributed to damping in the apparatus or sample assembly. The resolution limit for Q^{-1} of $\sim$10^{-3} is limited by a small amount of frictional sliding at the interface between the sample and the sample grips.

Data Analysis

On the basis of the equation of motion for a torsion pendulum, G_1, G_2, and Q^{-1} were determined from δ and ω ($= 2\pi f$) for rectangular samples of width a, thickness b, and length L using the following relationships (Nowick and Berry 1972; Xu 1997):

$$G_1 = \left(\frac{L}{ab^3/3}\right)(2mr^2\omega^2)\left(1 - \frac{\delta^2}{4\pi^2}\right), \qquad (10.2)$$

$$G_2 = \left(\frac{L}{ab^3/3}\right)(2mr^2\omega^2)\left(\frac{\delta}{\pi}\right) \tag{10.3}$$

and

$$Q^{-1} = \frac{G_2}{G_1} \approx \frac{\delta}{\pi}, \tag{10.4}$$

where m is the mass of each of two inertial weights fastened to the sample at a radial distance r from the torsion axis. The quantity $2mr^2$ is the mass moment of inertia, and the term $(3L/ab^3)$ is the area moment of inertia for a rectangular sample.

Experimental Results

The dependence of the real part of the complex modulus and of the internal friction on temperature are shown in figures 10.2 and 10.3 for samples of olivine plus 0.4 volume percent melt and olivine plus 6 volume percent basalt, respectively. On the basis of a series of calibration tests using alumina single crystal as a standard, the uncertainty in the modulus is <30 percent, whereas the uncertainty in the

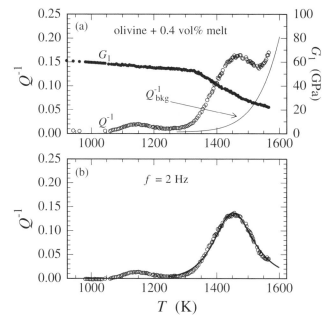

Figure 10.2 (a) Values of Q^{-1} and G_1 versus temperature for a sample of olivine plus 0.4 volume percent melt. Results were obtained at 5 K intervals from data similar to those shown in figure 10.1. Values of Q^{-1} were normalized to $f = 2$ Hz. (b) Values of Q^{-1} normalized to $f = 2$ Hz versus temperature after background damping has been subtracted. The solid curve was obtained by fitting the experimental data between 1,250 and 1,575 K to equations (10.6) and (10.8).

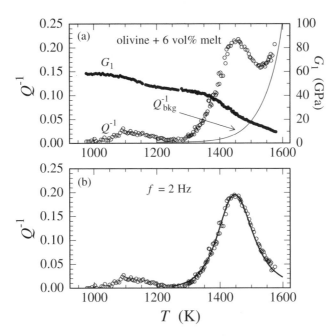

Figure 10.3 (a) Values of Q^{-1} and G_1 versus temperature for a sample of olivine plus 6 volume percent basalt. Results were obtained at 5 K intervals from data similar to those shown in figure 10.1. Values of Q^{-1} were normalized to $f = 2$ Hz. (b) Values of Q^{-1} normalized to $f = 2$ Hz versus temperature after background damping has been subtracted. The solid curve was obtained by fitting the experimental data between 1,250 and 1,575 K to equations (10.6) and (10.8).

internal friction is <10 percent. In repeat experiments on multiple samples of olivine plus melt/basalt of identical composition, scatter in the modulus and internal friction data was smaller than these uncertainties. The raw attenuation data, presented in figures 10.2a and 10.3a, exhibit two peaks, one centered near 1,150 K and the other near 1,450 K (Xu and Kohstedt 1994, 1997). These peaks (anelastic deformation) are superimposed on high-temperature background damping (viscous deformation) that increases systematically with increasing temperature. Because it affects the position, height, width, and shape of an internal friction peak, the high-temperature background damping must be subtracted from the raw spectra to isolate the attenuation (Debye) peaks. Therefore, on the basis of the model recently proposed by Bunton and Cooper (Bunton 2001:45), the high-temperature background attenuation (damping), Q_{bkg}^{-1}, was determined by fitting the experimental Q^{-1} versus T data away from the attenuation peaks to the relationship

$$Q_{bkg}^{-1} = Q_o^{-1} + \frac{2.2 + (G_1/\eta\omega)^{1/2}}{2.2 + (\eta\omega/G_1)^{1/2}}, \quad (10.5a)$$

where the zero offset or apparatus resolution Q_o^{-1} is estimated from the internal friction at low temperatures. The shear viscosity η is given by

$$\eta = \eta_o \exp\left(\frac{E_\eta}{RT}\right), \tag{10.5b}$$

where η_o is a materials parameter, E_η is the activation energy for the viscous damping process, R is the gas constant, and T is absolute temperature. Nonlinear least-squares fits of the high-temperature background data to equations (10.5a) and (10.5b) yields $E_\eta = 615 \pm 50$ kJ/mol for the sample with 0.4 volume percent melt and $E_\eta = 650 \pm 50$ kJ/mol for the sample with 6 volume percent melt with corresponding values of $\eta_o = 2.3 \times 10^{-8}$ and 1.8×10^{-9} Pa·s. These values of viscosity are in very good agreement with those of Bunton and Cooper (Bunton 2001) for melt-free samples. The errors associated with fitting the background data are rather large because, as is apparent in figures 10.2a and 10.3a, few of the data points are far enough away from the attenuation peaks to permit clear isolation of the background damping. Therefore, the constraint that the Debye peaks must be approximately symmetric about the peak temperature was used to help define the background damping, as described in the next paragraph.

The attenuation peaks are plotted in figures 10.2b and 10.3b after removal of the background damping. The attenuation peaks near 1,150 K in figures 10.2 and 10.3 are associated with the glass transition of the melt phase (Xu 1997) and will not be discussed further in this chapter. To analyze the dynamic anelastic response of our samples represented by the attenuation peaks near 1,450 K in terms of a relaxation process, Q^{-1} and G_1 are expressed in terms of a relaxation time τ combined with the relaxed and unrelaxed shear moduli, G_U and G_R, based on the model for a standard anelastic solid (Nowick and Berry 1972:41–63):

$$Q^{-1} \approx \frac{G_U - G_R}{G_R} \cdot \frac{\omega\tau}{1 + \omega^2\tau^2}. \tag{10.6}$$

This Debye equation for Q^{-1} has a peak at $\omega\tau = 1$. Likewise, G_1 decreases from G_U for $\omega\tau \gg 1$ to G_R for $\omega\tau \ll 1$ according to the relation

$$G_1 = G_U - \frac{G_U - G_R}{1 + \omega^2\tau^2}. \tag{10.7}$$

The relaxation time can be expressed by an Arrhenius equation of the form

$$\tau = \tau_o \exp\left(\frac{E_\tau}{RT}\right), \tag{10.8}$$

where E_τ is an activation energy characteristic of the relaxation process and τ_o is a materials parameter. Because ω is approximately constant in each of our experiments, the condition $\omega\tau \gg 1$ occurs at temperatures below the Debye peak where τ is large, while $\omega\tau \ll 1$ occurs at temperatures above the Debye peak where the relaxation time for the attenuation process is short. On the basis of equations

(10.6), (10.7), and (10.8), the anticipated dependence of Q^{-1} and G_1 on temperature is illustrated in figure 10.4 for the case in which ω is constant and both G_U and G_R are independent of temperature.

The attenuation peaks in figures 10.2b and 10.3b were fit to equation (10.6) with τ given by equation (10.8), taking into account the experimentally determined decrease in ω with increasing temperature, characteristic of a free-oscillation apparatus, as reflected in the dependence of G_1 on temperature in figures 10.2a and 10.3a. The activation energy is $E_\tau = 310 \pm 50$ kJ/mol for the sample with 0.4 volume percent melt and $E_\tau = 350 \pm 5\,0$ kJ/mol for the sample with 6 volume percent melt.

Comparison with Published Results

Results determined for our samples with 0.4 volume percent melt for the high-temperature background damping and for G_1 are in good agreement with those reported by Jackson (2000) and Gribb and Cooper (1998) on fine-grained, melt-

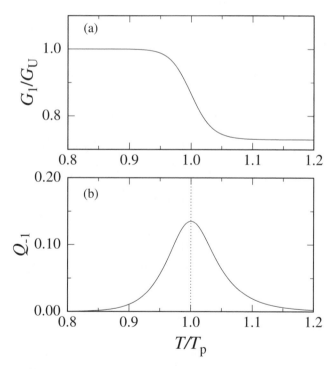

Figure 10.4 (a) Shear modulus normalized to unrelaxed value of shear modulus versus temperature normalized to the temperature at the attenuation peak T_p based on the model for a standard anelastic solid summarized in equations (10.7) and (10.8). (b) Q^{-1} versus normalized temperature for a standard anelastic solid based on equations (10.6) and (10.8). In both (a) and (b), values of $E = 300$ kJ/mol and $\tau_0 = 10^{-14}$ s were used. In addition, G_U, G_R, and ω were taken to be independent of temperature.

free samples. At a frequency of 1 Hz and 1,573 K, Jackson (2000) finds $Q^{-1} \approx$ 0.07 and $G_1 \approx 40$ GPa for a sample with a grain size of ~30 μm, compared with values at 2 Hz of $Q^{-1} \approx 0.14$ and $G_1 \approx 22$ GPa for our sample with a grain size of ~20 μm. At a frequency of 1 Hz and 1,523 K, Gribb and Cooper (1998) obtained $Q^{-1} \approx 0.1$ and $G_1 \approx 38$ GPa for a sample with a grain size of ~3 μm, compared with our values at 2 Hz of $Q^{-1} \approx 0.06$ and $G_1 \approx 27$ GPa. All three studies reveal a significant increase in $(\partial G_1/\partial T)_\omega$ with increasing temperature above ~1,275 K, correlated with the rapid rise in the high-temperature background damping. Below ~1,275 K, deformation in our samples is reversible (recoverable) and linear. Above ~1,275 K, deformation remains linear but some of the strain is irreversible; in this regime, plastic (viscoelastic) deformation of the sample begins to dominate as the high-temperature background damping increases systematically with increasing temperature.

For our samples with 6 volume percent melt, results for Q^{-1} measured for the high-temperature background damping and for G_1 are in good agreement with those reported by Gribb and Cooper (2000) on fine-grained samples with ~5 volume percent melt. At a frequency of 1 Hz and 1,523 K, Gribb and Cooper (2000) find $Q^{-1} \approx 0.25$ and $G_1 \gtrsim 20$ GPa for a sample with a grain size of ~3 μm, compared with values at 2 Hz of $Q^{-1} \approx 0.14$ and $G_1 \approx 22$ GPa for our sample with a grain size of ~20 μm. It should be emphasized that the results from the experimental studies of Jackson and coworkers, Cooper and coworkers, and the present investigation are all in very good agreement with one another.

Discussion of the Attenuation Results

High-temperature background damping has historically been interpreted in terms of a broad spectrum of relaxation times (compliances). For damping due to dislocation motion, this effect is due to a distribution of dislocation line lengths, whereas for damping due to grain-boundary sliding, this effect results from a distribution of grain size or step spacing along grain boundaries (Nowick and Berry 1972:454–461). Following the work of Raj and Ashley (1971) and Raj (1975), Gribb and Cooper (1998, 2000) and Bunton and Cooper (Bunton 2001) have recently presented a model that quantitatively explains the observed high-temperature background damping described by equation (10.5) for a deforming material in terms of diffusional dissipation of stress concentrations at triple junctions that arise due to sliding on grain boundaries with negligible viscous resistance. The results of our study are consistent with this interpretation in the sense that deformation above 1,275 K is clearly viscous; that is, a significant component of the deformation is irreversible (i.e., plastic).

Attenuation peaks have previously been observed in some (Xu and Kohlstedt 1994, 1997) but not all (e.g., Tan et al. 1997; Jackson 2000; Gribb and Cooper 1998, 2000) studies of anelastic and plastic deformation of fine-grained aggregates of olivine or olivine plus melt at low frequencies, possibly because the grain sizes of the samples used in other studies were somewhat smaller and the frequencies somewhat lower than those used in our experiments. Consistent with this deduction, Bunton and Cooper (Bunton 2001) observed an attenuation "plateau" at the high-frequency (0.1–0.3 Hz) end of their attenuation spectra; this plateau is most pronounced for their (melt-free) samples with the largest grain size (16.9 μm) investigated at their lowest temperature (1,473 K). They suggest that this plateau may be the beginning of an attenuation peak.

Such attenuation peaks superimposed on a monotonically increasing high-temperature background are characteristic of those arising due to grain-boundary mechanisms (e.g., Nowick and Berry 1972:455). It should be emphasized that the attenuation peaks correspond to anelastic deformation (reversible but time dependent). In the context of the analysis of Gribb and Cooper (1998) and Bunton and Cooper (Bunton 2001), McMillan et al. (unpublished manuscript) suggested that an absorption peak should occur at a frequency characteristic of the detailed (periodic) grain-boundary structure (e.g., the distance between steps/ledges in grain boundaries). For isothermal experiments, they propose that a peak should occur at the transition from high-frequency behavior, controlled by elastic loading of triple junctions (with negligible grain-boundary sliding), to low-frequency behavior, controlled by loading at triple junctions caused by sliding on inviscid grain boundaries. For our experiments in which temperature is systematically varied at approximately constant frequency, this transition and the associated absorption peak should develop with increasing temperature. In this context, we propose that the high-temperature peak is due to sliding of grain boundaries with finite viscous strength combined with elastic loading at triple junctions. This proposed mechanism is supported by the observation that, with increasing temperature (decreasing relaxation time), the attenuation peak occurs just at the onset of the steep rise in background damping due to grain-boundary sliding. Such grain-boundary-sliding behavior has been reported for numerous materials (e.g., Kê 1947a, 1947b; Nowick and Berry 1972:438–440, 454–457; Mosher and Raj 1974: Mosher et al. 1976; Weiner et al. 1987).

As demonstrated in the plot of height of the attenuation peak, ΔQ^{-1}, versus melt fraction, ϕ, in figure 10.5, extrapolation of ΔQ^{-1} to $\phi = 0$ yields a peak height of ~0.1, suggesting that the same attenuation mechanism operates in melt-free samples. This conclusion is consistent with the presence of the onset of an attenuation peak reported by Bunton and Cooper (Bunton 2001), as described above. Because the peak height increases with increasing melt fraction, the presence of the melt phase must enhance the grain-boundary relaxation process, possibly by providing short-circuit diffusion paths at triple junctions that assist in the relaxation of stress concentrations near the triple junctions, much as envisioned by Cooper et al. (1989).

Other models for anelastic deformation of melt-bearing materials involve long-range migration of melt through the interconnected network of triple junctions (Cooper 1990). Because cyclic deformation in torsional experiments is pure shear, no driving force exists for such long-range melt migration. However, at a local scale, the melt-filled triple junctions are subjected to states of stress that depend on their orientations. The resulting local gradient in pore pressure is a

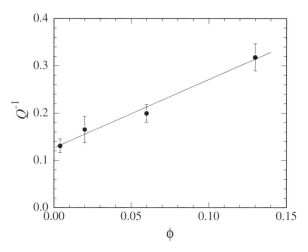

Figure 10.5 Height of attenuation peak versus melt fraction.

driving force for short-range movement of melt from one triple junction to the next, a process often referred to as "melt squirt". The viscosity of the melt phase μ is related to the relaxation time τ_{ms} for melt squirt through the expression (Mavko 1980)

$$\mu = \frac{\tau_{ms} K_m}{40} \left(\frac{R}{d}\right)^2,$$ (10.9)

where K_m is the bulk modulus of the melt phase, the tube (or triple junction) length is taken to be on the order of the grain size d, and R is a measure of the tube cross section, which in this geometry is approximately half the height of the triangle defined by a melt-filled triple junction. The attenuation peak occurs at $\omega\tau = 2\pi f\tau = 1$. For a typical value of $f = 2$ Hz in our experiments combined with $d = 20$ μm and $R = 1$ μm for our samples plus $K_m = 10$ GPa , equation (10.9) yields $\mu = 5 \times 10^4$ Pa·s, a value more than an order of magnitude larger than viscosity of basalt at 1,450 K. Thus, it is unlikely that the melt-squirt mechanism dominates attenuation in our experiments. This conclusion is supported by the observation that the activation energy determined for the attenuation peaks of 310 to 350 kJ/mol is significantly larger than that of 210 to 250 kJ/mol reported for the viscosity of basalt (Ryan and Blevins 1987; Murase and McBirney 1973). A more detailed discussion of this point and models involving melt squirt not only through triple junctions but also from grain boundaries are developed in Xu (1997).

If, as suggested previously, the attenuation peaks near 1,450 K are due to viscous sliding along grain boundaries coupled with the buildup of elastic stress at triple junctions, then the grain-boundary viscosity η_{gb} can be calculated from the grain-boundary sliding model of Mosher and Raj (1974) (see also, Mosher et al. 1976; Weiner et al. 1987), which yields

$$\eta_{gb} = \frac{2G_1 \tau h}{1.14(1 - v) d}, \qquad (10.10)$$

where h is the grain boundary thickness. For $h = 1$ nm (Ricoult and Kohlstedt 1983), η_{gb} is $\sim 3 \times 10^5$ Pa·s at 1,450 K. Combined with the average value for the activation energy determined previously for the attenuation peak, the temperature dependence of grain-boundary viscosity for dunite is given by $\eta_{gb} = \eta^\circ_{gb} \exp(-E_{gb}/RT)$ with $\eta^\circ_{gb} = 4 \times 10^{-7}$ Pa·s and $E_{gb} = 330$ kJ/mol.

Extrapolation of recent experimental results obtained on fine-grained samples to the larger grain sizes appropriate for mantle rocks remains an important challenge. First, for the viscous damping giving rise to the high-temperature internal friction background, Jackson (2000) suggested that $Q_{bkg}^{-1} \propto 1/d$ and Bunton and Cooper (Bunton 2001) determined that $\eta \propto 1/d^{2.2 \pm 0.3}$. The expression from Bunton and Cooper for Q_{bkg}^{-1}, which is presented here as equation (10.5a), can be well approximated by $Q_{bkg}^{-1} \approx (G_1/\eta\omega)^{1/2}/2.2$ at high temperatures where background damping dominates attenuation. This expression, combined with the experimental observation just quoted that $\eta \propto 1/d^{2.2}$ yields $Q_{bkg}^{-1} \propto 1/d^{1.1}$, is in good agreement with the conclusion reached by Jackson (2000). By combining this result with our experimentally determined values for η and G_1, we calculate $Q_{bkg}^{-1} = 0.01$ and 0.02 for frequencies of 0.1 and 0.01 Hz, respectively, for a grain size of 3 mm and a temperature of 1,675 K. These values for Q_{bkg}^{-1} are in good agreement with those reported for attenuation at depths of 80 to 220 km in the upper mantle (Romanowicz and Durek 2000). Second, for the anelastic damping giving rise to the attenuation peaks, on the basis of equation (10.10), an attenuation maximum would be expected at ~ 0.2 Hz for a temperature of 1,675 K and a grain size of 3 mm. At this point, it should be noted that Gribb and Cooper (1998) and Bunton and Cooper (Bunton 2001) suggest that the subgrain spacing, which in steady-state deformation is a function of differential stress, is the critical microstructural parameter controlling anelastic and viscoelastic deformation resulting from sliding on viscid and inviscid interfaces, respectively. If this suggestion is correct, then the high-temperature background damping associated with sliding on inviscid boundaries and the frequency of the attenuation peak associated with sliding on viscid boundaries would be a factor of ~ 10 larger for a subgrain size of 0.3 mm.

Plastic Deformation

Overview

The plastic-deformation behavior of partially molten rocks with mantle compositions has been investigated for nearly two decades (Cooper and Kohlstedt 1984, 1986; Bussod and Christie 1991; Beeman and Kohlstedt 1993; Jin et al. 1994; Kohlstedt and Chopra 1994; Hirth and Kohlstedt 1995a, 1995b; Kohlstedt and Zimmerman 1996; Zimmerman 1999; Mei and Kohlstedt 2000a, 2000b). The experimentally determined creep results are usually analyzed to take into account the dependence of strain rate $\dot{\varepsilon}$ on differential stress σ, grain size, temperature, and pressure P. With the addition of melt, the constitutive equation describing steady-state deformation will have the generalized form

$$\dot{\varepsilon} = \dot{\varepsilon}(\sigma, d, \phi, T, P) \tag{10.11}$$

As discussed in the following text, the dependence of $\dot{\varepsilon}$ on σ and d is often described using power-law relationships, and the dependence of $\dot{\varepsilon}$ on T and P is analyzed using the exponential relationship common for thermally activated processes. The dependence of $\dot{\varepsilon}$ on ϕ remains a topic of active experimental and theoretical analysis.

To apply laboratory results to problems involving flow in the mantle, extrapolation to significantly lower strain rates is necessary. To deform samples under laboratory conditions (most notably times of $\sim 10^5$ s), experiments are generally conducted at differential stresses larger than those appropriate for flow in the bulk mantle (10 MPa versus 0.1 MPa). Laboratory and upper-mantle temperatures are usually similar to each other, whereas laboratory pressures are generally significantly smaller than mantle pressures. Extrapolation requires that the deformation mechanism is the same in the laboratory as it is in the Earth. Hence, experiments often explore deformation behavior in both the diffusional creep regime and the dislocation creep regime. Viscosity is then calculated as

$$\eta = \frac{\sigma}{\dot{\varepsilon}(\sigma)}. \tag{10.12}$$

While η is independent of stress in the diffusional creep regime (i.e., diffusional creep is Newtonian, $\dot{\varepsilon} \propto \sigma^1$), η increases with decreasing differential stress in the dislocation creep regime (i.e., dislocation creep is non-Newtonian, $\dot{\varepsilon} \propto \sigma^n$ with $3 < n < 5$).

Plastic Flow of Partially Molten Lherzolite and Olivine plus Basalt Aggregates

Recently, strain rate has been determined from laboratory experiments as a function of melt fraction ϕ under both anhydrous (dry) and hydrous (wet) conditions (Mei et al. 2002). The results can be expressed in terms of a flow law of the form

$$\dot{\varepsilon}^{\text{dry/wet}}(\phi) = \dot{\varepsilon}^{\text{dry/wet}}_{\phi=0} \mathcal{F}(\phi), \tag{10.13}$$

where $\mathcal{F}(\phi)$ contains the functional dependence of strain rate on melt fraction. That is to say, the effect of water on strain rate can be separated from the effect of melt, and the effect of melt on strain rate is the same under anhydrous and hydrous conditions. The latter point is a statement of the observation that the wetting behavior of basalt in a peridotite matrix is the same or very nearly the same under hydrous and anhydrous conditions. Empirically, the melt-fraction dependence can be described quite well by (Kelemen et al., 1997; Mei et al. 2002; Zimmerman and Kohlstedt, in press):

$$\mathscr{F}(\phi) = \exp(\alpha\phi). \tag{10.14}$$

To illustrate the dependence of strain rate on melt fraction, recent creep results for fine-grained (10–20 μm) lherzolite samples deformed under anhydrous conditions are presented in figure 10.6 (Zimmerman and Kohlstedt, in press). These samples composed of 61.8 volume percent olivine, 26.0 volume percent enstatite, 10.0 volume percent chrome-diopside, and 2.2 volume percent spinel were prepared by hot-pressing lherzolite powders obtained from a Damaping, China, mantle

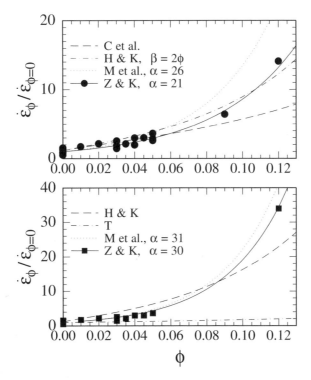

Figure 10.6 (a) Normalized strain rate versus melt fraction for samples of synthetic lherzolite deformed in the diffusional creep regime. C et al. = Cooper et al. (1989), equation (10.15); H & K = Hirth and Kohlstedt (1995a), equation (10.18); M et al. = Mei et al. (2002), equation (10.14); Z & K = Zimmerman and Kohlstedt (in press), equation (10.14). (b) Normalized strain rate versus melt fraction for samples of synthetic lherzolite deformed in the dislocation creep regime. H & K = Hirth and Kohlstedt (1995b), equation (10.21); T = Tharp (1983), equation (10.20); M et al., equation (10.14); Z & K, equation (10.14).

xenolith. In the diffusional creep regime (figure 10.6a), the experimental results are well fit to the exponential dependence in equations (10.13) and (10.14) with $\alpha = 21$. In the dislocation-creep regime, the strain rate is more sensitive to melt fraction such that $\alpha \approx 30$ yields the best fit to the data.

As also illustrated in figure 10.6, somewhat larger values of α are required to describe the dependence of strain rate on melt fraction for aggregates of olivine plus basalt. In the diffusion creep regime, $\alpha = 26$, while in the dislocation creep regime, $\alpha = 31$ (Mei et al. 2002).

Predicted Dependence of Viscosity on Melt Fraction

Theoretical analyses yield a less dramatic increase in strain rate with increasing melt fraction. In the diffusion-creep regime, Cooper and Kohlstedt (1986) and Cooper et al. (1989) considered the effect on strain rate of the local stress enhancement that arises due to replacement of a fraction of the grain-boundary contact $\Delta d'/d'$ by melt combined with the effect of rapid (short-circuit) diffusion through melt-filled triple junctions. In their analysis, these authors assumed that the melt distribution is governed by surface tension and that the dihedral angle θ is single-valued with $0 < \theta < 60°$, such that all of the melt is distributed in three- and four-grain junctions. Their model leads to the following relationship between strain rate and melt fraction:

$$\dot{\varepsilon}(\phi) = \dot{\varepsilon}\,^{\text{dry/wet}}_{\phi=0} \left(\frac{1}{1 - (\Delta d'/d')}\right)^2_{sc} \left(\frac{1}{1 - (\Delta d'/d')}\right)^2_{se} \qquad (10.15)$$

$$= \dot{\varepsilon}\,^{\text{dry/wet}}_{\phi=0} \left(\frac{1}{1 - \phi^{1/2}\mathcal{E}(\theta)}\right)^4,$$

where the subscripts sc and se indicate short-circuit transport and local stress enhancement, respectively, and

$$\mathcal{E}(\theta) = \frac{1.06 \sin(30° - (\theta/2)}{\left(\dfrac{1 + \cos\theta}{\sqrt{3}} - \sin\theta - \dfrac{\pi}{90}(30° - (\theta/2))\right)^{1/2}} \qquad (10.16)$$

Equation (10.15) underestimates the effect of melt fraction on strain rate, particularly at $\phi \gtrsim 0.05$. This equation breaks down at least in part because the melt distribution in partially molten mixtures of olivine plus basalt and partially molten lherzolite deviates significantly from the ideal microstructure used in the Cooper–Kohlstedt analysis (Cooper and Kohlstedt 1986; Cooper et al. 1989). Because of the anisotropy in solid-melt interfacial energy, melt is not confined to three- and four-grain junctions but also penetrates a significant fraction of the grain boundaries to separate neighboring grains. Hence, the partially molten samples contain short-circuit diffusion paths and stress-enhancement geometries not considered in the derivation of equation (10.15).

To take this effect into account in the diffusion creep regime, Hirth and Kohlstedt (1995a) quantified the fraction of grain boundaries completely wetted by melt β and determined that β increases approximately linearly with melt fraction:

$$\beta \propto \phi. \tag{10.17}$$

These authors argued that the effective diffusion path is shortened from $d' - \Delta d'$ to $(d' - \Delta d')(1 - \beta)$ due to wetted-grain boundaries and, hence, that equation (10.15) should be modified to

$$
\begin{aligned}
\dot{\varepsilon}(\phi) &= \dot{\varepsilon}\,_{\phi=0}^{\text{dry/wet}} \left(\frac{1}{(1 - \Delta d'/d')}\right)_{se}^2 \left(\frac{1}{(1 - \Delta d'/d')(1 - \beta)}\right)_{sc}^2 \\
&= \dot{\varepsilon}\,_{\phi=0}^{\text{dry/wet}} \left(\frac{1}{1 - \phi^{1/2}\mathscr{C}(\theta)}\right)_{se}^2 \left(\frac{1}{(1 - \phi^{1/2}\mathscr{C}(\theta))(1 - \beta)}\right)_{sc}^2 .
\end{aligned}
\tag{10.18}
$$

In the diffusion-creep regime, no correction was made to the stress-enhancement term because the grain-boundary area replaced by melt in partially molten samples is approximately the same as in an ideal system with isotropic solid-melt interfacial energy (Hirth and Kohlstedt 1995a). For aggregates of olivine plus basalt, Hirth and Kohlstedt (1995a) obtained a value of $\beta = 4.5\phi$ for anhydrous conditions, whereas Mei et al. (2002) determined a value of $\beta = 3\phi$ for hydrous conditions. For partially molten lherzolite samples, equation (10.18) fits the experimental data reasonably well using a value of $\beta = 2\phi$, as illustrated in figure 10.6a. Fewer grain-grain contacts are wetted in the lherzolite samples than in the dunite samples because, although melt often wets olivine-olivine grain boundaries, it does not tend to wet pyroxene-pyroxene grain boundaries or olivine-pyroxene phase boundaries.

On the basis of the premise that a large fraction of grain interfaces are wetted by melt, Paterson (2001) presented a theory for granular flow to describe the deformation of partially molten rocks. This model considers three regimes: one in which deformation rate is limited by transfer of melt by viscous flow, a second in which deformation rate is controlled by transfer of material away from points at which grains impinge on one another by diffusion through the melt, and a third in which deformation rate is governed by the rate of reaction at the melt-solid interface. Although the strain rate predicted for the diffusion-control regime for a rock with a grain size of 10 μm is in reasonable agreement with the laboratory results of Hirth and Kohlstedt (1995a), the dependence of strain rate on temperature and grain size is not in good agreement with values reported from experimental studies. On the basis of the model, strain rate is expected to be proportional to viscosity of the melt phase and inversely proportional to the square of the grain size. However, activation energies for viscous flow of Kilauea olivine tholeiite basalt (Ryan and Blevins 1987) and Columbia River basalt (Murase and McBirney

1973) are <250 kJ/mol, whereas the activation for creep of partially molten olivine-basalt aggregates are >300 kJ/mol (Cooper and Kohlstedt 1986; Hirth and Kohlstedt 1995a). In addition, the creep results yield an inverse cubed (rather than squared) dependence of strain rate on grain size. Hence, the granular flow model of Paterson (2001) has not been considered further in our analysis of flow of partially molten mantle rocks.

In the dislocation-creep regime, the model of Chen and Argon (1971) based on stress enhancement due to the presence of a weak second phase leads to the relationship

$$\dot{\varepsilon}(\phi) = \dot{\varepsilon}_{\phi=0}^{\text{dry/wet}} \left(\frac{1}{1 - 2\phi} \right) \left(\frac{1}{1 - \phi} \right)^{n-1}, \tag{10.19}$$

whereas that of Tharp (1983) for a small amount of porosity gives

$$\dot{\varepsilon}(\phi) = \dot{\varepsilon}_{\phi=0}^{\text{dry/wet}} \frac{1}{((1 - k\phi)^{2/3})^n} \tag{10.20}$$

with $0.98 \leq k \leq 2.26$. Equations (10.19) and (10.20) both markedly underestimate the effect of melt fraction on strain rate, even at relatively low-melt fractions.

To account for the discrepancy between the models represented by equations (10.19) and (10.20) and their experimental observations for dislocation creep, Hirth and Kohlstedt (1995b) used the stress-enhancement portion of equation (10.15) modified to take into account the fact that the stress exponent $n = 3.0$–3.5 for dislocation creep in olivine-rich rocks:

$$\dot{\varepsilon}(\phi) = \dot{\varepsilon}_{\phi=0}^{\text{dry/wet}} \left(\frac{1}{1 - \phi^{1/2}\mathcal{E}(\theta)} \right)_{se}^{2n}. \tag{10.21}$$

Again, as illustrated in figure 10.6b, this form of equation (10.15), as modified for dislocation creep in equation (10.21), provides a remarkably good fit to the data, in particular, given the geometric assumptions (e.g., hexagonal grains and melt restricted to triple junctions) made in the Cooper-Kohlstedt model (Cooper et al. 1989). Nonetheless, equation (10.21) overestimates the strain rate at low-melt fractions and underestimates the strain rate at higher-melt fractions. Some part of the weakening due to the presence of melt is likely due to the ease of grain-boundary sliding on those interfaces that are totally wetted by melt combined with melt-assisted relaxation of stresses at melt-filled triple junctions.

The term $\dot{\varepsilon}_{\phi=0}^{\text{dry/wet}}$ in this series of equations contains the dependence of strain rate on σ, d, T, P, and, in the case of hydrous (i.e., wet) conditions, OH concentration C_{OH}. In the usual power-law format,

$$\dot{\varepsilon}_{\phi=0}^{\text{dry}}(\sigma,T,P) = A_{\text{dry}} \frac{\sigma^{n_{\text{dry}}}}{d^{P_{\text{dry}}}} \exp\left(- \frac{E_{\text{dry}} + PV_{\text{dry}}}{RT} \right) \tag{10.22}$$

and

$$\dot{\varepsilon}^{\text{wet}}(\sigma, C_{\text{OH}}, T, P) = A_{\text{wet}} \frac{\sigma^{n_{\text{wet}}}}{d^{P_{\text{wet}}}} C_{\text{OH}}^{1} \exp\left(- \frac{E_{\text{wet}} + PV_{\text{wet}}}{RT} \right), \quad (10.23)$$

where $E_{\text{dry/wet}}$ is the activation energy and $V_{\text{dry/wet}}$ is the activation volume for creep under dry or wet conditions. The transition with increasing water concentration from flow described by equation (10.18) to flow described by equation (10.23) occurs at a OH concentration of $\sim$50 H/10^{6}Si (Mei and Kohlstedt 2000a, 2000b).

Deformation-Driven Melt Redistribution

Overview

Transport of melt from its source region at depth to the surface of the Earth occurs much more quickly than is possible by grain-scale porous flow (Spiegelman and Kenyon 1992). To accommodate rapid migration of melt, flow must become channelized shortly after it is formed. One possible mechanism for focusing melt into relatively narrow, melt-rich, high-permeability paths is by the formation of a reactive infiltration instabilities. Laboratory examples on the salt-water (Kelemen et al. 1995) and peridotite-basalt (Daines and Kohlstedt 1994) systems and mathematical models (Aharonov et al. 1997) based on compaction theory as introduced by McKenzie (1984) provide constraints on the scale and efficiency of the transition from porous to channelized flow.

The essential feature of a reactive-infiltration instability is the change in equilibrium composition of the melt phase relative to the composition of the host rock as melt rises buoyantly through the mantle. As basaltic melt flows upward, it becomes undersaturated in pyroxene due to changing P-T conditions. Melt therefore preferentially dissolves pyroxene from the host rock, locally producing regions of high-melt fraction, and hence high permeability. Positive feedback is thus established as more melt, again undersaturated in pyroxene, flows through the high-permeability channels. The result is a distinct fingering of melt-rich paths into the rock, as illustrated in figure 10.7 (Daines and Kohlstedt 1994).

In this section, we focus on a second method by which melt-rich, high-permeability channels form in a partially molten rock. In an article in 1989, Stevenson explored the possibility of melt segregation during pure-shear deformation. In brief, spatial variations in melt fraction produce spatial variations in mean pressure, because the stress that a region is capable of supporting increases with decreasing melt fraction. Melt then flows down the resulting (mean) pressure gradients from regions with lower-melt fraction (higher strength, higher mean pressure) to regions with higher-melt fraction (lower mean pressure). As with the reactive infiltration instability, positive feedback is established such that the melt fraction locally increases in regions of initially higher melt content. Although his analysis did not yield a characteristic scale length for this process, Stevenson

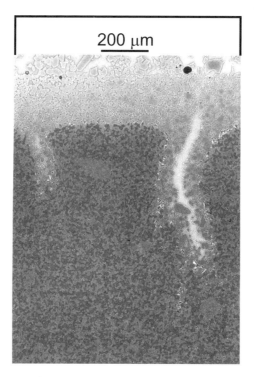

Figure 10.7 Backscattered electron image of melt migration couple. The lower part of the sample was composed of a 50:50 mixture of olivine plus pyroxene. Lighter gray mineral is olivine, and darker gray mineral is pyroxene. The upper portion was initially basalt undersaturated in pyroxene. Glass is white. After annealing at 1,573 K for 12 h, a region of olivine separated the basalt from the olivine plus pyroxene rock. Instability of the reaction front is indicated by the finger-like projections extending into the two-phase, olivine plus pyroxene portion of the sample. From Daines and Kohlstedt (1994).

concluded that band spacing must be larger than the grain spacing and possibly on the order of the compaction length.[1] Hall and Parmentier (2000) extended Stevenson's analysis by including the role of both water and melt in the localization of melt during deformation. They also conclude that the spacing between bands should be on the order of the compaction length. A complementary analysis for the growth of shear bands in a variable-viscosity porous medium undergoing simple-shear deformation (a deformation geometry analogous to the experiments described in the next subsection) also indicates that melt will segregate into melt-rich bands. This analysis predicts that the bands will be oriented in a direction antithetic to the shear direction and at an angle significantly less than 45° to the shear plane with a spacing of less than half the compaction length (M. Spiegelman, private communication).

Experimental Observations

Recent simple- and pure-shear experiments on samples of several solid plus melt systems reveal a dramatic reorganization of melt during plastic flow. In the olivine plus basalt system, melt becomes redistributed during deformation from randomly oriented triple-junction and grain-boundary pockets into highly aligned pockets (Kohlstedt and Zimmerman 1996; Daines and Kohlstedt 1997; Zimmerman et al. 1999). An example of this deformation-driven change in melt distribution is illustrated in figure 10.8 with optical micrographs of an undeformed and a deformed sample of olivine plus 5 volume percent basalt. The latter was deformed in simple shear to a shear strain of $\gamma \approx 1.$[2] Deformation aligns grain-scale melt pockets at ~25° to the shear plane and ~20° to the maximum principal stress, σ_1. In unsheared samples, the three-dimensional melt distribution forms an interconnected

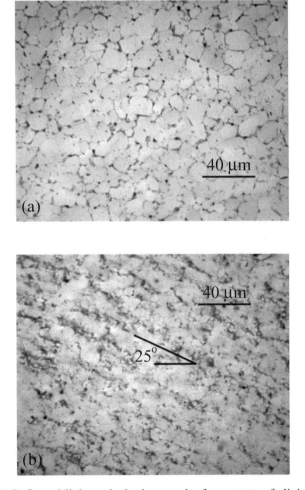

Figure 10.8 (a) Reflected-light optical micrograph of aggregate of olivine plus 7 volume percent basalt. Note the presence of basalt (the darker phase) not only in all of the triple junctions but also in many of the grain boundaries. (b) Reflected-light optical micrograph of aggregate of olivine plus 5 volume percent basalt sheared to a shear strain of ~1. As a result of deformation, a pronounced melt preferred orientation developed at an angle of ~25° to the shear direction. Olivine is lighter gray, and basalt is darker gray.

volume composed of planar and tubular sections, whereas in sheared samples, the three-dimensional melt distribution sculpts an interconnected volume approximating a series of parallel sheets separated by about the width of a single olivine grain (Scott and Kohlstedt, unpublished manuscript).

In systems composed of plagioclase plus basalt-derived melt (Ginsberg 2000), olivine plus chromite plus basalt (Holtzman et al., in press), and olivine plus FeS melt plus basalt (Hustoft, private communication), melt segregates during deformation into melt-rich zones that appear as bands in two dimensions. As illustrated in figure 10.9 for a sample of anorthite plus $\sim$5 volume percent basalt, melt-rich bands $\sim$20 μm wide are spaced $\sim$200 μm apart. The bands lie in approximately the same orientation as the grain-scale melt pockets in sheared samples of olivine plus basalt, specifically, at $\sim$10° to the shear plane and $\sim$35° to σ_1. The melt-rich bands contain $\sim$20 volume percent melt, and the regions between these bands contain $\lesssim$1 volume percent melt. Once established, the orientation of the melt-rich bands does not appear to evolve with increasing strain. Rather, the orientation seems to be controlled by the stress state. Hence, in a shear experiment, differential movement occurs between the solid and the melt-rich bands such that the melt acts as a porosity wave (magmon or soliton) moving through the solid (Stevenson and Scott 1987).

Theoretical Framework

On the basis of the results for partially molten samples of several different compositions, the transition from the first case in which melt pockets are locally reoriented during deformation to the second case in which melt is both reoriented

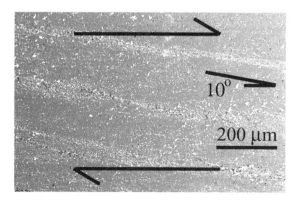

Figure 10.9 Reflected-light optical micrograph of sample of anorthite plus 5 volume percent MORB sheared to a shear strain of $\gamma \approx 2.5$. Deformation-driven melt segregation produced the three melt-rich bands at an angle of $\sim$10° to the shear direction. Anorthite is darker gray, and melt is lighter gray. A few relatively large clinopyroxene grains (brightest phase) that result from reaction between MORB and anorthite are scattered throughout the sample.

and segregated into melt-rich bands separated by melt-poor regions is well cor-related with the compaction length, δ_c, (Holtzman et al. 2002)

$$\delta_c = \sqrt{\kappa \frac{\eta}{\mu}}, \qquad (10.24)$$

where κ is permeability, η is sample viscosity, and μ is melt viscosity. In simple-shear experiments, if δ_c is larger than the thickness of the sample, $\mathscr{L}_{th}$ ($\lesssim 1$ mm), the melt pockets realign during deformation but melt-rich bands do not form (e.g., the olivine + basalt case). In contrast, in samples for which δ_c is approximately equal to or smaller than $\mathscr{L}_{th}$, not only does a strong melt preferred orientation develop but also melt-rich bands form (e.g., the plagioclase + basalt-derived melt, olivine + chromite + basalt and olivine + FeS melt + basalt cases) (Holtzman et al., in press). On the basis of measured values for κ, η, and μ for the olivine + basalt system $\delta_c \approx 10$ mm, that is, $\delta_c \gg \mathscr{L}_{th}$. However, for the plagioclase + basalt-derived melt, olivine + chromite + basalt and olivine + FeS + basalt samples, $\delta_c \lesssim 1$ mm, that is, $\delta_c \lesssim \mathscr{L}_{th}$. For the plagioclase + basalt-derived melt case, the compaction length is smaller than for olivine + basalt samples because plagioclase is less viscous than olivine while the melts have similar viscosities. For the olivine + chromite + basalt and olivine + FeS + basalt cases, the compaction length is smaller than for olivine + basalt samples because chromite and FeS melt reduce the permeability. Even though the wetting behavior is similar for basalt in olivine-rich samples with and without chromite or FeS, small grains of chromite and small spheres of FeS melt very frequently reside in the melt channels, thus significantly reducing permeability of the chromite- or FeS-bearing samples.

Although deformation-driven melt segregation produces melt-rich bands sepa-rated by approximately 100 µm in laboratory samples on fine-grained (~15 µm) synthetic rock samples, band spacing is predicted to be much larger under geologic conditions. On the basis of recent experiments, the spacing between melt-rich bands is $\sim \delta_c/10$. In equation (10.24) for compaction length, κ and η are both functions of ϕ [$\kappa \propto \phi^3$ and $\eta \propto \exp(\alpha\phi)$]; in addition, κ increases with increasing grain size ($\kappa \propto d^2$) and η increases with decreasing shear stress ($\eta \propto 1/\sigma^{2.5}$). Extrapolated to mantle stresses (~0.1 MPa) and grain sizes (~1 mm), $\delta_c \approx 0.1$– 1 km with a corresponding anticipated spacing of 10 to 100 m between melt-rich bands (Holtzman et al., in press).

The presence of melt-rich bands at depth in the mantle will have a pronounced effect on viscosity, permeability, and seismic waves properties. First, because vis-cosity is a sensitive function of melt fraction, deformation-driven segregation of melt into melt-rich bands will greatly decrease the viscosity in the bands while increasing the viscosity in regions between bands. For example, if the melt fraction in a region increases from 1 to 10 volume percent, the viscosity will decrease by a factor of ~20 in that region. The result will be the formation of shear zones in which deformation is spatially localized and directionally anisotropic. Second, with melt-rich bands present, permeability will be extremely anisotropic with en-hanced permeability in the melt-rich bands. The permeability in a band with 10 volume percent melt will be $\kappa \approx 10^{-10}$ m^2 compared to $\kappa < 10^{-12}$ m^2 normal to the bands. Third, seismic waves will be affected by the presence of melt-rich bands. Not only will the bands result in a decrease in wave speed, but they may also induce a pronounced polarization of compressional waves and splitting of shear waves.

Field evidence for deformation-driven melt segregation can be found in the form of plagioclase banding in the Oman ophiolite (Holtzman et al., in press) and the Lanzo peridotite massif (Boudier 1978). Tabular dunite reported in the Ingalls ophiolite (Kelemen et al. 2000), the Oman ophiolite (Kelemen et al. 2000) and the Bay of Islands ophiolite (Suhr, private communication, 2001) are almost certainly directly related to deformation-driven melt segregation, even though these features have generally been discussed in terms of a reactive infiltration instability (Ahronov et al. 1997). A recent detailed examination of an ancient shear zone in the Oman ophiolite demonstrated the presence of melt, providing another example of deformation-driven melt localization in the Earth (Dijkstra et al. 2002).

In summary, a small amount of melt affects the viscosity as well as seismic velocity and seismic attenuation in mantle rocks. At the same time, during deformation melt segregates into melt-rich bands separated by melt-depleted regions. Consequently, the permeability, viscosity, and seismic properties of the partially molten rock become highly anisotropic and spatially heterogeneous. In the melt-rich bands, the permeability increases and the viscosity decreases, while between the melt-rich bands the inverse holds. Experiments conducted with known boundary conditions provide a basis for understanding the driving force, kinetics, and mechanism by which deformation drives melt segregation. In addition, the experiments constrain the time and length scales over which deformation-driven melt segregation occurs in nature and its effect on physical properties of mantle and lower crustal rocks. A detailed understanding of the anisotropy in viscosity and permeability provides the foundation for understanding the dynamics of partially molten regions of the crust and mantle beneath ridges, in plumes, and at subduction zones.

Acknowledgments

This research was funded by the National Science Foundation through grants EAR-9815039, EAR-0126277, EAR-9906986, INT-0123224, and OCE-0002463 as well as NASA grant NAG5–10509. One of us (D.L.K.) was supported by the Alexander von Humboldt Foundation during the final stages of the writing of this chapter and wishes to thank Georg Dresen and Steve Mackwell for their intellectual support during his visit to their laboratories. Critical reviews by Reid Cooper, Georg Dresen, Greg Hirth, and Steve Mackwell led to significant improvements in the manuscript; we gratefully acknowledge their important insights and contributions.

Notes

1. Compaction theory describes the behavior of a melt phase migrating through a deformable partially molten rock (e.g., McKenzie 1984). A characteristic length scale, the compaction length (δ_c), is used to nondimensionalize the equations. Physically, δ_c is the distance over which melt flow and matrix deformation are coupled such that melt can migrate in response to deformation in the solid. Melt flow will eliminate pressure gradients in the solid over distances shorter than δ_c. In laboratory experiments, if δ_c is on the order of the sample thickness, the melt distribution will dynamically adjust to the pressure gradients produced during shear deformation.

2. As described by Zhang and Karato (1995) and Zimmerman et al. (1999), elliptically shaped samples with major and minor axes of ~8 and 6 mm, respectively, and a thickness of ≤ 1 mm were sheared between tungsten pistons cut at 45° to their cylindrical axes. These pistons plus the sample were enclosed in a nickel capsule. This capsule was part of a deformation column composed of a series of alumina and zirconia pistons inserted into an iron jacket, similar to the sample assemblies used in triaxial compression tests. The sample was then deformed in simple shear at a shear-strain rate of ~2×10^{-4}/s and a shear stress of ~100 MPa in a gas-medium high-pressure apparatus by compressing the deformation column at constant displacement rate at a temperature of 1,435 K and a confining pressure of 300 MPa.

References

Aharonov, E., M. Spiegelman, and P. B. Kelemen. 1997. Three-dimensional flow and reaction in porous media: Implications for Earth's mantle and sedimentary basins. *J. Geophys. Res.* 102:14821–14833.

Bagdassarov, N. 2000. Anelastic and viscoelastic behaviour of partially molten rocks and lavas. In N. Bagdassarov, D. LaPorte, and A. B. Thompson, eds., *Physics and Chemistry of Partially Molten Rocks*, pp. 29–65. Boston: Kluwer Academic Publishers.

Beeman, M. L. and D. L. Kohlstedt. 1993. Deformation of fine-grained aggregates of olivine plus melt at high temperatures and pressures. *J. Geophys. Res.* 98:6443–6452.

Boudier, F. 1978. Structure and petrology of the Lanzo peridotite massif (Piedmont Alps). *Geol. Soc. Am. Bull.* 89:1574–1591.

Bunton, J. H. 2001. *The Impact of Grain Size on the Shear Creep and Attenuation Behavior of Polycrystalline Olivine.* Ph.D. thesis, Department of Materials Science and Engineering, University of Wisconsin, Madison.

Bussod, G. Y. and J. M. Christie. 1991. Textural development and melt topology in spinel lherzolite experimentally deformed at hypersolidus conditions. *J. Petrol., Special Lherzolite Issue* 32:17–39.

Chan, W. W., I. S. Sacks, and R. Morrow. 1989. Anelasticity of the Iceland Plateau from surface wave analysis. *J. Geophys. Res.* 94:5675–5688.

Chen, I. W. and A. S. Argon. 1979. Steady state power-law creep in heterogeneous alloys with microstructures. *Acta Metall.* 27:785–791.

Cooper, R. F. 1990. Differential stress-induced melt migration: An experimental approach. *J. Geophys. Res.* 95:6979–6992.

Cooper, R. F. and D. L. Kohlstedt. 1984. Solution-precipitation enhanced diffusional creep of partially molten olivine basalt aggregates during hot-pressing. *Tectonophysics* 107:207–233.

Cooper, R. F. and D. L. Kohlstedt. 1986. Rheology and structure of olivine-basalt partial melts. *J. Geophys. Res.* 91:9315–9323.

Cooper, R. F., D. L. Kohlstedt, and K. Chyung. 1989. Solution-precipitation enhanced creep in solid-liquid aggregates which display a non-zero dihedral angle. *Acta Metall.* 37:1759–1771.

Daines, M. J. and D. L. Kohlstedt. 1994. The transition from porous to channelized flow due to melt/rock reaction during melt migration. *Geophys. Res. Lett.* 21:145–148.

Daines, M. J. and D. L. Kohlstead. 1997. Influence of deformation on melt topology in peridotites. *J. Geophys. Res.* 102:19257–10271.

Dijkstra, A. H., M. R. Drury, and R. M. Frijhogg. 2002. Microstructures and lattice fabrics in the Hilti mantle section (Oman Ophiolite): Evidence for shear localization and melt weakening in the crust-mantle transition zone? *J. Geophys. Res.* 107:2270, doi:10.1029/2001JB000458.

Forsyth, D. W. 1975. The early structural evolution and anisotropy of the oceanic upper mantle. *Geophys. J. R. Astron. Soc.* 43:103–162.

Ginsberg, S. B. 2000. *Deformation Experiments on Natural and Synthetic Diabasic Aggregates with Application to the Tectonics of Earth and Venus.* Ph.D. thesis. Department of Geology and Geophysics, University of Minnesota, Minneapolis.

Goetze, C. 1977. A brief summary of our present day understanding of the effect of volatiles and partial melt on the mechanical properties of the upper mantle. In M. H. Manghnani and S. Akimoto, eds., *High-Pressure Research: Applications in Geophysics*, pp. 3–23. New York: Academic Press.

Gribb, T. T. and R. F. Cooper. 1998. Low-frequency shear attenuation in polycrystalline olivine: Grain boundary diffusion and the physical significance of the Andrade model for viscoelastic rheology. *J. Geophys. Res.* 103:27267–27279.

Gribb, T. T. and R. F. Cooper. 2000. The effect of an equilibrated melt phase on the shear creep and attenuation behavior of polycrystalline olivine. *Geophys. Res. Lett.* 27:2341–2344.

Hall, C. E. and E. M. Parmentier. 2000. Spontaneous melt extraction from a deforming solid with viscosity variations due to water weakening. *Geophys. Res. Lett.* 27:9–12.

Hirth, G. and D. L. Kohlstedt. 1995a. Experimental constraints on the dynamics of the partially molten upper mantle: Deformation in the diffusion creep regime. *J. Geophys. Res.* 100:1981–2001.

Hirth, G. and D. L. Kohlstedt. 1995b. Experimental constraints on the dynamics of the partially molten upper mantle: 2. Deformation in the dislocation creep regime. *J. Geophys. Res.* 100: 15441–15449.

Holtzman, B. K., N. J. Groebner, M. E. Zimmerman, S. B. Ginsberg, and D. L. Kohlstedt. Deformation-driven melt segregation in partially molten rocks. *Geochem. Geophys. Geosys.*, in press.

Hwang, H. and B. J. Mitchell. 1987. Shear velocities, $Q\beta$, and the frequency dependence of $Q\beta$ in stable and tectonically active regions from surface wave observations. *Geophys. J. R. Astron. Soc.* 90:575–613.

Jackson, D. D. 1969. *Grain Boundary Relaxation and the Attenuation of Seismic Waves.* Ph.D. thesis. Massachusetts Institute of Technology, Cambridge.

Jackson, D. D. and D. L. Anderson. 1970. Physics mechanisms of seismic-wave attenuation. *Rev. Geophys. Space Phys.* 8:1–63.

Jackson, I. 1993. Progress in the experimental study of seismic wave attenuation. *Annu. Rev. Earth Planet. Sci.* 21:375–406.

Jackson, I. 2000. Laboratory measurement of seismic wave dispersion and attenuation: Recent progress. In S. Karato, A. M. Forte, R. C. Liebermann, G. Masters, and L. Stixrude, eds., *Earth's Deep Interior: Mineral Physics and Tomography, from the Atomic to the Global Scale*, pp. 269–285. Washington, DC: American Geophysical Union.

Jin, Z. M., H. W. Green II, and Y. Zhou. 1994. The rheology of pyrolite across its solidus. *EOS Trans. Am. Geophys. Union* 75:585.

Karato, S. and H. A. Spetzler. 1990. Defect microdynamics in minerals and solid state mechanisms of seismic wave attenuation and velocity dispersion in the mantle. *Rev. Geophys.* 28:399–421.

Kê, T.-S. 1947a. Experimental evidence of the viscous behavior of grain boundaries in metals. *Phys. Rev.* 71:533–546.

Kê, T.-S. 1947b. Stress relaxation across grain boundaries in metals. *Phys. Rev.* 72:41–46.

Kelemen, P. B., J. A. Whitehead, E. Aharonov, and K. A. Jordahl. 1995. Experiments on flow focusing in soluble porous media with applications to melt extraction from the mantle. *J. Geophys. Res.* 100:475–496.

Kelemen, P. B., G. Hirth, N. Shimizu, M. Spiegelman, and H.J.B. Dick. 1997. A review of melt migration processes in the adiabatically upwelling mantle beneath mid-oceanic spreading ridges. *Philos. Trans. R. Soc. Lond. A* 355:283–318.

Kelemen, P. B., M. Braun, and G. Hirth. 2000. Spatial distribution of melt conduits in the mantle beneath oceanic spreading ridges: Observations from the Ingalls and Oman ophiolites. *Geochem. Geophys. Geosys.* 1:1999GC000012.

Kohlstedt, D. L. and P. N. Chopra. 1994. Influence of basaltic melt on the creep of polycrystalline olivine under hydrous conditions. In M. P. Ryan, ed., *Magmatic Systems*, pp. 37–53. New York: Academic Press.

Kohlstedt, D. L. and M. E. Zimmerman. 1996. Rheology of partially molten mantle rocks. *Annu. Rev. Earth Planet. Sci.* 24:41–62.

Mavko, G. 1980. Velocity and attenuation in partially molten rocks. *J. Geophys. Res.* 85:5173–5189.

Mavko, G. and A. Nur. 1975. Melt squirt in the asthenosphere. *J. Geophys. Res.* 80:1444–1448.

McKenzie, D. P. 1984. The generation and compaction of partially molten rock. *J. Petrol.* 25:713–765.

McMillan, K. M., R. S. Lakes, R. F. Cooper, and T. Lee. Viscoelastic behavior of β-In$_3$Sn. Unpublished manuscript.

Mei, S. and D. L. Kohlstedt. 2000a. Influence of water on plastic deformation of olivine aggregates: 1. Diffusion creep regime. *J. Geophys. Res.* 105:21457–21469.

Mei, S. and D. L. Kohlstedt. 2000b. Influence of water on plastic deformation of olivine aggregates: 2. Dislocation creep regime. *J. Geophys. Res.* 105:21471–21481.

Mei, S., W. Bai, T. Hiruga, and D. L. Kohlstedt. 2002. Influence of melt on creep behavior of olivine-basalt aggregates under hydrous conditions. *Earth Planet. Sci. Lett.* 201:491–507.

Minster, J. B. and D. L. Anderson. 1980. Dislocation and nonelastic processes in the mantle. *J. Geophys. Res.* 85:6347–6352.

Minster, J. B. and D. L. Anderson. 1981. A model of dislocation-controlled rheology for the mantle. *Philos. Trans. R. Soc. London Ser. A* 299:319–356.

Mitchell, B. J. 1995. Anelastic structure and evolution of the continental crust and upper mantle from seismic surface wave attenuation. *Rev. Geophys.* 33:441–462.

Mosher, D. R. and R. Raj. 1974. Use of the internal friction technique to measure rates of grain boundary sliding. *Acta Metall.* 22:1469–1474.

Mosher, D. R., R. Raj, and R. Kossowsky. 1976. Measurement of viscosity of the grain-boundary phase in hot-pressed silicon nitride. *J. Mater. Sci.* 11:49–53.

Murase, T. and A. R. McBirney. 1973. Properties of some common igneous rocks and their melts at high temperatures. *GSA Bull.* 84:3565–3592.

Nowick, A. S. and B. S. Berry. 1972. *Anelastic Relaxation in Crystalline Solids*. New York: Academic Press.

Paterson, M. S. 2001. A granular flow theory for the deformation of partially molten rock. *Tectonophysics* 335:51–61.

Raj, R. 1975. Transient behavior of diffusion-induced creep and creep rupture. *Metall. Trans.* 6A:1499–1509.

Raj, R. and M. F. Ashby. 1971. On grain boundary sliding and diffusional creep. *Metall. Trans.* 2:1113–1127.

Romanowicz, B. and J. J. Durek. 2000. Seismological constraints on attenuation in the Earth: A review. In S. Karato, A. M. Forte, R. C. Liebermann, G. Masters, and L. Stixrude, eds., *Earth's Deep Interior. Mineral Physics and Tomography from the Atomic to the Global Scale*, pp. 161–180. Washington, DC: American Geophysical Union.

Ricoult, D. L. and D. L. Kohlstedt. 1983. Structural width of low-angle grain boundaries in olivine. *Phys. Chem. Miner.* 9:133–138.

Ryan, M. P. and J.Y.K. Blevins. 1987. *The Viscosity of Synthetic and Natural Silicate Melts and Glasses at High Temperatures and 1 bar (10^5 pascals) Pressure and at Higher Pressures*, USGS Bulletin 1764, pp. 455–457. Washington, DC: U.S. Government Printing Office.

Scott, T. S. and D. L. Kohlstedt. Lattice-Boltzmann determination of permeability in partially molten olivine-basalt aggregates. Unpublished manuscript.

Spiegelman, M. and P. Kenyon. 1992. The requirements for chemical disequilibrium during melt migration. *Earth Planet. Sci. Lett.* 109:611–620.

Stevenson, D. J. and D. R. Scott. 1987. Melt migration in deformable media. In D. E. Loper, ed., *Structure and Dynamics of Partially Solidified Systems,* vol. 125, pp. 401–415. Boston: Martinus Nijhoff Publishers.

Stevenson, D. J. 1989. Spontaneous small-scale melt segregation in partial melts undergoing deformation. *Geophys. Res. Lett.* 16:1067–1970.

Tan, B. H., I. Jackson, and J. D. FitzGerald. 1997. Shear wave dispersion and attenuation in fine-grained synthetic olivine aggregates: Preliminary results. *Geophys. Res. Lett.* 24:1055–1058.

Tharp, T. M. 1983. Analogies between the high-temperature deformation of polyphase rocks and the mechanical behavior of porous powder metal. *Tectonophysics* 96:T1–T11.

Versteeg, V. 1992. *Internal Friction in Lithium Aluminosilicate Glass-Ceramics*. M.S. thesis, Department of Materials Science and Engineering, Cornell University, Ithaca, NY.

Versteeg, V. A. and D. L. Kohlstedt. 1994. Internal friction in lithium aluminosilicate glass-ceramics. *J. Am. Ceram. Soc.* 77:1169–1177.

Walsh, J. B. 1969. New Analysis of attenuation in partially melted rock. *J. Geophys. Res..* 74:4333–4337.

Weiner, A. T., M. H. Manghnani, and R. Raj. 1987. Internal friction in tholeiitic basalts. *J. Geophys. Res.* 92:11635–11643.

William S. D., W. S. D. Wilcock, S. C. Solomon, G. M. Purdy, and D. R. Toomey. 1992. The seismic attenuation structure of a fast-spreading mid-ocean ridge. *Science* 258:1470–1474.

Xu, Y. 1997. *Anelasticity in Olivine-Basalt Aggregates*. Ph.D. thesis. Department of Geology and Geophysics, University of Minnesota, Minneapolis.

Xu, Y. and D. L. Kohlstedt. 1994. Attenuation at seismic frequencies in olivine-basalt aggregates. *EOS Trans. Am. Geophys. Union* 75:585.

Xu, Y. and D. L. Kohlstedt. 1997. Anelasticity in olivine-basalt aggregates: Application to seismic attenuation in the upper mantle. *EOS Trans. Am. Geophys. Union* 78:467.

Zhang, S. and S. Karato. 1995. Lattice preferred orientation of olivine aggregates deformed in simple shear. *Nature* 375:774–777.

Zimmerman, M. E. 1999. *The Structure and Rheology of Partially Molten Mantle Rocks*. Ph.D. Thesis. University of Minnesota, Minneapolis.

Zimmerman, M. E. and D. L. Kohlstedt. Rheology of lherzolite. *J. Petrol.*, in press.

Zimmerman, M. E., S. Zhang, D. L. Kohlstedt, and S. Karato. 1999. Melt distribution in mantle rocks deformed in shear. *Geophys. Res. Lett.* 26:1505–1508.

CHAPTER ELEVEN

Relations Among Porosity, Permeability, and Deformation in Rocks at High Temperatures

Brian Evans, Yves Bernabé, and Greg Hirth

Introduction

Fluid flow in rocks has far-reaching implications for seismogenic (Miller and Nur 2000; Rice 1992; Sibson 1981), sedimentary (Berner 1980; Etheridge et al. 1984), and metamorphic processes (Etheridge et al. 1984; Wood and Walther 1986). Because the physical properties of rocks strongly depend on porosity and pore geometry (e.g., Gangi 1979; Shankland et al. 1981; Simmons and Richter 1976; Walsh 1981) and because mechanical and thermal loads can substantially alter the pore geometry, mechanical forces directly affect transport properties. Conversely, rock strength is profoundly affected by pore fluids owing to diverse mechanical and chemical interactions (Bredehoeft and Hanshaw 1968; Carter et al. 1990; Fertl 1976; Fyfe et al. 1978; Ortoleva 1994; Ross and Lewis 1989; Sleep and Blanpied 1992; Walder and Nur 1984). Thus, one expects permeability and porosity to be dynamic quantities varying with time and space (Bredehoeft and Hanshaw 1968; Carter et al. 1990; Fertl 1976; Fyfe et al. 1978; Nur and Walder 1994; Ortoleva 1994; Ross and Lewis 1989; Sibson 1981; Sleep and Blanpied 1992; Walder and Nur 1984).

Similar effects are possible when the pore fluid is a silicate melt. For example, the formation of the oceanic crust, the migration of melt to hot spots and volcanic arcs, and the dynamics of flow beneath spreading centers and the mantle wedge of subduction zones all depend on the rheological and transport properties of partially molten ultramafic rocks. Each also requires an understanding of the relationships among deformation, melt topology, and melt migration (Daines and Kohlstedt 1997; Kelemen et al. 1997; Nicolas 1990; Phipps-Morgan 1987; Rubin 1998; Spiegelman 1996).

A brief discussion of some aspects of the mechanisms by which deformation and fluid flow interact follows. The basic ideas around which the discussion is organized are quite simple:

1. The mechanical and transport properties are linked by the mechanisms that change the geometry of the pore space (e.g., David et al. 1994; Evans et al. 1999; Ko et al. 1997; Wong et al. 1997; Zhu and Wong 1997).
2. The changing transport properties can be represented by an evolution curve in porosity-permeability space that is determined by the physical mechanisms that alter the pore structure, by the driving forces and thermodynamic conditions present, and by the initial porosity (Bernabé et al. 2003).
3. Providing that the initial geometry, driving forces, and mechanisms are known, qualitative descriptions of the changes in permeability can be given for rocks with aqueous fluids or melts as pore fluid (e.g., Bernabé et al. 2003; Renner et al., in press).

The discussion is based on experiments done on samples at the centimeter scale and thus is limited to grain-scale permeability processes. For more complete reviews on the issues of scaling to larger spatial dimensions, of incorporating dual-porosity structures, and of describing pressure-solution processes, see Gueguen et al. (1996), Holness (1997), and Jamtviet and Meakin (1999) and contributions within each.

The Evolution of the Relation between Pore Structure and Permeability (EPPR)

Permeability is a transport property relating the volume flux of a fluid, q, to the driving force for motion, namely, the gradient of fluid pressure, ∇P_f. In general, permeability is a tensor property, but when it may be regarded as isotropic, then Darcy's law gives:

$$q = \frac{\kappa}{\eta_f} \nabla P_f$$

where k is the permeability and η_f is the fluid viscosity (Gueguen and Palciauskas 1994). Permeability has units of length^2, emphasizing the fact that it is determined by aspects of the geometry of the pore space, including the total porosity, the sizes and size distributions of the pore space, their interconnections, surface roughness, and other geometric descriptors. (See table 11.1 for a list of symbols used.)

The relationship between porosity and permeability is quite complex and non-unique (Gueguen et al. 1996), as can be deduced from the fact that the permeability varies by at least five orders of magnitude from sandstone to shale, while porosity varies from 30–40 percent to 2–3 percent (Brace 1980, 1984). Despite that fact, classic models, such as the Kozeny-Carman relation, are often used to link permeability to some relevant measure of pore geometry. The simplest example is a

Table 11.1   List of Symbols Used

a	Exponent in power-law creep equation 11.3
A	Preexponential constant in equation 11.3
b	Grain size exponent in power-law creep equation 11.3
β_s	Storage capacity or storativity of a porous solid
β_f	Compressibility of fluid
d	Grain size
$\dot{\varepsilon}_v$	Volumetric strain rate
$\dot{\varepsilon}_{crit}$	Critical strain rate at which dilation hardening occurs
$f(\phi,\theta)$	Stress multiplier in compaction law 3
g	Hydraulic conductance of a varicose tube
g_o	Hydraulic conductance of a smooth tube of radius r_o
H_θ	Mass of mineral phase θ
η_f	Fluid viscosity
$\eta_{shear\,solid}$	Shear viscosity of solid
$\eta_{bulk\,solid}$	Bulk viscosity of porous solid
$k,\,k(t)$	Permeability
l	Length scale of consideration or sample length
L_c	Compaction length
m	Exponent relating $1/d$ to k
n	Exponent relating ϕ and k in equation 11.1
n'	Exponent relating ϕ_{eff} and k in equation 11.2
v_θ	Efficiency of change in effective porosity
P	Mean stress in solid (lithostatic pressure)
P_{eff}	Effective pressure (lithostatic pressure − pore fluid pressure)
∇P_f	Gradient of pore-fluid pressure
q	Volume flux of flowing fluid
Δr	Constriction height of a varicose tube
r_o	Reference radius of smooth tube
ϕ	Total porosity (or melt content)
ϕ_{eff}	Effective porosity
θ	Label for the identity of a mineralogical phase
ξ	Ratio of effective porosity to noneffective porosity

relation between k and the total porosity (ϕ), i.e., the fraction of the bulk volume that is pore space (Dullien 1979; Paterson 1983; Walsh and Brace 1984), such that:

$$k < \phi^n \tag{11.1}$$

where n is typically of the order 2 or 3. These models can be extended using spatial-correlation functions to characterize the pore structure (Berryman and Blair 1987; Blair et al. 1996), but other averaging schemes have also been proposed (Guéguen and Dienes 1989); some work uses effective medium theory (David et al. 1990; Doyen 1988) or renormalization group techniques (King 1989; Madden 1983). Recently, three techniques have emerged as ways to relate permeability to microstructure in materials with broad or skewed pore-size distributions: percolation theory (Katz and Thompson 1986); Monte Carlo simulations of fluid flow in realistic pore geometry (Auzerais et al. 1996) or in networks (Bernabé 1995;

Seeberger and Nur 1984); and lattice-gas cellular automata (Frisch et al. 1986; O'Connor and Fredrich 1999; van Genabeek and Rothman 1996).

When equation (11.1) is fit to data, the exponent, *n*, varies widely (see table 11.2 and David et al. 1994). Even within the same rock formation, the exponent may vary with porosity (Bourbie and Zinszner 1985; David et al. 1994). Part of the problem arises from the fact that total porosity is only one measure of pore geometry. Reflecting for a moment suggests that some parts of the pore space are more effective than others in allowing fluid transport (see figure 11.1 and Aharonov et al. 1997; Berryman and Blair 1987; Schwartz et al. 1989). For example, dead-end pores, pores that are completely disconnected, or those that lie between projecting rough points are not effective in aiding fluid transport.

To begin a more complex treatment of the pore geometry, we include a second geometric variable, the ratio of effective pore space to noneffective pore space (ξ) (Bernabé et al. 2003). In this description, permeability depends on these two independent variables: ξ and ϕ, respectively. The local slope of a curve in $\ln k - \ln \phi$ space, i.e., *n* in equation (11.1), is related to the way in which a process affects the effective pore space and, hence, to the mechanism of porosity change (figure 11.2). Thus, if the ratio of effective to noneffective porosity decreases, but the total porosity remains constant, then $\partial \ln k / \partial \ln \phi = -\infty$. Alternatively, if both effective and noneffective porosity change at the same rate, ξ remains constant, while total porosity decreases. For this process, the permeability could obey a Kozeny-Carman relation. Processes that increased or decreased the noneffective pore space but left the effective pore space unchanged would not change the permeability; and $\partial \ln k / \partial \ln \phi = 0$. Finally, a given process might change only the effective porosity, but not the noneffective porosity, in which case, $\partial \ln k / \partial \ln \phi < 0$.

Mechanisms of Permeability Evolution

The processes that change pore structure in rocks include elastic mechanisms (Brown 1987; Brown and Scholz 1986; Carlson and Gangi 1985; Gangi 1979; Tsang and Witherspoon 1981; Walsh 1981; Walsh and Grosenbaugh 1979), pore-producing fracturing caused by nonhydrostatic stresses (Atkinson 1984; Brown 1987; Brown and Scholz 1986; Carlson and Gangi 1985; Gangi 1979; Kranz et al. 1982; Martin 1972; Tsang and Witherspoon 1981; Walsh 1981; Walsh and Grosenbaugh 1979) or by thermal or pressure cycling (Carlson et al. 1990; Fredrich and Wong 1986; Heard and Page 1982; Wang et al. 1989). Both aqueous fluids and melts may increase porosity by infiltrating rocks to establish textural equilibrium (Riley and Kohlstedt 1991; Watson and Brenan 1987). Decreases in permeability might be caused by brittle pore collapse (Zhang et al. 1990, 1994a; Zhu and Wong 1997), pressure solution (Elliot 1973; Engelder 1979, 1982; Sprunt and Nur 1976, 1977a, 1977b), crack healing, crack sealing (Ramsay 1980; Roedder 1984), cementation of existing minerals, or precipitation of new minerals (Meyers 1974; Tada and Siever 1989). Positive and negative volume changes may occur during metamorphism or melting (Connolly et al. 1997; Walther and Orville 1982).

Table 11.2 Values of n for different processes (after Bernabé et al., in press)

Process	Material	Value of n (equation 11.1)
Plastic compaction	Synthetic aggregates	2.5–3 (n increasing with decreasing ϕ if disconnection occurs)
Sintering	Porous glass	4.5 for $\phi < 0.10$, disconnection at $\phi \approx 0.04$
Semibrittle compaction	Salt aggregates	5–7
Elastic compaction	Sandstones	1–25, depending on initial microstructure
Cataclastic compaction (hydrostatic)	Sandstones ($\phi > 0.30$)	≈ 20
	$0.30 > \phi > 0.15$	10–20
	$\phi < 0.15$	≈ 10
Cataclastic Compaction (triaxial)	Sandstones ($\phi > 0.30$)	5–10
	$0.0.30 > \phi > 0.15$	10–20
	$\phi < 0.15$	≈ 20
Dilatant microcracking	Dense rocks	7–8, n decreasing with increasing ϕ
Thermal microcracking	Dense rocks	5–7, $n \approx 1$ at very low ϕ
Dissolution	Sedimentary rocks	>20
Precipitation	Sedimentary rocks	≈ 8
Metamorphic reaction (roughening)	Porous glass Sedimentary rocks	>10, n decreasing with decreasing ϕ, $n \approx 2$ at $\phi < 0.10$

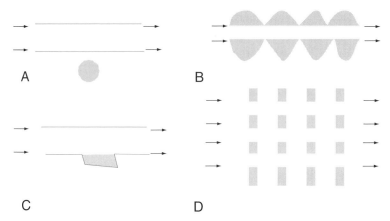

Figure 11.1 Schematic diagram to illustrate effective (lightly shaded) and noneffective (darkly shaded) pore space. Some portions of the pore space in complex geomaterials do not effectively contribute to fluid transport. Examples are (A) disconnected porosity, (B) roughness or undulations along pore walls, (C) dead-end porosity. (D) Some parts of the pore space may be effective for some pressure gradients, but not for others. Notice that rotating the direction of the pressure gradient will change the flow patterns in the pore space, allowing the darkly shaded pores to contribute to fluid flow. For complex networks of porosity with broad size distributions, separating the effective and noneffective pore space may be much more difficult.

In principle, the evolution of the relation between permeability and porosity (EPPR) can be quantitatively predicted, provided that the rate of change of porosity is uniquely determined by known variables that might include temperature, pressure, total porosity, local curvatures of the pore surfaces, and the values of the chemical and mechanical driving forces (for a brief discussion see Evans et al. 1999). In practice, exact forward quantifications are extremely difficult, especially given the complex nature of the pore spaces in rocks. Fortunately, some qualitative generalizations can be made on the basis of laboratory observations of several different mechanisms by which the permeability evolves, and we describe some of these in the following text.

Reactions and Permeability

Metamorphic reactions under hydrothermal conditions significantly affect permeability even on laboratory time-scales, and large permeability decreases may occur even when total porosity decreases are quite small. Granular mixtures of labradorite and quartz show two stages of densification when loaded non-hydrostatically and heated to 250–350°C in the presence of distilled water at 50 MPa (Aharonov et al. 1998; Blanpied et al. 1992; Karner et al. 1997; Karner and Schreiber 1993; Kranz and Blacic 1984; Moore et al. 1983; Morrow et al. 1981; Scholz et al. 1995; Tenthorey et al. 1998; Vaughan et al. 1986). Permeability can

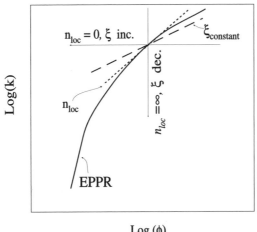

Log (φ)

Figure 11.2 Evolution of porosity-permeability relationship. In general, the permeability of a porous material will change if some physical process changes the geometry of the pore space; the evolution curve that tracks the change of permeability (k) with total porosity (ϕ) is labeled EPPR (heavy line). The EPPR could be concave, convex, or straight depending on the details of the process and the starting pore geometry. The local slope of the curve (dotted line labeled n_{loc}) is analogous to the exponent in equation 11.1. Both ϕ and the ratio of effective to noneffective porosity (ξ) influence k and n_{loc}. If effective pore space is exchanged for noneffective pore space, then the permeability will decrease even though the total porosity remains constant, and $n_{loc} = -\infty$. Similarly, if effective pore space remains constant while noneffective pore space decreases, ξ will increase, but k will remain unchanged, implying that $n_{loc} = 0$. Finally, if ξ is constant, but ϕ decreases, the EPPR will be a Kozeny-Carman relation (dashed line). (after Bernabé et al., in press).

continue to decrease even after compaction rates slow. Those studies showed that pore roughness and concomitant permeability decreases were directly related to the progress of the diagenetic reactions.

The roughness produced during diagenetic reactions is particularly important in sedimentary rocks (Aharonov et al. 1997; Berryman and Blair 1987; Schwartz et al. 1989), where the pore–solid interface may actually be self-affine, i.e., fractal (Thompson 1991). As the small crystals grow and roughen the pore space, non-effective porosity is created. For one recent example of the microstructural changes during diagenetic reactions during laboratory tests, see figure 11.3 and Mok et al. 2002. Although discussed most frequently for rocks of relatively low metamorphic grade, the roughening effect will, in fact, be evident whenever new, small crystals are formed within the pore space, even under high metamorphic grade or igneous conditions. For example, permeability decreases were inferred when chromite inclusions were formed in a partially molten peridotite system (cf. Holtzman et al., in press).

Using the effective porosity concept, Aharonov et al. (1998) analyzed permeability-reduction data in hydrothermally altered quartz and feldspar aggregates (Tenthorey et al. 1998) by assuming that the permeability, $k(t)$, is related to the effective porosity, ϕ_e, raised to a power, n'. Effective porosity is supposed to be related to the initial porosity by

$$k(t) < \phi_e^{n'} = \left(\phi + \sum v_\theta \int_{t=0}^{t} H_\theta \, dt \right)^{n'} \qquad (11.2)$$

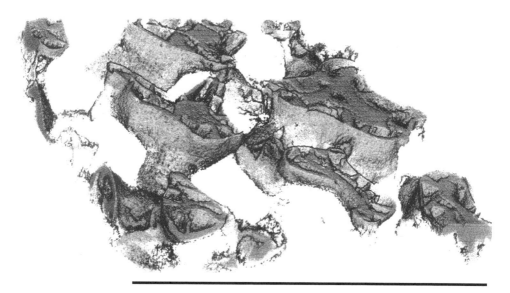

384 μm

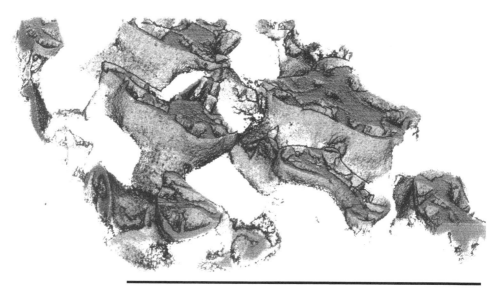

384 μm

where v_θ measures the efficiency in reducing the effective porosity for a given mass, H_θ of the precipitating mineral phase θ. The coefficient v_θ depends on the habit of the precipitating mineral as well as any preference in nucleation site. The exponent, n', is apparently 2 for a suite of sandstones with permeability ranging from 0.03×10^{-15} to $5,000 \times 10^{-15}$ m^2 (Aharonov et al. 1998). Much less is known about the relationship of permeability to effective porosity at low porosity where the pore space may change its connectivity.

The formulation used by Thompson, Aharonov, and coworkers (Aharonov et al. 1997, 1998; Thompson et al. 1987; Thompson 1991) is different from but complementary to the approach used here (Bernabé et al. 2003). To predict permeability using equation (11.2) requires independent determination of the effective porosity, which may then be substituted for the total porosity (Berryman and Blair 1987; Brown 1987; Sarkar and Prosperetti 1996). Actual determination of the effective porosity is not straightforward, and the effective dimensions might take different values when different transport properties are considered for the same pore space (Brown 1987). Blair et al. (1996) discussed the effect of roughness in sandstone, suggesting that permeability could be estimated to within a factor of 3 using image analysis, calculations of the two-point correlation function, and the Kozeny-Carman relation. As discussed previously, we take the total porosity as an independent variable, and use the local n to indicate changes in the partitioning of the effective pore space. Finally, note that we use the term "noneffective porosity" to include "rough" or "fractal" porosity, and the term "effective" pore space to include "Euclidean" pore space (Thompson 1991).

Some EPPR curves for diagenetic roughening in natural sandstones and shales (Pape et al. 1999, 2000) and in some synthetic sandstone experiments (Mok et al. 2002) are plotted in figure 11.4a; for more complete discussion see Bernabé et al. (2003). Characteristic features of the curves are an initial rapid drop in permeability of several orders of magnitude over a very small porosity range, followed by much more moderate decreases in permeability.

Figure 11.3 Laser scanning confocal microscope image of the pore structure of a synthetic sandstone sample (a) before and (b) after diagenetic reaction (after Mok et al. 2002). In these images, the pore structure is rendered as shaded. (a) The initial samples are sintered feldspar glass beads with grain size of about 90 μm. The total porosity is about 15 percent; all of it is connected into a framework around the edges of the glass spheres. The structure at the beginning of the experiment is smooth and clearly shows the indentation where the round glass beads reside. (b) Microstructure after alteration by aqueous pore fluid at 200°C, pore fluid pressure 20 MPa, confining pressure 50 MPa. Reaction of the glass with water produces mineralization in the pore structure that mimics diagenetic alteration. The effective porosity has been reduced by the production of dead-end porosity and roughened pore walls. Completely isolated porosity cannot be imaged by this technique, which depends on flooding the pore space with a dye-laden fluid, but such pores are probably also present. The bead pack suffers dramatic decreases in permeability as the pore space roughens (Mok et al. 2002).

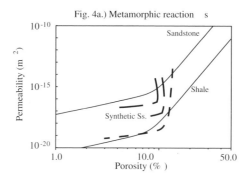

Fig. 4a.) Metamorphic reaction s

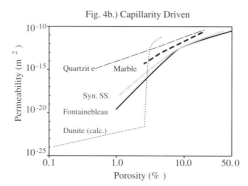

Fig. 4b.) Capillarity Driven

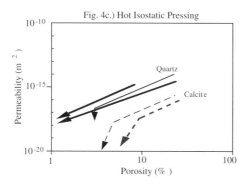

Fig. 4c.) Hot Isostatic Pressing

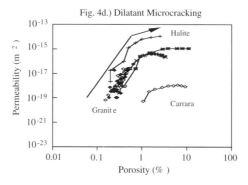

Fig. 4d.) Dilatant Microcracking

The effect of roughening on permeability was also demonstrated in simulations by Bernabé and Olson (2000), who used a lattice gas cellular automata to simulate fluid flow thorough a straight tube with a sinusoidally varying cross section, an object termed a varicose tube. The conductance of the varicose tube (g) normalized by the conductance (g_0) of a smooth tube of radius r_0, monotonically decreases from unity at $\Delta r/r_0 = 0$ to zero at $\Delta r/r_0 = 1$, where Δr is the height of the pore constriction. In the presence of roughness, a significant portion of the pore volume may be stagnant and hence does not effectively contribute to the flow. The simulations suggest that it is not necessary for the pore space to be fractal for a stagnant zone to develop.

Interface Tension or Capillarity Forces

Provided that the constituents of the solid phases are mobile, material redistribution via surface diffusion, lattice diffusion, or transport through fluid-saturated pores may be driven by surface or interphase tensions, often called capillarity forces (Blakely 1973). The pore morphology may range from tubules along grain edges to isolated fluid pockets, or sheet-like pores along grain faces, depending on fluid chemistry and interfacial energy (Beere 1975; Hickman and Evans 1987, 1992; Holness 1992, 1997; Laporte and Watson 1991; Lee et al. 1991; Olgaard and Fitzgerald 1993; Waff and Bulau 1979; Wark and Watson 1998; Watson and Brenan 1987). Intragranular porosity will evolve from cracks to inclusion planes or to a single pore depending on healing velocity (Roedder 1984), at a rate related to temperature (Bodnar and Sterner 1985, 1987; Brantley et al. 1990; Shelton and Orville 1980; Smith and Evans 1984; Wanamaker and Evans 1985), fluid chemistry (Brantley and Voight 1989), and crack size (Hickman and Evans 1987, 1995).

Great attention has been paid to determining the pore structure that results from capillarity forces during experiments on partially molten rocks (Cooper and Kohlstedt 1984a, 1984b; Hirth and Kohlstedt 1995b; Kohlstedt and Zimmerman 1996). At upper-mantle conditions in the absence of deformation, experimental studies on partially molten olivine aggregates suggest that basaltic melt is present at three and four grain junctions (Cooper and Kohlstedt 1984a; Toramaru and Fujii 1986; Vaughan et al. 1986; Waff and Faul 1992). Anisotropy of interfacial energy (Bussod and Chrisitie 1991; Cooper and Kohlstedt 1982; Cooper and Kohlstedt 1984a; Cooper and Kohlstedt 1984b; Hirth and Kohlstedt 1995a; Waff and Faul 1992) can result in melt topology that is more "sheet-" or "crack-" like than that

Figure 11.4 Evolution of porosity permeability relationships (EPPR) for (a) rocks and granular aggregates undergoing metamorphic reactions, (b) densification driven by capillarity forces (sintering or textural "equilibration"), (c) mechanical compaction under isostatic loading, (d) dilatant microcracking under triaxial loading. The sources of the data are given by Bernabé et al. (in press).

predicted assuming isotropic interfacial energies (Faul et al. 1994). Drury and Fitzgerald (1996) suggested that melt structures less than 10 nm thick might commonly be present along many grain boundaries. In anisotropic systems, relatively small melt fractions can dramatically affect elastic moduli and bulk viscosity (Faul et al. 1994; Schmeling 1985).

Numerous experimental, observational, and theoretical studies of melt transport in the mantle provide indirect constraints on the permeability of peridotite. Permeability has been estimated by monitoring the exchange of melt, driven by capillary force, between samples of different melt contents (Daines and Kohlstedt 1993; Riley and Kohlstedt 1991). Von Bargen and Waff (1986) predicted permeability on the basis of numerical calculations based on a pore structure in textural equilibrium, assuming isotropic interfacial energy and uniform grain sizes. It appears, however, that the interphase tension in the olivine/MORB system is anisotropic, and therefore melt may reside in disc-shaped inclusions as well as along grain-edge tubules. Depending on the exact shape of those discs, the permeability is expected to increase quite rapidly when the melt fraction increases to the point that the discs interconnect (see figure 11.4b and Faul 1997b, 2001). Other constraints on transport properties were obtained from analysis of melt topology in planar sections (Faul 1997a), from diffusion studies (Daines and Richter 1988; Watson and Wark 1997), and direct measurement of permeability in analog systems (Wark and Watson 1998).

The presence of a second phase might also change melt topology, even when the second-phase grain size is too large to be included in the pore space and cannot be considered to add roughness to the pore space. Because the energies of the interphase boundaries are different from the energies of the matrix-grain boundaries, the geometry of the pore space can be affected. In partially molten peridotite, observations of the microstructure in experimental charges suggests that melt connectivity is reduced in the presence of orthopyroxene (opx) or clinopyroxene (cpx) (Toramaru and Fujii 1986). Dihedral angles measured for opx-melt and cpx-melt were both greater than 60°, significantly greater than the critical value for connection assuming isotropic interfacial energy. This "pyroxene-effect" is apparently reduced in the presence of water (Fujii et al. 1986; von Bargen and Waff 1986). Toramaru and Fujii (1986) predicted that melt networks would become disconnected with opx contents between 25 and 60 percent, depending on the grain size of the orthopyroxene. On the basis of analyses of melt migration driven by capillary forces, Daines and Kohlstedt (1993) concluded that small amounts of orthopyroxene ($\leq$20 percent) did not significantly influence melt transport, although they did note numerous melt-free triple junctions in their samples.

Mechanical Compaction

When lithostatic loads are greater than the fluid pressure, a liquid-saturated rock can compact. Two processes must occur: the matrix must deform, and the fluid must be expelled. Thus, the rate of compaction will be influenced by both the viscosity of the pore fluid and the strength of the solid (Bercovici et al. 2001; McKenzie 1984; Scott and Stevenson 1989; Sleep 1974). When the pore fluid is relatively inviscid, then the matrix strength limits the rate of compaction, as might happen when wringing out a thick towel. When the pore fluid is quite viscous, and the solid phase is very weak, the egress of the fluid may limit compaction. Imagine throwing a sponge flat against a wall.

When the solid strength controls compaction, the kinetics will be determined by the type of deformation mechanism, e.g., brittle, plastic, creep, or solution-transfer (de Meer and Spiers 1999; Evans et al. 1999; Paterson 1995; Zhang et al. 1990, 1994; Zhu 1996; Zhu et al. 1999). The volumetric stain rate can be obtained using models for hot isostatic pressing (HIP) (Arzt 1982; Fischmeister and Arzt 1983; Helle et al. 1985). Thus, if the compaction rate is limited by power-law creep of the matrix, then the volumetric strain rate, $\dot{\varepsilon}_v$, is described by:

$$\dot{\varepsilon}_v = A \cdot f(\phi,\theta) \cdot d^{-a} \cdot P_{\text{eff}}^b \, \exp\left(\frac{Q}{RT}\right) \tag{11.3}$$

where A is a constant, $f(\phi,\theta)$ is a function of porosity/melt content (ϕ) and melt topology (e.g., dihedral angle θ), d is the grain size, P_{eff}, the effective pressure, is equal to the difference between the lithostatic pressure and the pore-fluid pressure, a and b are numerical exponents, Q is the activation energy, R is the gas constant, and T is the absolute temperature. The local stress enhancement, $f(\phi,\theta)$, depends on the coordination number of the solid grains as well as the total porosity. Since the coordination number increases with decreasing ϕ, the stress enhancement factor will also decrease (Helle et al. 1985).

Conversely, if the solid has no strength, then all the stresses are borne by the melt, and all of the effective pressure would be available to drive fluid flow. To first order, if the solid framework has permeability k, and the fluid flux is given by Darcy's law, then the volumetric strain rate would be

$$\dot{\varepsilon}_v = \frac{k}{\eta_f l} \cdot \frac{\partial P_f}{\partial x} = \frac{k}{\eta_f l^2} \cdot P_{\text{eff}}$$

where l is the length scale in question, and $\dfrac{\partial P_f}{\partial x}$ is approximately $\dfrac{P_{\text{eff}}}{l}$. A more detailed analysis would include the fact that k and the storage capacity, β_s, depend on porosity ϕ, which in turn varies strongly with P_{eff}. The transition between the solid-limited and the fluid-limited regimes would occur when:

$$k = \frac{\dot{\varepsilon}_v}{P_{\text{eff}}} \cdot \eta_f l^2 \tag{11.4}$$

Experimental examples of both solid-limited and fluid-limited cases exist. If the fluids are water or gas, i.e., relatively inviscid, then matrix strength limits compaction. In the case of calcite aggregates filled with gas, deformation at 400–650°C occurs by plastic flow, but the compaction is accompanied by a pore-disconnection process (Bernabé et al. 1982; Zhang et al. 1994). The evolution curve (figure 11.4c) first shows a decrease in permeability more or less in agreement with the Kozeny-Carman equation, followed by a more precipitous decrease of permeability at about 10 percent porosity. A critical porosity is reached at about 5 percent, at which time the permeability becomes too small to be measured.

The EPPR curves for creep compaction (figure 11.4c) may be partially rationalized by supposing that the ratio of ineffective porosity to total porosity increases as the total porosity decreases. Thus, n_{loc} becomes more negative as compaction proceeds, and permeability drops quite rapidly at low porosity. Notice the sharp contrast to permeability reductions in reacting systems (Johnson 1989), where the permeability decreases early in the compaction curve, even as total porosity remains relatively constant.

Indeed, microstructural observations in calcite compacting by creep show that, in addition to decreasing total porosity, the densification process produces disconnected pores that do not contribute to fluid transport (Bernabé et al. 1982; Zhang et al. 1994). Disconnected porosity is also predicted by wetting-angle measurements in dry carbonates (Bernabé et al. 1982; Holness and Graham 1995). Similar decreases of permeability have been observed in naturally cemented samples of Fontainebleau sandstone (Bourbie and Zinszner 1985) and in electrical conductivity of fluid-filled quartz powders, densified at high temperatures (Lockner and Evans 1995). In the latter case, it is not clear if continued densification will produce a percolation threshold, because very low porosities were not developed.

Using experimentally obtained pore statistics and a network model, Zhu (1996) showed that the actual decrease in permeability during densification by creep is greater than model predictions, unless the connectivity of the pore structure changes, as suggested by microstructure observations (Bernabé et al. 1982; Zhang et al. 1994b). Using a model of second-stage HIP, simplifying assumptions concerning grain-shape, packing, and dimensions, and experimentally obtained constants for the constitutive laws, Zhu et al. (1999) predicted the change in permeability with pore size as a function of time, P_{eff}, and T. To simulate changes in connectivity, they used rate equations taken from tube-ovulation experiments in calcite (Hickman and Evans 1987). The final model matched the permeability decreases well, but the absolute permeability was overestimated, sometimes by a large amount. One important, but unproven, assumption is that the HIP process and the tube-ovulation process are independent.

Melt Extraction in Partially Molten Olivine Rocks

The importance of the relationship between fluid transport and deformation is well-illustrated by considering the problem of melt migration. Analyses of geophysical data beneath oceanic spreading centers indicate that while melt is produced over a wide region in the upwelling mantle (Forsythe 1992; Toomey et al. 1998), the oceanic crust is largely produced in a very narrow region at the neo-volcanic zone (Macdonald 1982). Its thickness is largely independent of spreading rate. Several potential explanations for these observations have been explored theoretically, including: focusing of mantle upwelling owing to buoyancy-driven flow (Barnouin-Jha et al. 1997; Braun et al. 2000; Buck and Su 1989; Rabinowicz et al. 1987; Scott and Stevenson 1989), focusing of melt transport as a result of deformation-induced permeability structures; (Daines and Kohlstedt 1997; Phipps-Morgan 1987; Spiegelman 1996), dynamically induced pressure gradients (Phipps-Morgan 1987; Spiegelman 1996; Spiegelman and McKenzie 1987), or a reactive infiltration instability (Aharonov et al. 1995; Kelemen et al. 1997). Melt migration inherently involves coupling between melt transport and compaction (Ahern and Turcotte 1979; McKenzie 1984; Spiegelman and Elliot 1993). All of these models require an understanding of both the rheology and the permeability of partially molten rocks. It is important to understand the effect of melt on both shear and bulk viscosity.

Geochemical analyses of mid-ocean ridge basalt (MORB) and residual mantle peridotite also provide strong constraints on mechanisms of melt migration. Trace-element analyses of clinopyroxenes in abyssal peridotites indicate that melt is mobile at intergranular porosities as small as 0.5 percent during melting beneath oceanic-spreading centers (Johnson et al. 1990). Similarly, analyses of U-series disequilibria in MORB suggest that melt extraction occurs deep in the melting column by porous flow at intergranular porosities less than ~ 0.2–0.5 percent (Lundstrom et al. 1995; McKenzie 1985; Spiegelman and Elliot 1993). At the same time, MORB is not in chemical equilibrium with residual harzburgite or lherzolite (Kelemen et al. 1997) and the observation of U-series disequilibria suggests melt must migrate in the mantle at velocities of at least 1–10 m/yr (Kelemen et al. 1997; McKenzie 2000). These constraints, combined with geological observations from ophiolites, have motivated models in which melt migration from the mantle occurs in localized conduits that form by hydraulic fracture (Nicolas 1990; Rubin 1998), viscous instabilities (Hall and Parmentier 2000; Stevenson 1989), or the reactive infiltration instability (Kelemen et al. 1997). Again, the application of these models requires an understanding of both the permeability (including the influence of melt fraction and the presence of second solid phases) and rheology (including the effect of melt pressure and melt fraction on viscosity and possibly fracture toughness) of partially molten peridotite.

Many constraints on our understanding of the processes that occur in arc settings are also based on chemical analysis of magmatic rocks. Apparently, arc magmas are equilibrated with the mantle at moho conditions ($\sim 1000°C$ and ~ 1 GPa), rather than the source conditions in the mantle wedge (Debari et al. 1989; Tatsumi et al. 1983, 1986). Numerical models (Davies and Stevenson 1992; Kincaid and Sacks 1997; Spiegelman and McKenzie 1987) informing us of the spatial and temporal distribution of the temperature and pressure conditions in the subduction zone, combined with petrologic and petrophysical observations (Christensen and

Mooney 1995; Kelemen 1995; Rudnick 1995; Rudnick and Fountain 1995) constrain the composition of the solid and liquid phases. Equilibration of the melt at those conditions is taken as an indication that melt migrates through the wedge by reactive porous flow, i.e., migration that is slow enough to allow for continuous equilibration between melt and mantle host (Aharonov et al. 1995). Alternatively, equilibration might simply indicate significant ponding of the magma in the mantle wedge. To understand the implication of these geochemical and petrologic studies, it is important to further constrain the physics of magma migration.

Compaction of Partially Molten Peridotites Under Isotropic Lithostatic Loading

In some recent experiments under compressive lithostatic loading of drained, partially molten, synthetic peridotites (Renner et al., in press), the rate of melt extraction was measured as a function of melt content (see figure 11.5), melt viscosity, grain size, and effective pressure (P_{eff}). Samples containing 30–40 percent melt were subjected to a confining pressure of 300 MPa at 1,473 K. Three different melt compositions were used: melts fluxed with lithium silicate, formed from

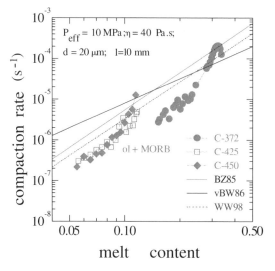

Figure 11.5 Normalized compaction rates for olivine/MORB aggregates compared with rates calculated assuming that compaction is limited by fluid transport. Compaction rates were normalized to an effective pressure of 10 MPa, a sample length of 10 mm, and a grain size of 20 μm. Two permeability models were used to model the olivine data: $k = (\phi^3 d^2)/C$ with $C = 100$ (Wark and Watson 1998); $k = (\phi^3 d^2)/C$ with $C = 1,250$ (von Bargen and Waff 1986); the C-value was determined from their figure 15 at a melt content of 5 percent. Additionally, we show the permeability-porosity relation for Fontainebleau sandstone determined by Bourbie and Zinszner (McKenzie 1984). (after Renner et al., in press).

MORB, or produced by melting albite glass. Melt pressure was controlled by connecting the sample to an argon reservoir; melt and confining pressures were maintained at $3 < P_{eff} < 50$ MPa. For comparison, densification and permeability measurements were also performed on porous olivine aggregates with argon as the pore fluid. Compaction rates always decreased with decreasing ϕ and increased with increasing P_{eff}, but they also showed an interesting relationship with melt viscosity. At the same P_{eff} and ϕ, olivine with MORB compacted fastest, aggregates made from olivine plus lithium silicate had intermediate rates, bracketed by the two olivine plus Ar experiments, and aggregates of olivine plus albite compacted the slowest.

Two regimes of melt segregation can be distinguished: one where solid compaction controls deformation, and a second where melt extraction is most important. In the olivine plus MORB systems, compaction was isotropic, i.e., the shape changes and melt content were homogeneous. When other variables are fixed, compaction rate is linearly proportional to P_{eff}. In the olivine/albite system compaction was not isotropic, and ϕ continuously increases from the drained interface to the far end of the sample. The exponent relating compaction rate to P_{eff} appears to be larger than 1, even though the olivine grain size is smaller than that in the olivine plus MORB system. The heterogeneity of melt in the compacting olivine/albite system indicated that compaction was controlled by melt extraction.

Using equation (11.4) and the volumetric strain rates determined in the deformation-limited regime, one may estimate lower bounds for the actual permeability, because fluid flow must be faster than solid compaction (figure 11.5). Such results agree with predictions based on pore structure at textural equilibrium (von Bargen and Waff 1986) for a grain size of 20 μm, and with measurement of permeability in fluid-rock systems (quartz + NaCl + H_2O) that exhibit similar dihedral angles (Wark and Watson 1998). Predictions of more rapid decreases in permeability at low porosity cannot be confirmed until the experiments are extended to lower melt contents.

Deformation of Partially Molten Rocks under Triaxial Stresses

In the diffusion creep regime, for melt fractions $\phi < \sim 0.05$, the influence of melt on strain rate is modest and consistent with models for diffusion creep enhanced by solution-precipitation (Cooper and Kohlstedt 1984a; Hirth and Kohlstedt 1995b). When $\phi > \sim 0.05$, the strain-rate enhancement is significantly greater than that predicted by theoretical models (Hirth and Kohlstedt 1995b). Microstructural observations of both statically annealed and deformed aggregates with $\phi > \sim 0.05$ demonstrate that a significant number of two-grain boundaries are "wetted," suggesting that the melt topology in the olivine-basalt system is affected by anisotropic interfacial energy (Hirth and Kohlstedt 1995b). Wetted boundaries may provide rapid-transport paths not accounted for in theories assuming isotropy.

The strength of partially molten olivine rock deforming by dislocation creep also depends on ϕ. In fine-grained rocks, melt has only a modest effect on strain rate for $\phi < 0.04$ (Hirth and Kohlstedt 1995b). However, when $\phi > 0.04$, creep rate is enhanced by more than an order of magnitude relative to melt-free aggregates, an effect not predicted by theories of the influence of weak secondary inclusions on dislocation creep (Chen and Argon 1979; Chopra and Paterson 1984; Tharp 1983; Tullis et al. 1991). The unexpectedly large effect of melt fraction on strain rate is apparently a result of enhanced grain-boundary sliding in the presence of melt (Hirth and Kohlstedt 1995b). Experiments on both melt-free and melt-added samples demonstrate that creep is limited by slip on the easiest system, [010](100). The effect of melt in coarser-grained aggregates (i.e., further from the transition between diffusion and dislocation creep) is not well constrained. However, limited data on coarse grained (>100 μm) dunites (Chopra and Paterson 1984) and lherzolites (Kohlstedt and Zimmerman 1996) suggest that the influence of melt is also modest when $\phi < 0.05$.

Dilation and Compaction during High-Temperature Deformation

Melt topology is also influenced by deformation. During triaxial deformation and undrained conditions, significant deviations from ideal melt topologies have been observed, even at low-melt fractions (Bussod and Chrisitie 1991; Daines and Kohlstedt 1997; Zimmerman et al. 1999). When partially molten olivine rocks are deformed in simple shear, melt-rich bands are formed and migrate through the sample (Holtzman et al., in press). The melt bands appear to result from localized melt transport as the aggregate is sheared. Thus, the segregation of melt that occurs during deformation results in the formation of high-permeability pathways and in the localization of deformation.

The characteristic length for coupling of melt flow and matrix deformation is given by the compaction length (L_c) (McKenzie 1984):

$$L_c = \sqrt{\frac{k\left(\eta_{\text{bulk solid}} + \frac{4}{3}\eta_{\text{shear solid}}\right)}{\eta_f}}$$

where $\eta_{\text{bulk solid}}$ is the bulk viscosity of a porous solid, $\eta_{\text{shear solid}}$ is the shear viscosity of the solid, and η_f and k are as before. This coupling is particularly clear in the simple-shear experiments (Holtzman et al., in press). By adding melts of differing viscosity, those authors demonstrated that the formation of the bands occurs when the compaction length L_c is nearly equivalent to the sample dimension, l.

A more subtle cause of topological changes might result from stress dependence of interfacial energy (Heidug and Leroy 1994; von Bargen and Waff 1986). Finally, deformation-induced "wetting" of two-grain boundaries has been suggested in both brine-saturated salt (Urai et al. 1986) and peridotite (Bai et al. 1997;

Jin et al. 1994). None of these stress effects has been thoroughly studied and much more work needs to be done to verify their importance. Deformation-induced changes in melt topology might also arise owing to the combined effects of anisotropic interfacial energy and the development of lattice-preferred orientations (LPO; Cooper and Kohlstedt 1982; Waff and Faul 1992). LPO can form by grain rotation during plastic flow, or by orientation of elongate crystals during flow of a crystal (Nicolas 1992).

Deformation of partially molten granite and aplites at laboratory strain rates and temperatures up to 1,100°C occurs with brittle fracture of the solid grains, but deformation and fluid flow are still intimately linked (Arzi 1978; Rutter and Neumann 1995; van der Molen and Paterson 1979). The initial failure of the partially molten granite (Rutter and Neumann 1995) obeyed a yield-cap model similar to that observed in water-saturated soils and porous rocks deformed at low temperatures (Curran and Carroll 1979; Wong 1990; Zhu and Wong 1997). Below a critical effective mean stress, the strength of partially molten granitic rock is positively correlated to effective mean stress. Above that critical value, strength decreases with increasing mean stress, a phenomenon known as shear-enhanced compaction. The maximum strength for initial yield occurs at the critical mean stress when the sample neither dilates nor compacts.

The effect of increasing melt content on strength of granite may also be linked to fluid flow. On the basis of experiments on partially molten granites, Arzi (1978) and van der Molen and Paterson (1979) proposed that, above a critical ϕ, ≈ 25 percent, the aggregate strength dropped rapidly from that characteristic of the solid phase to that approaching the viscosity of the melt. Rutter and Neumann (1995) questioned the existence of this rheologically critical melt fraction (RCMF). One major difference between the studies was that in the work of Arzi (1978) and van der Molen and Paterson (1979), water was added to produce melt, whereas Rutter and Neumann (1995) did not. Thus, it seems likely that the viscosity of the melts produced differ greatly.

The discrepancy in the studies may be resolved if one considers the influence of fluid transport during deformation. If the partially molten granites dilate during deformation via brittle fracture or by any other mechanism, the strength may increase, an effect known as dilation hardening (Brace and Martin 1968). Thus, the RCMF may not be a universal constant, but may be related to volumetric strain, drainage conditions, and melt viscosity (Renner et al. 2000).

The strain rate at which fluid pressure can be maintained constant within a rock volume is approximately (Fischer and Paterson 1989)

$$\dot{\varepsilon}_{\text{crit}} \propto k/(\eta_f \beta_s) \tag{11.5}$$

Although approximate, equation (11.5) predicts dilatancy hardening and compaction weakening in low-temperature rocks quite well. By assuming that k and β_s are related to ϕ, and that k is related to d, the RCMF is given as

$$\phi_{\text{crit}} \propto (\dot{\varepsilon} \eta_f \beta_f / d^m)^{1/(n-1)} \tag{11.6}$$

where β_f is the melt compressibility, d the grain size, and m and n relate permeability with grain size and ϕ, respectively. Applying this estimate to deformation of partially molten granites resolves the apparent conflicts between the three studies.

Finally, for a rock with an unreactive pore fluid failing by brittle processes, drained to an external reservoir at fixed pressure, peak fracture strength is affected by strain rate as well as by the effective pressure (Paterson 1978). Above a critical strain rate, the permeability is inadequate to maintain constant pore pressure even under drained conditions. Olgaard et al. (1995) termed this intermediate condition "transiently drained." Once again, it is clear that pore pressure, strength, dilation, compaction, and failure mode are strongly interrelated (Rice and Simons 1976; Rudnicki and Rice 1975). If dilation processes can occur at hypersolidus conditions, they might exhibit strong rate dependence, possibly extending the range of strain rates over which transiently drained conditions exist. Because the transport and mechanical properties of partially molten materials are closely linked, we believe that it will be extremely important to explore the influence of melt pressure on the physical properties of partially molten rocks.

Conclusion

The interaction between fluid transport and mechanical behavior is important for understanding a large number of processes at both passive and active margins, including sediment compaction and fluid transport at accretionary wedges, slip along mature faults, and melt migration in the mantle wedge along subduction zones. On the basis of experiments in centimeter-scale samples, we expect permeability changes to follow distinct evolution curves that depend on the mechanism altering the pore space, as well as the starting pore structure. The kinetics of the changes depend on the magnitude of the driving forces, temperature, lithostatic pressure, pore-fluid pressure, and chemical activities and fugacities of the phases present. Metamorphic and diagenetic reactions cause rapid decreases in permeability, followed by more gradual decreases at lower porosity. The rapid decreases result from the extreme roughening of the pore surface and disconnection of the pore space. In contrast to the reaction mechanism, compaction of the pore space by creep processes causes the permeability to decrease relatively slowly at first, followed by more rapid decreases at lower porosity. Brittle fracture of a low porosity rock causes permeability to follow an inverse pattern: initially, permeability increases quickly as a correlated pore structure is developed, followed later by a reduced rate of increase.

Compaction rates in experiments using partially molten peridotite can be limited either by fluid egress or by compaction of the solid matrix (Renner et al., in press) as predicted earlier by mechanical analyses (Bercovici et al. 2001; McKenzie 1984; Scott and Stevenson 1989). Because of the porosity range tested, the experiments do not confirm or deny the two-stage permeability behavior at low porosity predicted by Faul (1997b, 2001). When even a small amount of melt is present, the strength of peridotite rocks is decreased by at least a factor of 10, both in the dislocation-creep and diffusion-creep fields (Kohlstedt and Zimmerman 1996). Partially molten granites (Rutter and Neumann 1995) and peridotite (Holtzman et al., in press) show large decreases in strength when the melt content rises to 0.05 or greater. In both cases, deformation is coupled with fluid flow, even though the deformation mechanisms differ. The rheologically critical melt fraction may be a function of the permeability and melt viscosity, rather than a fixed numerical value (Renner et al. 2000). For partially molten granites, the relation between the stress to cause initial yield and the mean pressure seems to be described by a yield cap (Rutter and Neumann 1995) and is similar in many respects to flow in granular materials at lower temperatures.

Acknowledgments

We gratefully acknowledge helpful comments by David Kohlstedt, Ernie Rutter, Teng-fong Wong, and an anonymous reviewer; funding was provided by National Science Foundation grants OCE 0095936 (B.E.), NSF OCE 0099316 (G.H.), and DOE-FG0297ER14706 (Y.B.).

References

Aharonov, A., D. H. Rothman, and A. H. Thompson. 1997. Transport properties and diagenesis in sedimentary rocks: The role of microscale geometry. *Geology* 25:547.

Aharonov, E., J. A. Whitehead, P. B. Kelemen, and M. Spiegelman. 1995. Channeling instability of upwelling melt in the mantle. *J. Geophys. Res.* 100:20433–20450.

Aharonov, E., E. Tenthorey, and C. H. Scholz. 1998. Precipitation sealing and diagenesis: Theoretical analysis. *J. Geophys. Res.* 103:23969–23981.

Ahern, J. L. and D. L. Turcotte. 1979. Magma migration beneath an ocean ridge. *Earth Planet Sci. Lett.* 45:115–122.

Arzi, A. A. 1978. Critical phenomena in the rheology of partially melted rocks. *Tectonophysics* 44:1783–1184.

Arzt, E. 1982. The influence of an increasing particle coordination on the densification of spherical powders. *Acta Metall.* 30:1883–1890.

Atkinson, B. K. 1984. Sub-critical crack growth in geological materials. *J. Geophys. Res.* 89:4077–4113.

Auzerais, F. M., J. Dunsmuir, B. B. Ferréol, N. Martys, J. Olson, T. S. Ramakrishnan, D. H. Rothman, and L. M. Schwartz. 1996. Transport in sandstone: A study based on three-dimensional microtomography. *Geophys. Res. Lett.* 23:705–708.

Bai, Q., Z. M. Jin, and H. W. Green. 1997. Experimental investigation of partially molten peridote at upper mantle pressure and temperature. In M. Holness, ed., *Deformation Enhanced Fluid Transport in the Earth's Crust and Mantle*. London: Chapman and Hall.

Barnouin-Jha, K., E. M. Parmentier, and D. Sparks. 1997. Buoyant mantle upwelling and crustal production at oceanic spreading centers: On-axis segmentation and off-axis melting. *J. Geophys. Res.* 102:11979–911989.

Beere, W. 1975. A unifying theory of the stability of penetration liquid phases and sintering pores. *Acta Metall.* 23:131–138.

Bercovici, D. Y., Y. Ricard, and G. Schubert. 2001. A two-phase model for compaction and damage 1. General theory. *J. Geophys. Res.* 106:8887–8906.

Bernabé, Y. 1995. The transport properties of networks of cracks and pores. *J. Geophys. Res.* 100:4231–4241.

Bernabé, Y., W. F. Brace, and B. Evans. 1982. Permeability, porosity, and pore geometry of hot-pressed calcite. *Mech. Mater.* 1:173–183.

Bernabé, Y., U. Mok, and B. Evans. 2003. Permeability-porosity relationships in rocks subjected to various evolutionary processes. *Pure Appl. Geophys.* 160:937–960.

Bernabé, Y. and J. F. Olson. 2000. The hydraulic conductance of a capillary with a sinusoidally varying cross-section. *Geophys. Res. Lett.* 27:245–248.

Berner, R. A. 1980. *Early Diagenesis: A Theoretical Approach.* Princeton, NJ: Princeton University Press.

Berryman, J. G. and S. C. Blair. 1987. Kozeny-Carman relations and image processing methods for estimating Darcy's constant. *J. Appl. Phys.* 62:2221–2228.

Blair, S. C., P. A. Berge, and J. G. Berryman. 1996. Using two-point correlation functions to characterize microgeometry and estimate permeabilities of sandstones and porous glass. *J. Geophys. Res.* 101:20359–320375.

Blakely, J. M. 1973. *Introduction to the Properties of Crystal Surfaces.* Oxford, UK: Pergamon Press.

Blanpied, M. L., D. A. Lockner, and J. D. Byerlee. 1992. An earthquake mechanism based on rapid sealing of faults. *Nature* 358:574–576.

Bodnar, R. J., and S. M. Sterner. 1985. Synthetic fluid inclusions in natural quartz. II. Application to PVT studies. *Geochim. Cosmochim. Acta* 49:1855–1859.

Bodnar, R. J. and S. M. Sterner. 1987. Synthetic fluid inclusions. In G. C. Ulmer and H. L. Barnes, eds., *Hydrothermal Experimental Techniques*, pp. 423–457. New York: John Wiley.

Bourbie, T. and B. Zinszner. 1985. Hydraulic and acoustic properties as a function of porosity in Fontainebleau Sandstone. *J. Geophys. Res.* 90:11524–511532.

Brace, W. F. 1980. Permeability of crystalline and argillaceous rocks. *Int. J. Rock Mech. Min. Sci. Geomech. Abstr.* 17:241–251.

Brace, W. F. 1984. Permeability of crystalline rocks: New in situ measurements. *J. Geophys. Res.* 89:4327–4430.

Brace, W. F. and R. J. Martin. 1968. A test of the effective stress for crystalline rocks of low porosity. *Int. J. Rock Mech. Min. Sci. Geomech. Abstr.* 5:415–426.

Brantley, S. L., B. Evans, S. H. Hickman, and D. A. Crerar. 1990. Healing of microcracks in quartz: Implications for fluid-flow. *Geology* 18(2):136–139.

Brantley, S. L. and D. Voight. 1989. *Effects of Fluid Chemistry on Quartz Microcrack Healing.* Paper presented at the Water-Rock Interaction WRI-6: Proceedings of 6th International Symposium on Water Rock Interaction. Malvern, Rotterdam.

Braun, M. G., G. Hirth, and E. M. Parmentier. 2000. The effects of deep damp melting on mantle flow and melt generation beneath mid-ocean ridges. *Earth Planet. Sci. Lett.* 176:339–356.

Bredehoeft, J. D. and B. Hanshaw. 1968. On the maintenance of anomalous fluid pressures. I. Thick sedimentary sequences. *Geol. Soc. Am. Bull.* 79:1097–1106.

Brown, S. R. 1987. Fluid flow through rock joints: The effect of surface roughness. *J. Geophys. Res.* 92:1337–1347.

Brown, S. R. and C. H. Scholz. 1986. Closure of rock joints. *J. Geophys. Res.* 91:4939–4948.

Buck, W. R. and W. Su. 1989. Focussed mantle upwelling below mid-ocean ridges due to feedback between viscosity and melting. *Geophys. Res. Lett.* 16:641–644.

Bussod, G. Y. and J. M. Chrisitie. 1991. Textural development and melt topology in spinel lherzolite experimentally deformed at hypersolidus conditions. *J. Petrol.* (Special Lherzolite Issue):17–39.

Carlson, R. L. and A. F. Gangi. 1985. Effect of cracks on the pressure dependence of P wave velocities in crystalline rocks. *J. Geophys. Res.* 90:8675–8684.

Carlson, S. R., M. Wu, and H. F. Wang. 1990. Micromechanical modeling of thermal cracking in granite. In A. G. Duba, W. B. Durham, J. W. Handin, and H. F. Wang, eds., *The Brittle-Ductile Transition in Rocks: The Heard Volume.* Geophysics Monograph Ser. 56, pp. 37–48. Washington, DC: American Geophysical Union.

Carter, N. L., A. K. Kronenberg, J. V. Ross, and D. V. Wiltschko. (1990). Controls of Fluids on Deformation of Rocks. In R. J. Knipe and E. H. Rutter, eds., *Deformation Mechanisms, Rheology and Tectonics*. Geological Society Special Publication 54.

Chen, I. W. and A. S. Argon. 1979. Steady state power-law creep in heterogeneous alloys with coarse microstructures. *Acta Metall.* 27:785–791.

Chopra, P. and M. S. Paterson. 1984. The role of water in the deformation of dunite. *J. Geophys. Res.* 100:9761–9788.

Christensen, N. I. and W. D. Mooney. 1995. Seismic velocity structure and composition of the continental crust: A global view. *J. Geophys. Res.* 100:9761–9788.

Connolly, J.A.D., M. B. Holness, D. C. Rubie, and T. Rushmer. 1997. Reaction-induced microcracking: An experimental investigation of a mechanism for enhancing anatectic melt extraction. *Geology* 25:591–594.

Cooper, R. F. and D. L. Kohlstedt. 1982. Interfacial energies in the olivine-basalt system. In S. Akimota and M. H. Manghnani, eds., *High-Pressure Research in Geophysics*, vol. 12. Tokyo, Japan: Center for Academic Publications.

Cooper, R. F. and D. L. Kohlstedt. 1984a. Sintering of olivine and olivine-basalt aggregates. *Phys. Chem. Miner.* 11:5–16.

Cooper, R. F. and D. L. Kohlstedt. 1984b. Solution-precipitation enhanced diffusional creep of partially molten olivine basalt aggregates during hot-pressing. *Tectonophysics* 107:207–233.

Curran, J. H. and M. M. Carroll. 1979. Shear stress enhancement of void compaction. *J. Geophys. Res.* 84:1105–1112.

Daines, M. and F. M. Richter. 1988. An experimental method for directly determining the interconnectivity of melt in a partially molten system. *Geophys. Res. Lett.* 15:1459–1462.

Daines, M. J. and D. L. Kohlstedt. 1993. A laboratory study of melt migration. *Philos. Trans. R. Soc. Lond. A* 342:43–52.

Daines, M. J. and D. L. Kohlstedt. 1997. Influence of deformation on melt topology in peridotites. *J. Geophys. Res.* 102(5):10257–210271.

David, C., Y. Guéguen, and G. Pampoukis. 1990. Effective medium theory and network theory applied to the transport properties of rock. *J. Geophys. Res.* 95:6993–7006.

David, C., T.-f. Wong, W. Zhu, and J. Zhang. 1994. Laboratory measurement of compaction-induced permeability change in porous rocks: Implications for the generation and maintenance of pore pressure excess in the crust. *Pure Appl. Geophys.* 143:425–456.

Davies, J. H. and D. J. Stevenson. 1992. Physical model of source region of subduction zone volcanics. *J. Geophys. Res.* 97:2037–2070.

de Meer, S. and C. J. Spiers. (1999). On mechanisms and kinetics of creep by intergranular pressure solution. In B. Jamtviet and P. Meakin, eds., *Growth, Dissolution, and Pattern Formation in Geosystems*, pp. 345–366. Dordrecht, The Netherlands: Kluwer Academic Publishers.

Debari, S. M., R. G. Coleman, and R. A. Page. 1989. Examination of the deep levels of an island arc: Evidence from the Tonsina ultramafic-mafic assemblage, Tonsina, Alaska. *J. Geophys. Res.* 94:4373–4391.

Doyen, P. M. 1988. Permeability, conductivity, and pore geometry of sandstone. *J. Geophys. Res.* 93:7729–7740.

Drury, M. R. and J. D. FitzGerald. 1996. Grain boundary melt films in an experimentally deformed olivine-orthopyroxene rock: Implication for melt distribution in upper mantle rocks. *Geophys. Res. Lett.* 23:701–704.

Dullien, F.A.L. 1979. *Porous Media, Fluid Transport and Pore Structure*. New York: Academic Press.

Elliot, D. 1973. Diffusion flow law in metamorphic rocks. *Bull. Geol. Soc. Am.* 84:2645–2664.

Engelder, T. 1979. Mechanisms for strain within the Upper Devonian clastic sequence of the Appalachian plateau, Western New York. *Am J. Sci.* 279:527–542.

Engelder, T. 1982. A natural example of the simultaneous operation of free-face dissolution and pressure solution. *Geochim. Cosmochim. Acta* 46:69–74.

Etheridge, M. A., V. J. Wall, S. F. Cox, and R. H. Vernon. 1984. High fluid pressures during regional metamorphism and deformation: Implications for mass transport and deformation mechanisms. *J. Geophys. Res.* 89:4344–4358.

Evans, B., Y. Bernabé, and W. Zhu. 1999. Evolution of Pore Structure and Permeability of Rocks during Laboratory Experiments. In B. Jamtveit and P. Meakin, eds., *Growth, Dissolution, and Pattern Formation in Geosystems,* pp. 327–344. Dordrecht, The Netherlands: Kluwer Academic Publishers.

Faul, U. 1997a. Permeability of partially molten upper mantle rocks from experiments and percolation theory. *J. Geophys. Res.* 102:10299–210311.

Faul, U. H. 1997b. Permeability of partially molten upper mantle rocks from experiments and percolation theory. *J. Geophys. Res.* 102(B5):10299–10311.

Faul, U. H. 2001. Melt retention and segregation beneath mid-ocean ridges. *Nature* 410(6831):920–923.

Faul, U. H., D. R. Toomey, and H. S. Waff. 1994. Intergranular basaltic melt is distributed in thin, elongated inclusions. *Geophys. Res. Lett.* 21:29–32.

Fertl, W. H. 1976. *Abnormal Formation Pressures: Implications to Exploration, Drilling and Production of Oil and Gas Resources,* 2. Amsterdam: Elsevier.

Fischer, G. J. and M. S. Paterson. 1989. Dilatancy during rock deformation at high temperatures and pressures. *J. Geophys. Res.* 94:17607–617617.

Fischmeister, H. and E. Arzt. 1983. Densification of powders by particle deformation. *Powder Metallurgy* 26:82–88.

Forsythe, D. W. 1992. Geophysical constraints on mantle flow and melt generation beneath mid-ocean ridges. In J. Phipps-Morgan, D. K. Blackman, and J. M. Sinton, eds., *Mantle Flow and Melt Generation Beneath Mid-Ocean Ridges.* Geophysics Monograph 71, pp. 1–66. Washington, DC: American Geophysical Union.

Fredrich, J. and T.-F. Wong. 1986. Micromechanics of thermally induced cracking in three crustal rocks. *J. Geophys. Res.* 91:12743–12764.

Frisch, U., B. Hasslacher, and Y. Pomeau. 1986. Lattice-gas automata for the Navier-Stokes equations. *Phys. Rev. Lett.* 56:1505–1508.

Fujii, N., K. Osamura, and E. Takahashi. 1986. Effect of water saturation on the distribution of partial melt in the olivine-pyroxene-plagioclase system. *J. Geophys. Res.* 91:9253–9259.

Fyfe, W. S., N. J. Price, and A. B. Thompson. (1978). Fluids in the earth's crust. In *Developments in Geochemistry 1.* New York: Elsevier Scientific.

Gangi, A. F. 1979. Variation of whole and fractured porous rock permeability with confining pressure. *Int. J. Rock Mech. Min. Sci. Geomech. Abstr.* 15:249–257.

Guéguen, Y. and J. Dienes. 1989. Transport properties of rocks from statistics and percolation. *Math. Geol.* 21:1–13.

Guéguen, Y., P. Gavrilenko, and M. Le Ravelec. 1996. Scales of rock permeability. *Surv. Geophys.* 17:245–263.

Guéguen, Y. and V. Palciauskas. 1994. *Introduction to the Physics of Rocks.* Princeton, NJ: Princeton University Press.

Hall, C. and E. M. Parmentier. 2000. Spontaneous melt localization in a deforming solid with viscosity variations due to water weakening. *Geophys. Res. Lett.* 27:9–12.

Heard, H. C. and L. Page. 1982. Elastic moduli, thermal expansion and inferred permeability of two granites to 350°C and 55 MPa. *J. Geophys. Res.* 87:9340–9348.

Heidug, W. K. and Y. M. Leroy. 1994. Geometrical evolution of stressed and curved solid-fluid phase boundaries: 1. Transformation kinetics. *J. Geophys. Res.* 99:505–516.

Helle, A. S., K. E. Easterling, and M. F. Ashby. 1985. Hot-isostatic pressing diagrams: New developments. *Act Metall.* 33:2163–2174.

Hickman, S. H. and B. Evans. 1987. Influence of geometry upon crack healing rate in calcite. *Phys. Chem. Miner.* 15(1):91–102.

Hickman, S. H. and B. Evans. 1992. Growth of grain contacts in halite by solution-transfer; implications for diagenesis, lithification, and strength recovery. In B. Evans and T.-F. Wong, eds., *Fault Mechanics and Transport Properties of Rocks: A Festschrift in Honor of W. F. Brace*, pp. 253–280. London: Academic Press.

Hickman, S. H. and B. Evans. 1995. Kinetics of pressure solution at halite-silica interfaces and intergranular clay films. *J. Geophys. Res.* 100(B7):13113–113132.

Hirth, G. and D. L. Kohlstedt. 1995a. Experimental constraints on the dynamics of the partially molten upper mantle: 2. Deformation in the dislocation creep regime. *J. Geophys. Res.* 100: 15441–15449.

Hirth, G. and D. L. Kohlstedt. 1995b. Experimental constraints on the dynamics of the partially molten upper mantle: Deformation in the diffusion creep regime. *J. Geophys. Res.* 100:1981–2001.

Holness, M. B. 1992. Equilibrium dihedral angles in the system quartz-CO_2-H_2O-NaCl at 800°C and 1–15 kbar: The effects of pressure and fluid composition on the permeability of quartzites. *Earth Planet. Sci. Lett.* 114:171–184.

Holness, M. B., ed. 1997. *Deformation-Enhanced Fluid Transport in the Earth's Crust and Mantle*, vol. 8. London: Chapman & Hall.

Holness, M. B. and C. M. Graham. 1995. P-T-X effects on equilibrium carbonate-H_2O-CO_2-NaCl dihedral angles: Constraints on carbonate permeability and the role of deformation during fluid infiltration. *Contrib. Mineral. Petrol.* 119(2–3):301–313.

Holtzman, B. K., N. J. Groebner, M. E. Zimmerman, S. B. Ginsberg, and D. L. Kohlstedt. Deformation-driven melt segregation in partially molten rocks. *Geochem. Geophys. Geosyst.*, in press.

Jamtveit, B. and P. Meakin, eds. 1999. *Growth, Dissolution and Pattern Formation in Geosystems*. Dordrecht, The Netherlands: Kluwer Academic Publishers.

Jin, Z. M., H. W. Green, and Y. Zhou. 1994. Melt topology during dynamic partial melting of mantle peridotite. *Nature* 372:164–167.

Johnson, D. L. and L. M. Schwartz. 1989. Unified theory of geometrical effects in transport properties of porous media. *Log Analyst* 30:89.

Johnson, K.T.M., H.J.B. Dick, and N. Shimizu. 1990. Melting in the oceanic upper mantle: An ion microprobe study of diopsides in abyssal peridotites. *J. Geophys. Res.* 95:2661–2628.

Karner, S. L., C. Marone, and B. Evans. 1997. Laboratory study of fault healing and lithification in simulated fault gouge under hydrothermal conditions. *Tectonophysics* 277(1–3):41–55.

Karner, S. L. and B. C. Schreiber. 1993. Experimental simulation of plagioclase diagenesis at P-T conditions of 3.5 km burial depth. *Pure Appl. Geophys.* 141:221–247.

Katz, A. J. and A. H. Thompson. 1986. Quantitative prediction of permeability in porous rock. *Phys. Rev. B* 34:8179–8181.

Kelemen, P. B. 1995. Genesis of high Mg# andesites and the continental crust. *Contrib. Mineral Petrol.* 120:1–19.

Kelemen, P. B., G. Hirth, N. Shimizu, M. Spiegelman, and H.J.B. Dick. 1997. A review of melt migration processes in the adiabatically upwelling mantle beneath oceanic spreading ridges. *Philos. Trans. R. Soc. Lond. A* 355:283–318.

Kincaid, C. and I. S. Sacks. 1997. Thermal and dynamical evolution of the upper mantle in subduction zones. *J. Geophys. Res.* 102:12295–212315.

King, P. R. 1989. The use of renormalization for calculating effective permeability. *Transport Porous Media* 4:37–58.

Ko, S. C., D. L. Olgaard, and T. F. Wong. 1997. Generation and maintenance of pore pressure excess in a dehydrating system: 1. Experimental and microstructural observations. *J. Geophys. Res.* 102(B1):825–839.

Kohlstedt, D. L. and M. E. Zimmerman. 1996. Rheology of partially molten mantle rocks. *Ann. Rev. Earth Planet. Sci.* 24:41–62.

Kranz, R. L. and J. D. Blacic. 1984. Permeability changes during time-dependent deformation of silicate rock. *Geophys. Res. Lett.* 9:1–4.

Kranz, R. L., W. J. Harris, and N. L. Carter. 1982. Static fatigue of granite at 200°C. *Geophys. Res. Lett.* 9:1–4.

Laporte, D. and E. B. Watson. 1991. Direct observation of near-equilibrium pore geometry in synthetic quartzites at 600–800°C and 2–10.5 kbar. *J. Geophys. Res.* 99:873–878.

Lee, V. W., S. J. Mackwell, and S. L. Brantley. 1991. The effect of fluid chemistry on wetting textures in novaculite. *J. Geophys. Res.* 96(B6):10023–10037.

Lockner, D. and B. Evans. 1995. Densification of quartz powder and reduction of conductivity at 700-degrees-C. *J. Geophys. Res.* 100(B7):13081–13092.

Lundstrom, C. C., J. Gill, and Q. Williams. 1995. Mantle melting and basalt extraction by equilibrium porous flow. *Science* 270:1958–1961.

Macdonald, K. C. 1982. Mid-ocean ridges: Fine scale tectonic, volcanic, and hydrothermal processes within the plate boundary zone. *Ann. Rev. Earth Planet. Sci.* 10:155–190.

Madden, T. R. 1983. Microcrack connectivity in rocks: A renormalization group approach to the critical phenomena of conduction and failure in crystalline rocks. *J. Geophys. Res.* 88:585–592.

Martin, R. J. 1972. Time-dependent crack growth in quartz and its application to the creep of rocks. *J. Geophys. Res.* 77:1406–1419.

McKenzie, D. 1984. The generation and compaction of partially molten rock. *J. Petrol.* 25:713–765.

McKenzie, D. 1985. ^{230}Th-^{238}U disequilibrium and the melting processes beneath ridge axes. *Earth Plan. Sci. Lett.* 72:149–157.

McKenzie, D. 2000. Constraints on melt generation and transport from U-series activity ratios. *Chem. Geol.* 162:81–94.

Meyers, W. J. 1974. Carbonate cement stratigraphy of the Lake Valley Formation (Mississippian) Sacramento Mountains, New Mexico. *Sediment. Pet.* 44:837–861.

Miller, S. A. and A. Nur. 2000. Permeability as a toggle switch in fluid-controlled crustal processes. *Earth Planet. Sci. Lett.* 183:133–146.

Mok, U., Y. Bernabé, and B. Evans. 2002. Permeability, porosity, and pore geometry of chemically altered porous silica glass. *J. Geophys. Res.* 107:10.1029/2001JB000247.

Moore, D. E., C. A. Morrow, and J. D. Byerlee. 1983. Chemical reactions accompanying fluid flow through granite held in a temperature gradient. *Geochim. Cosmochim. Acta* 47:445–453.

Morrow, C. A., D. A. Lockner, D. E. Moore, and J. D. Byerlee. 1981. Permeability of granite in a temperature gradient. *J. Geophys. Res.* 86:3002–3008.

Nicolas, A. 1990. Melt extraction from mantle peridotites: Hydrofracturing and porous flow, with consequences for oceanic ridge activity. In M. P. Ryan, ed., *Magma Transport and Storage*, pp. 159–173. Chichester, UK: John Wiley & Sons.

Nicolas, A. 1992. Kinematics in magmatic rocks. *J. Petrol.* 33:891–915.

Nur, A., and J. Walder. (1994). Hydraulic pulses in the Earth's crust. In B. Evans and T.-F. Wong, eds., *Fault Mechanics and Transport Properties of Rocks*, pp. 461–473. San Diego, CA: Academic Press.

O'Connor, R. M. and J. T. Fredrich. 1999. Microscale flow modelling in geologic materials. *Phys. Chem. Earth (A)* 24(7):611–616.

Olgaard, D. L. and J. D. Fitzgerald. 1993. Evolution of pore microstructures during healing of grain-boundaries in synthetic calcite rocks. *Contrib. Mineral. Petrol.* 115(2):138–154.

Olgaard, D. L., S.-C. Ko, and T.-F. Wong. 1995. Deformation and pore pressure in dehydrating gypsum under transiently drained conditions. *Tectonophysics* 68:131–146.

Ortoleva, P. J. 1994. *Geochemical Self-Organization*, vol. 23. New York: Oxford University Press.

Pape, H., C. Clauser, and J. Iffland. 1999. Permeability prediction based on fractal pore-space geometry. *Geophysics* 64:1447–1460.

Pape, H., C. Clauser, and J. Iffland. 2000. Variation of permeability with porosity in sandstone diagenesis interpreted with a fractal pore space model. *Pure Appl. Geophys.* 157:603–619.

Paterson, M. S. 1978. *Experimental Rock Deformation: The Brittle Field*. New York: Springer.

Paterson, M. S. 1983. The equivalent channel model for permeability and resistivity in fluid saturated rock: A reappraisal. *Mech. Mater.* 2:345–352.

Paterson, M. S. 1995. A theory for granular flow accommodated by material transfer via an intergranular fluid. *Tectonophysics* 245:133–151.

Phipps-Morgan, J. 1987. Melt migration beneath mid-ocean spreading centers. *Geophys. Res. Lett.* 144:1238–1241.

Rabinowicz, M., G. Ceuleneer, and A. Nicolas. 1987. Melt segregation and flow in mantle diapirs beneath spreading centers: Evidence from the Oman Ophiolite. *J. Geophys. Res.* 92: 3475–3486.

Ramsay, J. G. 1980. The crack-seal mechanism of rock deformation. *Nature* 284:135–139.

Renner, J., B. Evans, and G. Hirth. 2000. On the rheologically critical melt percentage. *Earth Planet. Sci. Lett.* 181:585–594.

Renner, J., K. Viskupic, G. Hirth, and B. Evans. Melt segregation from partially molten peridotites. *Geochem. Geophys. Geosystems*, in press.

Rice, J. R. (1992). Fault stress states, pore pressure distributions, and the weakness of the San Andreas Fault. In B. Evans and T.-F. Wong, eds., *Fault Mechanics and Transport Properties in Rocks: A Festschrift in Honor of W. F. Brace*, pp. 457–503. London: Academic Press.

Rice, J. R. and D. Simons. 1976. The stabilization of spreading shear faults by coupled deformation-diffusion effects in fluid-infiltrated porous materials. *J. Geophys. Res.* 81:5322–5334.

Riley, G. N., Jr., and D. L. Kohlstedt. 1991. Kinetics of melt migration in upper mantle-type rocks. *Earth Planet. Sci. Lett.* 105(4):500–521.

Roedder, E. 1984. *Fluid Inclusions*, vol. 12. Washington, DC: Mineralogical Society of America.

Ross, J. and P. D. Lewis. 1989. Brittle-ductile transition: Semi-brittle behavior. *Tectonophysics* 167:75–79.

Rubin, A. M. 1998. Dike ascent in partially molten rock. *J. Geophys. Res.* 103:20901–920919.

Rudnick, R. L. 1995. Making continental crust. *Nature* 378:571–578.

Rudnick, R. L. and D. M. Fountain. 1995. Nature and composition of the continental crust: A lower crustal perspective. *Rev. Geophys.* 33:267–309.

Rudnicki, J. W. and J. R. Rice. 1975. Conditions for the localization of deformation in pressure sensitive dilatant materials. *J. Mech. Phys. Solids* 23:271–394.

Rutter, E. H. and D.H.K. Neumann. 1995. Experimental deformation of partially molten Westerly granite under fluid-absent conditions, with implications for the extraction of granitic magmas. *J. Geophys. Res.* 100:15697–615715.

Sarkar, K. and A. Prosperetti. 1996. Effective boundary conditions for Stokes flow over a rough surface. *J. Fluid Mech.* 316:223–240.

Schmeling, H. 1985. Numerical models on the influence of partial melt on elastic, anelastic and electric properties of rocks. Part I: Elasticity and anelasticity. *Phys. Earth Planet. Sci.* 41:34–57.

Scholz, C. H., A. Leger, and S. L. Karner. 1995. Experimental diagenesis: Exploratory results. *Geophys. Res. Lett.* 22:719–722.

Schwartz, L. M., P. N. Sen, and D. L. Johnson. 1989. Influence of rough surfaces on electrolytic conduction in porous media. *Phys. Rev. B* 40: 450–2458.

Scott, D. R. and D. J. Stevenson. 1989. A self-consistent model of melting, magma migration and buoyancy-driven circulation beneath mid-ocean ridges. *J. Geophys. Res.* 94:2973–2988.

Seeberger, D. A. and A. Nur. 1984. A pore space model for rock permeability and bulk modulus. *J. Geophys. Res.* 89:527–536.

Shankland, T. J., R. J. O'Connell, and H. S. Waff. 1981. Geophysical constraints on partial melt in the upper mantle. *Rev. Geophys. Space Phys.* 19:394–406.

Shelton, K. L. and P. Orville. 1980. Formation of synthetic fluid inclusions in natural quartz. *Am Mineral.* 65:1233–1236.

Sibson, R. 1981. Fluid flow accompanying faulting: Field evidence and models. In D. W. Simpson and P. G. Richards, eds., *Earthquake Prediction: An International Review, Maurice Ewing Ser.*, vol. 4, pp. 593–604. Washington, DC: American Geophysical Union.

Simmons, G. and D. Richter. 1976. Microcracks in rocks. In R. G. J. Strens, ed., *The Physics and Chemistry of Minerals and Rocks*, pp. 105–137. New York: Wiley-Interscience.

Sleep, N. H. 1974. Segregation of a magma from a mostly crystalline mush. *Bull. Geol. Soc. Am.* 85:1225–1232.

Sleep, N. H. and M. L. Blanpied. 1992. Creep, compaction and the weak rheology of major faults. *Nature* 359:687–692.

Smith, D. L. and B. Evans. 1984. Diffusional Crack Healing in Quartz. *J. Geophys. Res.* 89(B6):4125–4135.

Spiegelman, M. 1996. Geochemical consequences of melt transport in 2-D models for melt extraction at mid-ocean ridges and island arcs. *Earth Plan. Sci. Letters* 139:115–132.

Spiegelman, M. and T. Elliot. 1993. Consequences of melt transport for uranium series disequilibrium. *Earth Plan. Sci. Lett.* 118:1–20.

Spiegelman, M. and D. McKenzie. 1987. Simple 2-D models for melt extraction at mid-ocean ridges and island arcs. *Earth Planet. Sci. Lett.* 83:137–152.

Sprunt, E. S. and A. Nur. 1976. Reduction of porosity by pressure solution: Experimental verification. *Geology* 4:463–466.

Sprunt, E. S. and A. Nur. 1977a. Destruction of porosity through pressure solution: Experimental verification. *Geophysics* 42:726–741.

Sprunt, E. S. and A. Nur. 1977b. Experimental study of the effects of stress on solution rate. *J. Geophys. Res.* 82:3013–3022.

Stevenson, D. J. 1989. Spontaneous small-scale melt segregation in partial melts undergoing deformation. *Geophys. Res. Lett.* 16:1067–1070.

Tada, R. and R. Siever. 1989. Pressure solution during diagenesis. *Annu. Rev. Earth Planet. Sci.* 17:89–118.

Tatsumi, Y., D. L. Hamilton, and R. W. Nesbitt. 1986. Chemical characteristics of fluid phase released from a subducted lithosphere and origin of arc magmas: Evidence from high-pressure experiments and natural rocks. *J. Vol. Geotherm. Res.* 29:293–309.

Tatsumi, Y., M. Sakuyama, H. Fukuyama, and I. Kushiro. 1983. Generation of arc basalt magmas and thermal structure of the mantle wedge in subduction zones. *J. Geophys. Res.* 88:5815–5825.

Tenthorey, E., E. Aharonov, and C. H. Scholz. 1998. Precipitation sealing and diagenesis: Experiments. *J. Geophys. Res.* 103:23951–23968.

Tharp, T. M. 1983. Analogies between the high-temperature deformation of polyphase rocks and the mechanical behavior of porous powder metal. *Tectonophysics* 96:T1–T11.

Thompson, A. B., A. J. Katz, and C. Krohn. 1987. The microgeometry and transport of sedimentary rock. *Adv. Phys.* 36:625–694.

Thompson, A. H. 1991. Fractals in rock physics. *Annu. Rev. Earth Planet. Sci.* 19:237–262.

Toomey, D. R., S. C. Wilcock, S. C. Solomon, W. C. Hammond, and J. A. Orcutt. 1998. Mantle seismic structure beneath the MELT region of the East Pacific Rise from P and S wave tomography. *Science* 280:1224–1227.

Toramaru, A. and N. Fujii. 1986. Connectivity of melt phase in a partially molten peridotite. *J. Geophys. Res.* 91:9239–9252.

Tsang, Y. W. and P. A. Witherspoon. 1981. Hydromechanical behavior of a deformable rock fracture subject to normal stress. *J. Geophys. Res.* 86:9287–9298.

Tullis, T. E., F. Horowitz, and J. Tullis. 1991. Flow laws of polyphase aggregates from end member flow laws. *J. Geophys. Res.* 96:8081–8096.

Urai, J. L., C. J. Spiers, H. J. Zwart, and G. S. Lister. 1986. Weakening of rock salt by water during long-term creep. *Nature* 324(6097):554–557.

van der Molen, I. and M. S. Paterson. 1979. Experimental deformation of partially-melted granite. *Contrib. Mineral. Petrol.* 70:299–318.

van Genabeek, O. and D. H. Rothman. 1996. Macroscopic manifestations of microscopic flows through porous media: Phenomenology from simulation. *Annu. Rev. Earth Planet. Sci.* 24:63–87.

Vaughan, P. J., D. E. Moore, C. A. Morrow, and J. D. Byerlee. 1986. Role of cracks in progressive permeability reduction during flow of heated aqueous fluids through granite. *J. Geophys. Res.* 91:7517–7530.

von Bargen, N. and H. S. Waff. 1986. Permeabilities, interfacial areas and curvatures of partially molten systems: Results of numerical computations of equilibrium microstructures. *J. Geophys. Res.* 91:9261–9276.

Waff, H. S. and J. R. Bulau. 1979. Equilibrium fluid distribution in an ultramafic partial melt under hydrostatic stress conditions. *J. Geophys. Res.* 84:6109–6114.

Waff, H. S. and U. H. Faul. 1992. Effects of crystalline anisotropy on fluid distribution in ultramafic partial melts. *J. Geophys. Res.* 97(B6):9003–9014.

Walder, J. and A. Nur. 1984. Porosity reduction and crustal pore pressure development. *J. Geophys. Res.* 89:11539–511548.

Walsh, J. B. 1981. The effect of pore pressure and confining pressure on fracture permeability. *Int. J. Rock Mech. Min. Sci. Geomech. Abstr.* 18:429–435.

Walsh, J. B. and W. F. Brace. 1984. The effect of pressure on porosity and the transport properties of rock. *J. Geophys. Res.* 89:9425–9431.

Walsh, J. B. and M. A. Grosenbaugh. 1979. A new model for analyzing the effect of fractures on compressibility. *J. Geophys. Res.* 84:3532–3536.

Walther, J. V. and P. Orville. 1982. Volatile production and transport in regional metamorphism. *Contrib. Mineral. Petrol.* 79:252–257.

Wanamaker, B. J. and B. Evans. (1985). Experimental diffusional crack healing in olivine. In R. N. Schock, ed., *Point Defects in Minerals*, Geophysical Monograph 31, pp. 194–210. Washington, DC: American Geophysical Union.

Wang, H., B. P. Boner, S. R. Carlson, B. J. Kowallis, and H. C. Heard. 1989. Thermal stress cracking in granite. *J. Geophys. Res.* 94:1745–1758.

Wark, D. A. and E. B. Watson. 1998. Grain-scale permeabilities of texturally equilibrated, monominerallic rocks. *Earth Planet. Sci. Lett.* 164:591–605.

Watson, E. B. and J. M. Brenan. 1987. Fluids in the lithosphere: 1. Experimentally-determined wetting characteristics of CO_2-H_2O fluids and their implications for fluid transport, host-rock physical properties, and fluid inclusion formation. *Earth Planet. Sci. Lett.* 85:497–515.

Watson, E. B. and D. A. Wark. 1997. Diffusion of dissolved SiO_2 in H_2O at 1 GPa with implications for mass transport in the crust and upper mantle. *Contrib. Mineral. Petrol.* 130:66–80.

Wong, T.-f. 1990. Mechanical compaction and the brittle-ductile transition in porous sandstones. In R. J. Knipe and E. H. Rutter, eds., *Deformation Mechanisms, Rheology and Tectonics,* Geological Society Special Publication 54, pp. 111–122. London: Geological Society.

Wong, T. F., S. C. Ko, and D. L. Olgaard. 1997. Generation and maintenance of pore pressure excess in a dehydrating system: 2. Theoretical analysis. *J. Geophys. Res.* 102(B1):841–852.

Wood, G. J. and J. V. Walther. 1986. Fluid flow during metamorphism and its implications for fluid-rock ratios. In J. V. Walther and B. J. Wood, eds., *Fluid Rock Interactions During Metamorphism*, pp. 89–108. New York: Springer-Verlag.

Zhang, J., T.-f. Wong, and D. M. Davis. 1990. Micromechanics of pressure-induced grain crushing in porous rocks. *J. Geophys. Res.* 95:341–352.

Zhang, S., S. F. Cox, and M. S. Paterson. 1994. Porosity and permeability evolution during hot isostatic pressing of calcite aggregates. *J. Geophys. Res.* 99:15741–15760.

Zhang, S., S. F. Cox, and M. S. Paterson. 1994a. The influence of room temperature deformation on porosity and permeability in calcite aggregates. *J. Geophys. Res.* 99(8):15761–15775.

Zhang, S., M. S. Paterson, and S. F. Cox. 1994b. Porosity and permeability evolution during hot isostatic pressing of calcite aggregates *J. Geophys. Res.* 99(8):15741–715760.

Zhu, W. 1996. *Effects of Stress, Cementation and Hot Pressing on Permeability: Experimental Observations and Network Modeling.* Unpublished Ph.D. Thesis. State University of New York, Stony Brook.

Zhu, W. and T.-F. Wong. 1997. The transition from brittle faulting to cataclastic flow: Permeability evolution. *J. Geophys. Res.* 102: 3027–3041.

Zhu, W. L., B. Evans, and Y. Bernabé. 1999. Densification and permeability reduction in hot-pressed calcite: A kinetic model. *J. Geophys. Res.* 104(B11):25501–25511.

Zimmerman, M. E., S. Zhang, D. L. Kohlstedt, and S.-I. Karato. 1999. Melt distribution in mantle rocks deformed in shear. *Geophys. Res. Lett.* 26(10):1505–1508.

INDEX

Page numbers in *italics* indicate that the information will be found in illustrations or tables.